SpringerWienNewYork

Valery A. Menshikov • Anatoly N. Perminov
Yuri M. Urlichich

Global Aerospace Monitoring and Disaster Management

SpringerWienNewYork

Valery A. Menshikov
A.A. Maksimov Space Systems
Research Institute
M. K. Tichonravova Str. 27
141092 Yubileynyy City
Russia
sergeycherkas@mail.ru

Anatoly N. Perminov
JSC "Russian Space Systems"
Aviamotornaya Str. 53
111250 Moscow
Russia

Yuri M. Urlichich
JSC "Russian Space Systems"
Aviamotornaya Str. 53
111250 Moscow
Russia

Softcover re-print of the Hardcover 1st edition 2012
SpringerWienNewYork is a part of Springer Science+Business Media
springer.at

Typesetting: SPi, Pondicherry, India

Printed on acid-free and chlorine-free bleached paper
SPIN: 80018698

With 184 Figures

ISBN 978-3-7091-1934-1 ISBN 978-3-7091-0810-9 (eBook)
DOI 10.1007/978-3-7091-0810-9
SpringerWienNewYork

Preface

This publication is devoted to the prospects of using aerospace monitoring potential for new approach to disaster management matter: early warning about cataclasmic natural events and man-made catastrophes. The International Global Monitoring Aerospace System (IGMASS) is actively promoted on the international level within recent 3 years initiative of Russian scientific and public organizations to create a system for early warning of the world society about global threats, including those of space origin. On base of the Project, implemented under the auspices of the International Academy of Astronautics, there is the conception of so-called disaster precursors, manifested as geosphere anomalies, identification with the use of special equipment of space, air and ground base, and forecasting of geological or meteorological character catastrophic phenomena on this base. Furthermore, among the prospect IGMASS issues there are early warning about near-Earth objects (meteor and asteroid threats), as well as dangerous situations in the near-Earth space, caused by "space debris" phenomenon. Finally, IGMASS will be able to provide a solution of a wide range of humanitarian issues of mankind, such as: protection and preservation of cultural values, eradication of illiteracy (distance learning) and disease control (telemedicine), providing conditions for gradual formation of global common "information security field".

The authors analyze the nature of disasters and man-made emergencies, opportunities for their effective forecasting with the use of aerospace monitoring, the key trends of the up-to-date space technologies, which will be able to reduce global risks and threats to the world society, ensuring sustainable development of civilization.

The book is intended to introduce the IGMASS concept to the wider scientific community and popularize it as an advanced Russian initiative to integrate the efforts of the international community on the peaceful using of outer space for the benefit and prosperity of all people on the Globe.

Valery A. Menshikov
Anatoly N. Perminov
Yuri M. Urlichich

Contents

Introduction

The current epoch of general globalization and integration calls for a new policy in space exploration. It is to differ from the policy that humanity had for half a century of the space era by a radically changed paradigm of space activities – a gradual transition from chasing the "space leader" in order to ensure expansion and supremacy in space to joint space activities aimed at satisfying the real needs of the international community in general and, first and foremost, promoting security and sustainable social development. The paradigm is consonant to the ideas of the classics of space exploration (Konstantin E. Tsiolkovsky, Alexander L. Chizhevsky, and Vladimir I. Vernadsky), as well as the adepts, who advocate *the idea of cosmic consciousness* (the Roerich family). In this case philosophic aspects of space activities in this century and the formation of an adequate concept make us think the current state of the planet, which tired of wars, technogenic and social upheavals and dishonest experiments with its residents, is close to the fatal borderline that is fraught with inevitable perish following a planetary catastrophe. Prof. Vernadsky has suggested the "noosphere" of the Earth is already overfilled with negative images of violence, lack of spirituality, death and destruction, while geo-pathogenic zones in various parts of the planet alarm the people with fractured the Earth crust triggering the Earthquakes, volcano eruptions, tsunami and landslides. Current global climate changes, which are difficult to forecast and comprehend, lead to catastrophic changes in the use of land, decrease crops and destroy them by natural calamities and weather anomalies. In the emerging conditions the humanity should as never focus on how to secure itself and save the planet. Space activities should play a keynote role in this matter Figs. I1 and I2.

Natural and man-made disasters continue to be among the main threats to sustainable development of mankind, inflicting huge losses on the states and the world. This is confirmed by disastrous-recent earthquakes in Haiti, Chile, Turkey, Japan and other countries, killing hundreds of thousands of people. The April (2010) eruption of the Icelandic Eyjafjallajökull volcano - paralyzed air communications throughout Europe for 10 days and caused economic damage counting hundreds of millions US dollars. Around the same time in the Gulf of Mexico, there

Fig. I1 Verdnardskiy Noosphere art concept

Fig. I2 Major natural and man-made disasters of 2010

was the largest in recent years man-made catastrophe, which rightly called the "oil Chernobyl", damage which amounts to many billions of dollars, not to mention the irreparable environmental loss. Japanese tragic events on March 2011 (several earthquakes, tsunami and caused by them destroying on nuclear power stations) add this "sad list" of natural and man-made disasters. Moreover, the problem of preventing asteroid danger moves to the foreground as it is fraught with devastating calamities of planetary scope. Meantime, according to some estimates, each million dollars invested into risk prevention and mitigation of the effects of such calamities can pay back seven times more, which makes efficient forecasts extremely important and vital Fig. I3.

Despite considerable progress achieved by the international community in space technologies to monitor natural and man-made calamities, so far it does not have global character and lacks coordination at the organizational, technical and information levels. The creation of a viable international mechanism for efficient forecasting and early warning against dangerous natural and man-made phenomena that pose planetary-scope danger is high on the agenda. It is time to seriously state that modern and maximum efficient warning against impending emergencies of space, natural or artificial origin can be provided only on the basis of large-scale international projects with complex, coordinated, and rational use of the scientific and technical potential of all countries of the world.

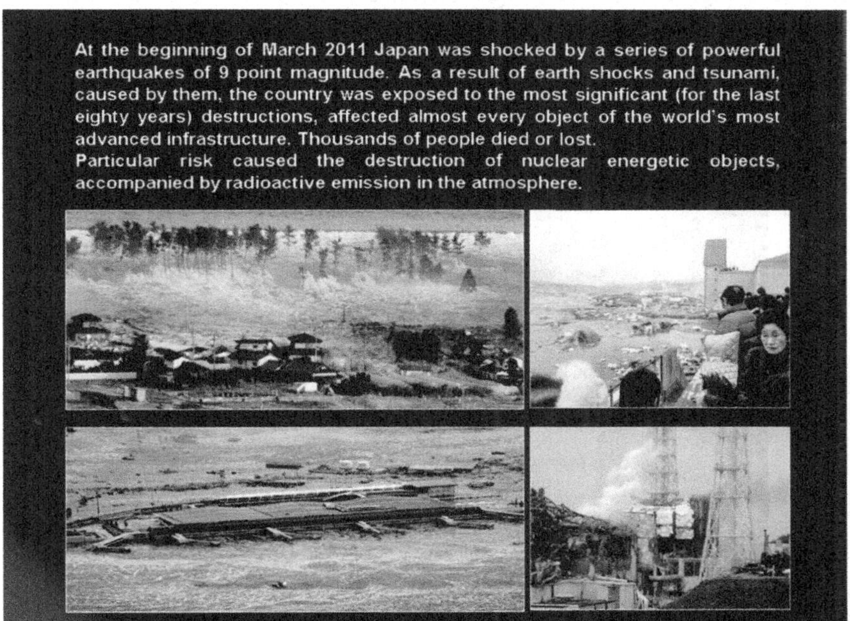

Fig. I3 Dramatic events in Japan (March, 2011)

The first chapter of the book is devoted to natural calamities, their classification, analyzes the devastating aftermath, including numerous casualties. The authors review the problems of exposing and diagnosing natural calamities of geological origin – earthquakes, volcano eruptions, and resulting tsunami as the most dangerous natural events. Today there are several hundred earthquake precursors related to the conduct of various geophysical fields at various stages of the Earth crust movement. They can be all detected by special equipment that covers practically the whole forecasting range from long-term (months) to short-term (hours).

Geophysical precursors are divided into seismic, hydro geodynamic, deformational, geochemical, thermal, gravitational, and electromagnetic. However, despite the big number none of the parameters can precisely indicate the time, place and force of the upcoming calamity. Various precursors manifest themselves differently in various seismic regions due to the sophisticated object of the research – seismic hotbed, its emergence and expansion conditions, as well as due to major influence of disturbances that can hardly be excluded from the analysis. Therefore, seismic forecast, like weather forecast, is probabilistic in nature. It is to be specially noted that reports about reliable precursors of such natural calamities have been few and non-systemic so far and, consequently, they make it difficult and often impossible to retrospectively assess their basic statistical characteristics – the probability of any false alarm or the average time from precursor to calamity Figs. I4 and I5.

The first chapter also analyses forecasting methods for natural calamities of geological origin with the use of spacecraft, including devices operating micro and nanotechnologies. Russia is a major space power and occupies proper positions in

Fig. I4 Two of the book authors (right to left): prof. Anatoly A. Perminov, Chairman of IGMASS Executive Committee and prof. Yuri M. Urlichich – Designer General of "GLONASS" Global Space Navigation System on one of scientific forums in Moscow

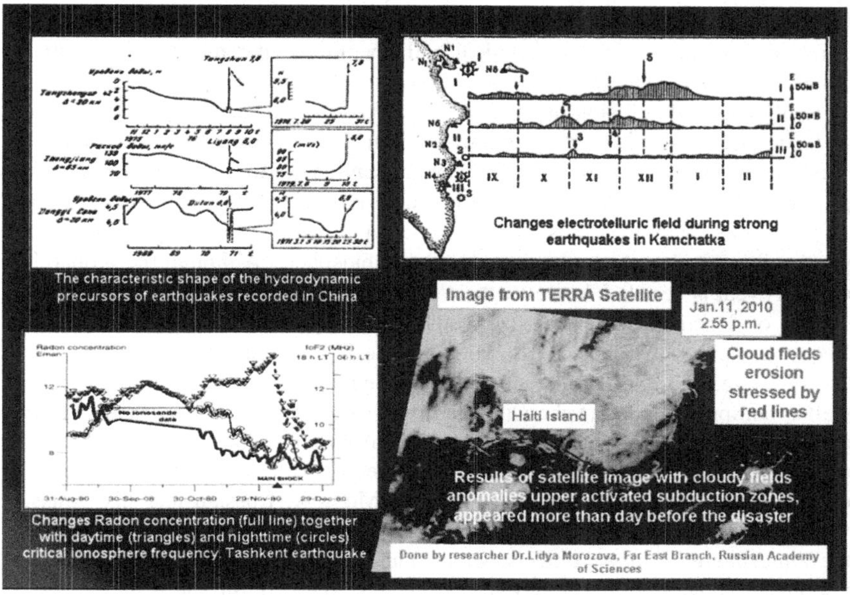

Fig. I5 Some precursors of coming earthquakes

remote observation of the Earth and in design of corresponding devices. Forecasts of threats emerging both on the planet and originating in space considerably depend on the applied use of existing scientific, technical and technological designs to detect the signs of impending danger on the basis of a complex analysis of parameters of various abnormal physical phenomena that precede natural calamities and subsequent technogenic emergencies. It has been confidently proved today that space equipment reliably detects and measures such precursors to forecast the place, time and force of a seismic event. Corresponding research is underway in many countries, including Russia.

The second chapter of the book analyses another group of modern civilization threats caused by anthropogenic factors, i.e. human activities to design complicated and hazardous industrial infrastructure, which will inevitably trigger accidents and catastrophes fraught with aftermath of the same grave scope as from natural calamities. Among the small number of precursors of man-made accidents and catastrophes there are those which are not directly linked to subjective human activities: reliability of complicated technical systems and remote control to maintain safety; natural impact (meteorological, seismic, etc.) that triggers load above projected limits, synergic failures and accidents occurring because of a long exposure of technical facilities to the environment (e.g. long operation in unfavorable weather conditions or seismically active regions) that results in fatigue and wear-and-tear of equipment fraught with catastrophic consequences. Thus, the task of forecasting man-made accidents and disasters is closely linked to global uninterrupted monitoring of technical and operational parameters of potentially dangerous systems and the territories of their location Fig. I6.

Real danger of collision of the Earth with the asteroids and long-period comets and, consequently, the necessity to prevent such a turn of developments is no longer doubted and is considered at the level of the United Nations as a vital and global problem. The most danger for the civilization is posed by "space strangers" with the size of more than hundreds meters – collision with such object could cause practically complete destruction of the planet biosphere. Although the probability of such event is very low however there is also the danger of collision with smaller objects of dozens of meters in size whose trajectory crosses the Earth orbit more frequently and which can cause major trouble if they enter the atmosphere at a certain angle. The study of impact craters on the Earth and other celestial bodies, as well as computer modeling show the energy released in a collision of the Earth with an asteroid of 50–100 m in diameter is comparable to the energy of a thermonuclear explosion of several dozen megaton in TNT equivalent. The explosion can fully destroy such mega cities, as New York or Moscow. Russian and U.S. missile warning space-deployed systems annually register close to a dozen of entries into the atmosphere of major objects, which luckily cease to exist several thousand kilometers above the surface. From 1975 to 1992 the U.S. missile warning system

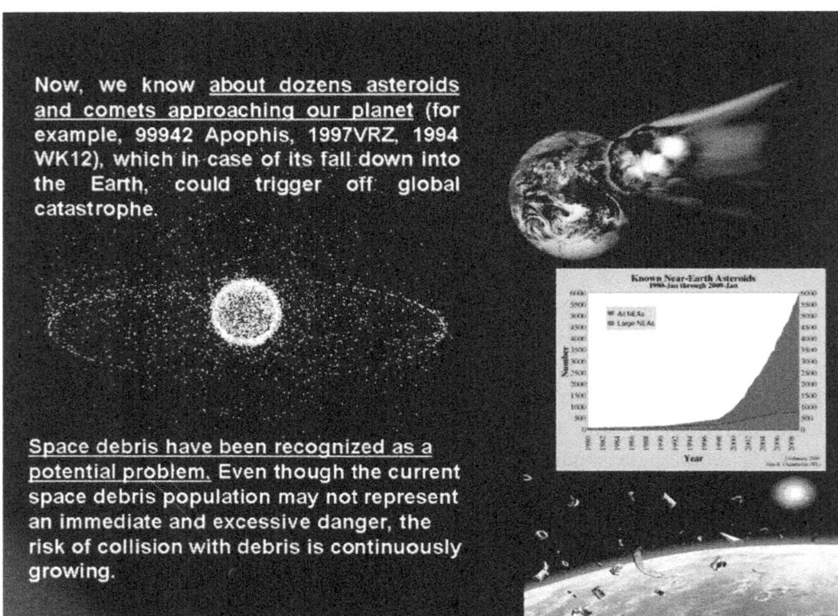

Fig. I6 Global risks and threats from outer space

reported 126 such entries, some of which released the energy equivalent to a million tons of TNT.

Experts say the Earth encounters asteroids of several to dozens meters in size every 10 years on average. Until now the events took place mostly above uninhabited regions of the planet however the growing density of the population increases the probability of catastrophic consequences from a falling asteroid. The permanently growing number of such objects can expand the scope of such catastrophes as they may be destroyed even by a small asteroid with most devastating consequences Fig. I7.

A number of researchers believe it is possible to prevent such a turn of developments by creating a planetary defense system from asteroids and comets. Up-to-date technological development of Russia and other world industrialized nations allow to begin designing such a system. Practically all its basic components or prototypes have been created and underwent live testing. In particular, they include many samples of missiles, nuclear weapons, communication, navigation, control facilities, etc. Today there is a unique possibility to use the mostly military infrastructure not for destruction, but for protection of humanity from "traveling" space objects. The third chapter is fully devoted to that interesting and important issue.

Despite considerable progress achieved by the international community in space technologies to monitor natural and man-made calamities, so far it does not have global character and lacks coordination at the organizational, technical and information levels. The creation of a viable international mechanism for efficient forecasting and early warning against dangerous natural and technogenic phenomena that pose a planetary-scope danger is high on the agenda. It is time to seriously state that modern and maximum efficient warning against impending emergencies of space, natural or man-made origin can be provided only on the basis of large-scale international projects with complex, coordinated, and rational use of the scientific

Fig. I7 VIP participants of the first international specialized symposium "Space & Security of Humanity" (Limassol, Cyprus) from Kazakhstan, Kyrgyzstan, Romania, France, Russia and Ukraine

and technical potential of all countries of the world. In 2007 the Maximov Space Research Institute, which is branch of the biggest Russian space holding – Khrunichev State Research and Production Space Center – proposed the Russia-patented idea of the creation of the International Global Monitoring Aerospace System, which is the result of over a decade-long design of efficient and systemic technologies of space monitoring.

The fourth chapter of the book introduces the reader with the IGMASS project. IGMASS' authors and supporters believe the system has to develop into a major organizational and technical structure that would integrate its own specialized space segment, which comprises small and micro satellites with the latest equipment to detect early signs of natural calamities, as well as ground and airborne facilities of monitoring and remote Earth observation space systems, communications and retranslation lines, meteorological and navigational provision together with corresponding infrastructure. The system will be created to provide early warning to the international community about impending natural calamities and technogenic emergencies, to efficiently eliminate their aftermath, upgrade and integrate navigational, telecom and information resources of the planet in the interests of repelling global threats and resolving humanitarian problems of the mankind Fig. I8.

Priority IGMASS missions include space monitoring of the lithosphere, atmosphere and ionosphere of the Earth and the near-the Earth space to detect early

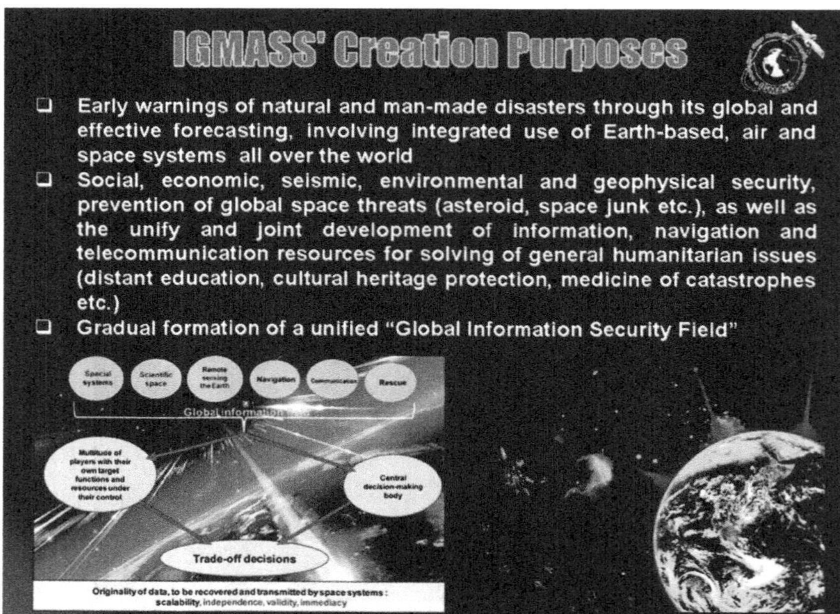

Fig. I8 The IGMASS Project keynote missions

signs of dangerous natural events, collect, review, process, interpret, store, and provide monitoring data and real-time information about exposed both natural and unnatural threats to affected states as well as specialized UN bodies concerned. With time it is planned to use the IGMASS technical and organizational potential for navigation and telecom services provision to customers in the world to promote emergency operations, disaster medicine, humanitarian relief, remote training of specialists in various spheres, and in more distant future – for efficient warning against asteroid and meteorite threats and abnormal phenomena with the aim of a staged creation of a common and planetary "global security information space."

Meanwhile, the IGMASS project is no alternative to the current efforts of the international community to monitor natural calamities and disasters. It is planned to engage in its framework the capabilities and potential of all available international, regional and national projects in the sphere of remote Earth observation and disaster prevention, such as UN-SPIDER program, the Global The Earth Observation System of Systems (GEOSS), the European Global Monitoring for Environment and Security (GMES), the Sentinel Asia natural disaster monitoring system for Asia and the Pacific Rim, the International Charter "Space and Major Disasters" (Disaster Charter), the Ukrainian "Ionosat" project to monitor natural and technogenic calamities, etc. Finally, it will provide collection, processing and compiling of reliable warning information about potentially dangerous processes and events to promote timely steps by the countries concerned and the world community to prevent them or ease its aftermaths.

A certain technical and technological base has been created already for the formation of IGMASS own orbital constellation on the basis of small and micro spacecraft with various payloads to detect and report early signs of natural and man-made disasters. For example, it may include spacecraft platforms with a mass of up to 400 kg deployed in sun-synchronous and geostationary orbits. The current state of vehicle design operating geophysical forecast technologies and of sensors for space-rocket hardware in general give grounds to predict that spacecraft with required parameters can appear in the coming years.

The fifth chapter of the book is devoted to prospects of the formation of a common security information space with the use of the IGMASS navigational and telecommunication resources (efficient warning about planetary threats also from space, maintaining space ecology, solution of humanitarian problems: preservation of cultural values, eradication of illiteracy, remote training, disaster medicine, etc.). The authors draw the attention of the readers to the absence of alternatives to space systems as the efficient instruments capable of receiving and distributing data adequate to the burden of problems, risks and threats, which humanity carries to the postindustrial stage of development – information society that turns information into an economic category and identifies all types of national and planetary resources. The intellectual foundation is created to resolve problems of the fight against natural calamities that are described in this publication. It also considers the special IGMASS distant training subsystem in the sphere of monitoring and forecasting natural and technogenic disasters that expands the possibilities of mastering corresponding educational and re-training programmes to prepare specialists for the operation of the system and raise the level of education at space training facilities located far away from major administrative, industrial and cultural centers. IGMASS information capabilities can be engaged for the solution of urgent tasks in emergency situations, in particular, for disaster and emergency medical assistance that needs first to locate people trapped in extreme situations, provide urgent medical assistance to them, and then remotely observes their state of health Fig. I9.

The book was prepared by well-known authors: prof. Anatoliy N. Perminov – PhD Tech., Chairman of IGMASS Executive Committee, Deputy Director General/ Designer General, JSC "Rossiyskiye Kosmicheskiye Sistemy" (Russian Space Systems), prof. Valery A. Menshikov – PhD Tech., Director General of the IGMASS Project, Designer General of the Multifunctional space system of the Union State, and prof. Yury M. Urlichich, PhD Tech., Director General and Designer General of JSC "Rossiyskiye Kosmicheskiye Sistemy" (Russian Space Systems). Other contributors comprise: prof: Yefim M. Malitikov – PhD Economy, PhD Tech., president of the "ZNANIE" International Association; prof. Vladimir A. Alexeyev, PhD Physics and Math's., Leading Expert of The Troitsk Institute of Innovations and Thermonuclear Researches (Russian Academy of Sciences); prof. Vladimir G. Degtyar, PhD Tech. and Corresponding Member of the Russian Academy of Sciences; as well as a team of experts from Maximov Space Systems

Fig. I9 One of the book authors prof.Valeriy Menshikov – Director General of the IGMASS Project and ideologist of the system – speaking at the Plenary Meeting of the "Russian Space Week" in Madrid, Spain (May, 2011)

Research Institute, dealing with IGMASS issues and problems of spacecraft use for natural calamity forecasting: prof Anatoly I. Rembeza – PhD Tech.; Alexander V. Rad'kov – PhD Tech., Sergey R. Lysyy – PhD Tech.

On behalf of editors we're much grateful personally to Mr. Andrey A. Kachekan and Mrs.Svetlana S. Kovalenko, who had taken all routine technical jobs for editing and preparing the book. The authors of the publication will be luck for any remarks, proposals and ideas of the readers to be submitted by tel.: +7 495 785-79-29 or via e-mails: IGMASS@mail.ru, info@igmass.com.

Natural Calamities and Their Forecasting 1

Man and the Environment (habitat) comprise an integrated system consisting of numerous inter-related elements, possessing specific traits, and streamlined within certain boundaries. The interaction between Man and the Environment depends on various factors that affect both of them and may be positive and negative at the same time. The negative influence of habitat on Man and his activities is exerted mostly by natural calamities and hazards [1].

Natural calamities and hazards comprise natural geophysical phenomena of extreme character that disturb normal human activities, damage and destroy material values, and result in loss of human life. Natural calamities are usually divided into two main groups depending on their origin: geological and meteorological (Fig. 1.1). They can be both independent and mutually dependent. Thus, major dependence exists between earthquakes and tsunamis, volcanic eruptions and fires. Floods and landslides can be triggered by tropical cyclones and earthquakes (e.g., as a result of river blocking). Earthquakes and eruptions are mutually dependent most of all, as tremors may be caused by volcanic eruptions and vice versa – rapid movement of underground mass results in eruptions. At the same time many natural calamities emerge because of negative anthropogenic impact (forest and peatland fires, production of explosions in mountains to build dams, and pits that trigger landslides, avalanches, etc.).

The statistics of the breakdown (Fig. 1.2) of natural calamities and their geographic origin (Fig. 1.3) over 35 years of the past century produce some interesting examples. For example, typhoons, storms, earthquakes, floods, and droughts comprised 88% of all natural calamities and over half of them were devastating phenomena of meteorological origin. None of the world regions were free of the impact of the elements; however Asia and its southeastern part, including Oceania, occupy leading positions with nearly half of all disasters.

The number of natural calamities has also been on the rise in Russia of late. In the 10 years from 1990 to 1999 the Russian Emergencies Ministry registered 2,877 such destructive events related to dangerous natural processes (Fig. 1.4). The average annual number of such events in the last decade of the twentieth century clearly showed a growing tendency and comprised close to 300. This growth was

V.A. Menshikov et al., *Global Aerospace Monitoring and Disaster Management*,
DOI 10.1007/978-3-7091-0810-9_1, © Springer-Verlag/Wien 2012

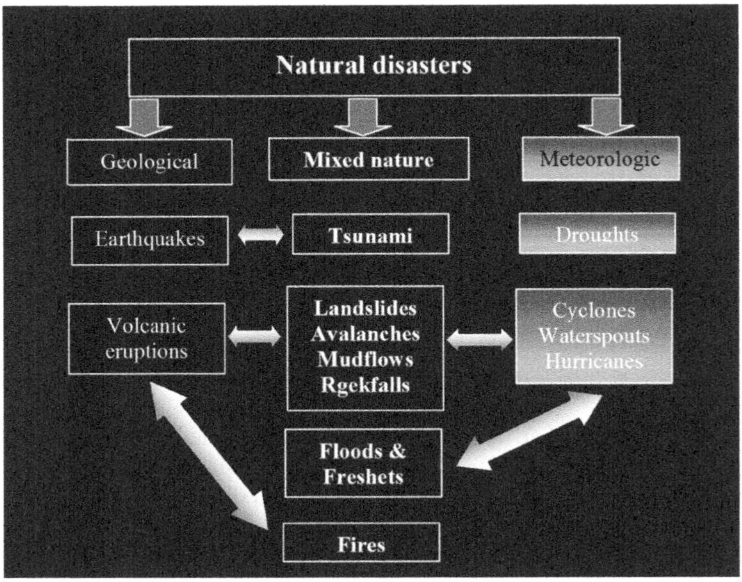

Fig. 1.1 Natural calamity classification (one of the approaches)

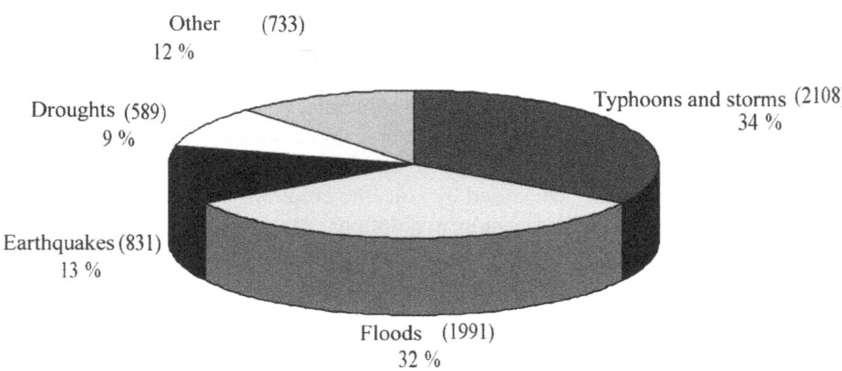

Fig. 1.2 Disaster statistics (1965–1999)

especially fast in 1997–1998 likely because of a critical rise in annual average air temperatures during that period [2].

Natural hazards of atmospheric origin were most frequent in Russia (Fig. 1.5) – storms, hurricanes, tornados, and squalls (28% of total natural emergencies). Earthquakes and floods follow with 24% and 19%, respectively. The share of major forest fires is tremendous in Russia and comprises 25% of total natural calamities [3].

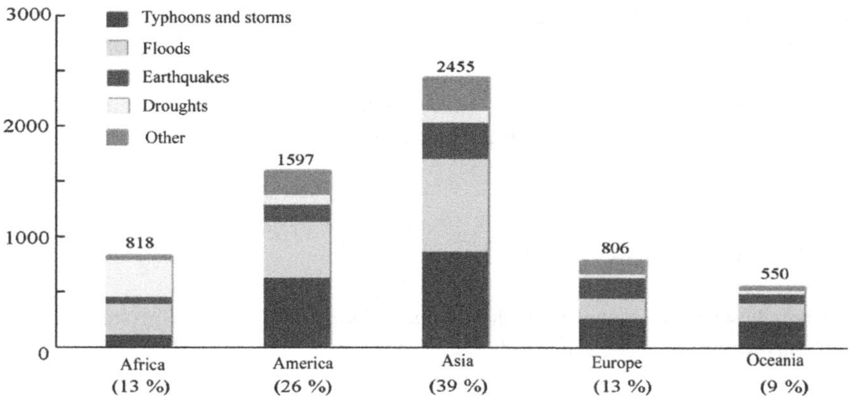

Fig. 1.3 Geography of natural calamities (1965–1999)

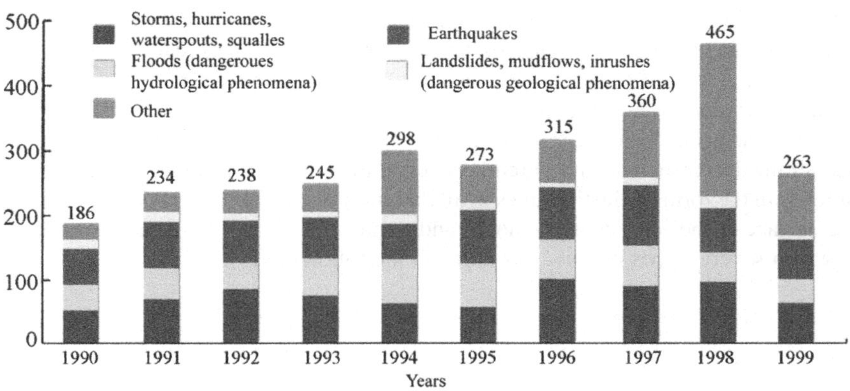

Fig. 1.4 Natural calamity statistics in Russia (1965–1999)

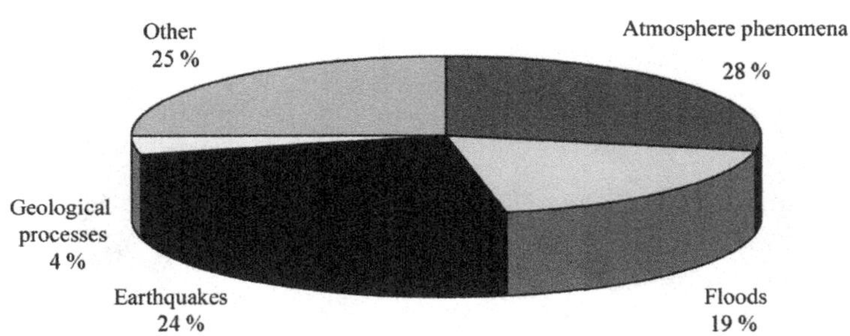

Fig. 1.5 Natural calamities in Russia (1965–1999)

1.1 Natural Calamities of Geological Origin

Geological natural calamities mostly comprise earthquakes and volcanic eruptions. However, sometimes tsunamis are added to this group, as they originate from tectonic processes, as well as earthquake-triggered landslides and avalanches.

1.1.1 Earthquakes

Earthquakes are lithosphere developments accompanied by underground shocks and surface tremors of natural (tectonic processes and lava discharge during volcanic eruptions) or man-made (explosions, artificial lake filling, collapsing underground driftage) origin. Earthquakes are thus divided into endogenous (related to deep mantle processes) and exogenous (caused by human activities). Close to a million earthquakes annually occur on the earth, but most are too weak to be noticed. Major earthquakes capable of inflicting considerable damage take place once in 2 weeks on average. As most of them occur under the ocean bed, they do not result in catastrophic consequences for humanity (unless they cause a tsunami) (Fig. 1.6).

It is common knowledge that elastic strain accumulates in the deep mantle for various reasons and should it surpass the resistance level of the rock it ruptures it, the strain decreases and the potential energy transforms into kinetic energy and spreads in the form of elastic waves in all directions from the earthquake focus up to the surface of the earth causing shocks and tremors. The earth's crustal movements result in several types of waves as noted in the following [4].

Fig. 1.6 Underwater seismic events

Primary waves (P) are compression and discharge waves that travel alternately with a velocity (in hard rock) of several kilometers per second. P-type waves originate as the reaction of the medium to changing volumes. They can travel through solids, liquids, and gases.

Secondary waves (S) are a result of the reaction of the environment to changing form and consequently cannot travel through liquids and gases. With S-type waves the ground is displaced perpendicularly to the direction of propagation.

Surface or Rayleigh waves (L) originate under special conditions on the convergent boundary of two media that differ in aggregate state (liquid and gas, solid and gas, etc.) and are triggered by tremors from the focus reaching the boundary. They travel more slowly than P or S-type waves and quickly subside although they may be highly destructive in the epicenter.

The type and travel velocity of elastic waves that trigger earthquakes depend on the physical characteristics of lithospheric rock. Elastic features include volumetric strain, which is resistance to compression without changing form, and shear, which is resistance to strain force. Thus, elastic wave velocity growth is directly proportional to the square root of the elasticity and density parameters of the medium.

Basic earthquake characteristics (Fig. 1.7) comprise origin depth, magnitude, and force on the surface of the earth. Earthquakes usually originate at a depth of

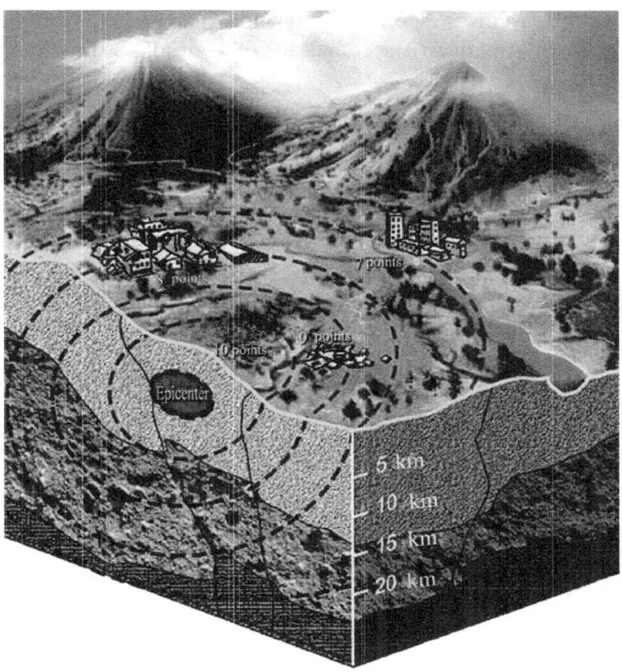

Fig. 1.7 Nature of typical earthquake

10–30 km, although the focus can sometimes be deeper. The magnitude characterizes the general force of the earthquake and is a logarithm of maximum earth displacement amplitude measured in microns by a seismogram at a distance of 100 km from the epicenter. Most commonly magnitude (M) is measured from 0 to 9. A one-point increase means tenfold growth of earth tremor amplitude (or ground displacement) and 30-fold increase in earthquake energy. It is of interest that strong and weak earthquakes differ not so much because of strain, which is constant and equal mostly to 103 erg/cm^3, but because of the focus depth.

Earthquake intensity, which is its outer effect or manifestation on the earth's surface, is measured in points. It depends on the focus depth, magnitude, distance from the epicenter, geological framework of the bedrock, and other factors. To measure seismic intensity the 12-point Medvedev-Sponheuer-Karnik scale is used which offers the following degrees: I–III points – light, IV–V – moderate and fairly strong, VI–VII – strong (dilapidated houses collapse), VIII – damaging (solid houses partially collapse, damage to chimneys), IX – destructive (most buildings collapse), X – devastating (bridges collapse, landslides triggered), XI – catastrophic (all buildings destroyed, landscape changes), XII – very catastrophic (upheaval of the landscape).

Earthquake epicenters spread unevenly on the surface of the earth (Fig. 1.8). The majority (close to 90% of all registered quakes in the past 100 years) were located along two narrow seismic belts that imborder the planet. The Pacific belt stretches along the eastern coast of Asia to the north and east of Australia and along the eastern coast of America (it accounts for 68% of all earthquakes, notably in

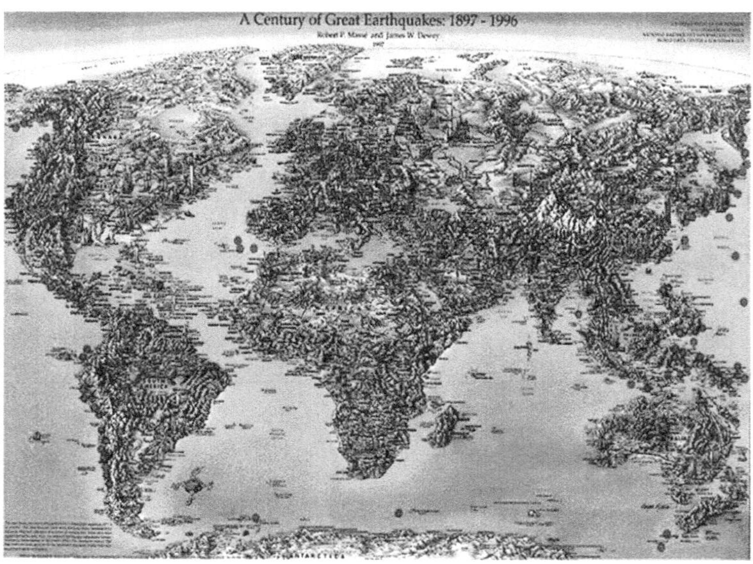

Fig. 1.8 Seismically dangerous regions on the earth

Japan and the Philippines). The Mediterranean belt covers Cape Verde Islands, Portugal, the Mediterranean and Black Seas, Asia Minor, the Himalayas, and Indonesia with a side branch towards Central China. Earthquake hotbeds also exist in the middle of the Atlantic Ocean from Spitsbergen via Iceland to Bouvet Island in the south Atlantic. Small belts stretch from the Nile River through the area of the Great Lakes of Africa; along the Urals, etc. As a rule, earthquakes occur in areas with the acutest highland displacements, i.e., where mountains alternate with trenches. Major plains (East-European, West-Siberian, etc.) are aseismic and the reason is tectonic strain. Areas with mixed terrain emerge due to major manifestations of tectonic strain, while plains are platforms are stable areas. It should be noted that seismically dangerous areas usually territorially coincide with volcanic areas because of common processes of earth magma movement.

According to statistics, on average one catastrophic earthquake with a toll of thousands of lives occurs annually on the earth. The history of mankind includes numerous earthquakes with catastrophic consequences (Table 1.1). Thus, for example, the 1990's earthquake in the northern Iranian province of Gilan killed over 50,000 people, the 2008 earthquake in the southern Chinese province of Sichuan killed from 70,000 to 90,000 people, and the 2010 quake in Haiti had a toll of hundreds of thousands of lives. In the meantime, the earth's seismic activity has been intensifying every year in the past two decades. The reasons behind this are an issue for a separate analysis. The authors cited this gloomy statistics with the sole aim of substantiating the necessity for urgent measures to forecast earthquakes and consequently resulting devastating natural and technogenic calamities.

1.1.2 Volcanic Eruptions

Volcanoes are geological structures on the surface of the earth's crust where magma discharges to the surface in the form of lava, volcanic gases, stones (volcanic bombs), and pyroclastic flows. Two types of volcanoes are distinguished by the type of eruption: Hawaiian which discharge liquid basalt lava and often create lava lakes; and hydro explosive when eruptions occur in shallow ocean and sea waters and produce huge volumes of steam due to the contact of red-hot magma with sea water.

Volcanoes are divided into active, dormant, and dead by the level of volcanic activity. Active are volcanoes that erupted during the historic period of time or in the Holocene. The notion of active is rather inaccurate as some scientists refer volcanoes with active steam vents to the active type, while others – to dead. Volcanoes of the dormant type are inactive volcanoes where eruptions are possible, while dead volcanoes have a low chance of eruptions. However, volcanologists differ in defining the notion of an active volcano, as active periods can alternate with quiet times (from several months to several thousand years), while many volcanoes were active dozens of thousands of years ago and are inactive at present.

There are two types of volcanic eruptions: central vent and fissure. Central-vent eruptions well up lava, scoria, ashes, and gases to the surface from a shallow

Table 1.1 Most devastating earthquakes in the history of mankind

January 23, 1556 – Gansu and Shaanxi, China – death toll of 830,000, a record high in human history.

1693 – Sicily, Italy, 60–100,000 people killed.

1737 – Calcutta, India – 300,000 people killed.

1755 – Lisbon — from 60,000 to 100,000 people killed, city fully destroyed.

1783 – Calabria, Italy – from 30,000 to 60,000 people killed.

1887 – Verny (now Alma Ata), Starshy Zhuz, Russian empire – material damage estimated at 2.5 million rubles; 1799 stone and 839 wooden buildings destroyed.

1896 – Sanriku, Japan – earthquake focus was under the seabed. A giant wave washed away 27,000 people and 10,600 buildings.

April 4, 1905 – Kangra, India. Magnitude 8.7 Richter scale. 19,000 people killed.

April 18, 1906 – San Francisco, USA. 7,000 people killed, 10 km^2 of the city destroyed.

December 28, 1908 – Sicily, Italy. 83,000 people killed, Messina ruined.

December 16, 1920 – Gansu, China. 20,000 people killed.

September 1, 1923 – Great Kanto earthquake – Tokyo and Yokohama, Japan (8.3 Richter scale) – 143,000 people killed and close to a million remained homeless.

October 6, 1939 – Inner Taurus, Turkey. 32,000 people killed.

1948 – Ashkhabad, Turkmenia, USSR, Ashkhabad quake killed 110,000 people.

August 5, 1949 – Ecuador. 10,000 people killed.

August 15, 1950 – Assam, India. Thousands killed.

February 29, 1960 – Agadir, Morocco. 12,000–15,000 people killed.

May 21, 1960 – Great Chile quake. Close to 10,000 killed, cities of Concepcion, Valdivia, Puerto Montt ruined.

July 26, 1963 – Skopje, Yugoslavia. Close to 2,000 killed, most of the city ruined.

May 31, 1970 – Peru. 63,000 people killed, 600,000 left homeless.

February 4, 1976 – Guatemala. Over 20,000 people killed, over one million left homeless.

July 28, 1976 – Taishan, northeastern China, Taishan earthquake (8.2 Richter scale) – over 655,000 killed.

October 10, 1980 – El Asnam, Algeria. Magnitude 7.7 points. 17,000–20,000 killed.

September 18, 1985 – Mexico City, Mexico. Magnitude 8.2 Richter scale – over 7,500 people killed.

December 7, 1988 – Spitak earthquake, Armenia, USSR – cities Spitak, Leninakan, and numerous settlements destroyed, 40,000–45,000 people killed.

June 21, 1990 – Gilan quake, Iran. Magnitude 8 Richter scale. 50,000 people killed and one million left homeless.

May 28, 1995 – Neftegorsk, northeastern Sakhalin (magnitude – 7.5) 1,841 people killed.

August 17, 1999 – Izmit earthquake, Turkey. Magnitude – 7.6. 17,217 people killed, 43,959 injured, close to 500,000 left homeless.

December 26, 2004 – Indian Ocean earthquake and resulting tsunami killed 225,000–250,000 people.

May 12, 2008 – Sichuan quake in central China killed close to 70,000 people.

January 12, 2010 – Haiti earthquake. Magnitude 7.0. 220,000 people killed, 300 thousand injured, 1.1 million left homeless.

February 27, 2010 – Santiago, Chile. Magnitude 8.8. Close to 800 people killed, over 1.5 million houses damaged by tremors and tsunami.

March 11, 2011 – Tohoku, Japan. Magnitude 9.0. The Japanese National Police Agency has confirmed 15,760 deaths, 5,927 injured, and 4,282 people missing across eighteen prefectures, as well as over 125,000 buildings damaged or destroyed.

magma hotbed that existed before the eruption through a single vent or a cluster of centrally placed vents, fed by a single pipe-like supply channel. In fissure eruptions magma pressure creates new volcanic channels and fissures.

An especially dangerous type of central-vent eruptions are directed volcanic blasts. They often last for several dozens of minutes and throw out gas-saturated red-hot mass that flows dozens of kilometers and destroys everything in its path. Such blasts were registered three times on Kamchatka over the last century: in 1907 in the Ksudach caldera, in 1956 – in Bezymyanny volcano, and in 1964 – in Shiveluch volcano. On May 20, 1980 such a blast destroyed the top of the St. Helens volcano in the west of the United States.

Volcanic eruptions are usually preceded by small tremors caused by the magma rising from depths of 100–120 km.

Volcanic eruptions are geological emergencies that can result in natural calamities, as eruptions may last from several hours to numerous years. Human history knows several major eruptions with catastrophic consequences. Besides the widely known tragedy of the ancient Roman city of Pompeii that was buried by ashes and lava from the eruption of Vesuvius some 3,600 years ago, 1,500 years before that a volcanic eruption on the Island of Santorini in the Aegean Sea completely destroyed the Minoan civilization [5].

Such a grandiose geological event as the Santorini eruption was definitely accompanied by underground tremors. The rattle of volcanic blasts and tremors was followed by a black ash cloud that covered ancient Santorini. Scientists estimate the eruption discharged 60 cubic kilometers of volcanic ash into the atmosphere and temperatures fell on the whole planet as a result. The ashes partially helped preserve Minoan-era buildings in the same way as ashes from Vesuvius covered Pompeii 1,500 years later and preserved it to the present day. Ashes fell during three consequent eruptions at Santorini and the ash layer became dozens of meters thick.

There is evidence that the underground tremors, blasts, and the caldera cave-in that accompanied the Santorini eruption in the fourteenth century B.C. caused a giant tsunami. On Anafi Island 25 km east of Santorini a 5-m thick tephra layer was discovered in a valley at a height of 250 m above sea level. It mostly consisted of pumice stone that usually originates in water. Experts believe the pumice was deposited in the sea during the Minoan-era eruption at Santorini and was then washed ashore by the tsunami. The tsunami caused by the Santorini disaster was definitely the strongest.

The Santorini catastrophe not only caused tremendous damage to the Kiklad Islands and Crete, but also killed a lot of people. There were casualties on other islands in the Aegean Sea as well. There is a hypothesis that the Santorini disaster influenced not only the development of Mediterranean civilizations, but also triggered mass resettlement of peoples that inhabited the southern Urals and adjacent areas. A sharp and prolonged fall in temperature made them look for new and more comfortable living habitats which they found in India. It is believed to be the reason why the Mahabharata of Vyasa epic pays so much attention to the cold, absent sunrises, and long nights: Roaring gale-force winds blew and stones fell instead of

Fig. 1.9 Vesuvius eruption

rain. Birds began to wheel, making circles from left to right. Great rivers turned back on themselves, mist covered all cardinal points, and fireballs fell on the Earth from the sky throwing red-hot coals all around. The Sun's disc was veiled and clouds fully blocked its rays in the east. Both the Moon and the Sun had an ominous tricolor halo with sharp black edges and red tints resembling hot ashes (Fig. 1.9).

The Santorini disaster provides indirect evidence to scientists, while the strongest volcanic eruption registered by modern science occurred on Krakatoa – a small volcanic island located in the strait between Java and Sumatra. It began on May 20, 1883 [6]. Before that Krakatoa was inactive for two centuries and was believed to be dead. However, it unexpectedly welled out a column of black smoke and volcanic ashes 11 km high. Tremors were felt at great distances from the volcano, up to Jakarta, while people in settlements on the coast of the Sunda Strait heard powerful explosions. Then silence followed for 3 weeks, but from the middle of June the volcano resumed its eruption with new force. In August in Krakatoa there were three craters instead of one and they all threw out ashes and volcanic gases. The size of the island increased from 12 to 30 km^2. On August 26, 1883 a scary rumble came from the volcano and became so loud by night that people on Java could not sleep. Close to 10 h on August 27 an incredible explosion occurred. It threw volcanic gases, sand, and debris 30 km high, while ashes went up as high as 70 km. The explosion was heard over a territory equal to 1/13th of the Globe, for example, three and a half thousand kilometers away on Sri Lanka and in central Australia, as well as on Rodriguez Island in the eastern Indian Ocean 5,000 km away from Krakatoa. The explosive wave broke glass, tore off doors, and cracked wall plaster in houses on Java (150 km from the volcano). Damage was reported in several old buildings (800 km away) [7] (Fig. 1.10).

Fig. 1.10 Eruption of Krakatoa volcano

An hour after the explosion Jakarta, which is 200 km away from the volcano, was in darkness as ash clouds fully blocked off the sun. Tropical forests were completely destroyed on the coast of the Sunda Strait, while land was covered with dirt, ashes, pieces of lava, and uprooted trees. Dead bodies of humans and animals were scattered everywhere. The ocean surface around Krakatoa was covered with a carpet of pumice that was so thick that ships failed in their attempts to cross the floating barrier. Pieces of pumice were found on that day in the coastal waters off Australia and the Maldives (Fig. 1.11).

However, the "seaquake" caused by the powerful explosion inflicted the most damage. A resulting 40-m high giant tsunami hit the islands of Sumatra and Java. It razed to the ground close to 300 towns and settlements with surrounding fields and plantations. Over 6,000 fishing vessels sank. The wave washed the Dutch gunboat Barrow 3 km ashore into a forest. Thirty-six thousand people were killed and hundreds of thousands were made homeless. The population of Sebesi Island located 20 km from the volcano was fully exterminated. The tsunami was 15 m high even 90 km away from the volcano. It was 5 m high when it reached Sri Lanka. The wave also reached the coast of Australia, Africa, and South America. Even vessels in the English Channel were affected by it.

Fig. 1.11 Krakatoa eruption

The Krakatoa volcano completely disappeared after the catastrophe. Only a small part of one of the three craters remained above water to demonstrate a classical cut of the lava cone. The rest of the volcano developed into a depression 7 km in diameter and 300 m deep.

However, the ruined volcano did not cease its activity. It resumed it half a century later and in 1952 a new and young volcanic cone emerged out of the water and began to grow and expand due to small but frequent eruptions to dominate the strait. Today the volcanic island is 250 m high and 1 km long and continues to grow. It is called Anak Krakatoa (Child of Krakatoa) (Fig. 1.12).

1.1.3 Tsunami

Tsunami (translated from the Japanese as "big harbor wave") is a long and disastrous wave caused mostly by tectonic displacements on the ocean bed. A tsunami may follow underwater volcanic eruptions and major landfalls into the ocean. Tsunamis travel in deep ocean waters at a speed of above 1,000 km/h and the distance between subsequent wave ridges can exceed several 100 km. Therefore, they are practically harmless for vessels in the high sea. However, when a tsunami reaches shallow coastal waters its speed falls sharply and its height rises quickly. It is there that the tsunami becomes devastating for infrastructure and a lethal danger for people. It is there that the wave can rise 30–50 m high and acquire major devastating force. A tsunami is especially dangerous for residential settlements and constructions in harbors and bays that are wide open to the ocean and arrow-headed in shape towards land (Fig. 1.13).

Such a disaster occurred on the night of August 26–27, 1883 when the Krakatoa volcano exploded between the islands of Java and Sumatra. Thirty minutes later a 40-m high tsunami caused by the eruption hit the coasts of Java and Sumatra.

Fig. 1.12 Krakatoa's child

Thirteen years later three tsunamis devastated the eastern coast of Japan on June 1896 between latitudes 30°N and 40°N and destroyed practically all coastal settlements (100,000 houses were ruined and washed away and 27,000 people were killed) [8] (Table 1.2).

Tsunamis, like earthquakes, have an intensity scale:

I – very weak tsunami when the wave is registered only by mareographs;

II – weak tsunami capable of flooding plain coast;

III – medium force tsunami. Plain coast flooded, light vessels may be washed ashore. Current may temporarily reverse in cone-shaped river entries. Insignificant damage inflicted to port structures;

IV – strong tsunami. Coast flooded, coastal buildings and erections damaged. Big sailing ships and small motorboats washed ashore and then back to the sea. Coast polluted with debris and garbage;

V – very strong tsunami. Coastal territories flooded, breakwaters and harbor bars damaged. Bigger ships washed ashore. Major damage inflicted to inner parts of the coast. High storm surges in river entries. Human casualties;

VI – catastrophic tsunami. Complete devastation of the coast and adjacent territories. Land flooded to a major distance from the sea. Big vessels damaged. Numerous casualties.

Close to 85% of all tsunamis are triggered by underwater earthquakes in the Pacific Rim. However, it should be noted that tsunamis as natural calamities occur more rarely than earthquakes. The deadly wave emerges only in the case where an underwater earthquake creates a sharp vertical displacement of the seabed: a part depresses and a part alleviates. As a result, the water surface begins a vertical wave-like movement trying to restore its initial level (average sea level) and triggering

Fig. 1.13 Tsunami on the Thailand seashore

a series of waves. Earthquakes with a shallow focus usually trigger tsunamis. Detailed analysis of tsunami origins showed that maximum wave amplitude is registered when rock displacements occur at a depth of 10 km.

Tsunamis can be triggered also by landslides. Thus, on July 9, 1958 an Alaskan earthquake in Lituya Bay resulted in a gigantic landslide. Ice and rock fell into the ocean from a height of 1,100 m triggering a wave over 500 m high at the opposite coast. In Indonesia, which has major shelf sedimentation, landslide-triggered tsunamis are especially dangerous as they occur regularly and cause local waves over 20 m high (Fig. 1.14).

Close to 5% of all tsunamis are caused by volcanic eruptions. A classic example is offered by the tsunami caused by the Krakatoa eruption in 1883. Huge tsunamis from the Krakatoa volcano hit practically all harbors in the world and destroyed a total of 5,000 vessels.

Tsunamis may also be caused by falling meteorites as their fast entry speed into the atmosphere of the earth discharges immense kinetic energy that triggers huge

Table 1.2 Major tsunamis in the past 50 years

5.11.1952.	Severo-Kurilsk (USSR). Caused by powerful earthquake (magnitude estimated at 8.3–9 points by various sources) in the Pacific Ocean 130 km off the Kamchatka coast. Three waves 15–18 m high (according to various estimates) destroyed the city of Severo-Kurilsk and damaged several other residential settlements. The death toll exceeded 2,000 people, according to official estimates.
9.03.1957	Alaska (USA). Caused by an earthquake with a 9.1 magnitude on the Andreanof Islands (Alaska) that triggered two waves 15 and 8 m high, respectively. Over 300 people killed.
9.07.1958	Lituya Bay (southwestern Alaska, USA). The earthquake that occurred north of the Bay (Fairweather fault) triggered a major landslide on the mountain dominating Lituya Bay (close to 300 million cubic meters of ground, stones, and ice). The mass fell from a height of 1,100 m, covered the northern part of the bay, and triggered a huge 524-m high wave traveling at a speed of 160 km/h.
28.03.1964	Alaska (USA). The strongest earthquake in Alaska (magnitude 9.2) occurred in the Prince Williams Strait and caused a tsunami consisting of several waves with the highest at 67 m. The disaster (mostly tsunami) killed from 120 to 150 people, according to various estimates.
17.07.1998	Papua New Guinea. An earthquake with 8.1 magnitude that occurred on the northwestern coast of New Guinea Island caused a major underwater landslide that triggered a tsunami which killed over 2,000 people.
06.09.2004	Coast of Japan. Two strong earthquakes (6.8 and 7.3 magnitude, respectively) occurred 110 km away from the coast of the Kii Peninsula and 130 km from the coast of the Kochi Prefecture and triggered a tsunami with 1-m high waves. Several dozen people were injured.
26.12.2004	South-East Asia. At 00:58 h the second strongest of all registered earthquakes (9.3 magnitude) triggered the biggest of all known tsunamis that affected several countries in South-East Asia (Indonesia – 180,000 people, Sri Lanka – 31,000–39,000 people, Thailand – over 5,000 people, India – 4,000 people, etc.), as well as Somalia (Africa). The total death toll exceeded 235,000 people.
02.04.2007	Solomon Islands archipelago. Caused by an eight-point magnitude earthquake in the southern part of the Pacific Ocean several-meter high waves reached New Guinea. Fifty-two people were killed by the tsunami.

waves after splashdown. Geological evidence (bottom sediments) of a catastrophic tsunami caused by a big falling meteorite 65 million years ago was found in the territory of Texas (USA).

In comprehending tsunami danger it may be unclear why the deadly several-meter high waves trigger catastrophic consequences, while waves of the same height triggered by a sea storm result in no casualties or damage? There are several factors that cause the devastating consequences of a tsunami.

Generally speaking, the height of a tsunami at the coast is no determinant factor. Depending on the coastal seabed configuration the phenomenon of a tsunami may occur without any wave in the usual sense of the word, but as a series of high and low tides that can also result in casualties and damage.

A storm rolls only the water's surface, while a tsunami moves the whole water body. A tsunami hits the coast with a bigger water mass which results in grave consequences.

Fig. 1.14 Giant tsunami

A tsunami's speed at the coast exceeds the speed of wind-induced waves and has higher kinetic energy. As a rule, tsunamis consist of several waves rather than a single wave. The first wave may not be the biggest and waters the land surface to decrease resistance for follow-up waves.

During a storm the roll of the sea increases gradually and people usually have time to shelter in a safe place before big waves arrive. Tsunamis come unexpectedly. Its force may increase in harbors where swell-induced waves are weak and allow one to erect houses at the seafront.

The population is usually unaware of the pending hazard as there are either no tsunami warning systems or they do not always operate. Thus, during the 2004 tsunami the sea retreated from the coast, but many locals remained on the beach out of curiosity or in an attempt to catch the fish that remained ashore. In addition, after the first wave many survivors returned home to estimate the damage or find relatives and were unaware of the follow-up waves.

1.2 Meteorological Natural Calamities

Meteorological natural calamities are conditioned by complex processes in the atmosphere and hydrosphere of the planet, including the influence of space radiation and gravitation fields of the Earth and the Moon, as well as numerous geophysical factors of various origins. Meteorological calamities often surpass earthquakes, volcanic eruptions, and tsunamis in terms of their devastating consequences.

1.2.1 Cyclones, Tornados, Squalls, Typhoons, and Hurricanes

Atmospheric whirl (tropical cyclones, tornados, squalls, typhoons, and hurricanes) is a high-speed movement of air mass. Usually cyclones originate only in low altitudes close to the equator between 5° and 20° of the Northern and Southern hemispheres. The heating of air over a major territory decreases its density and, consequently, the atmospheric pressure. Air saturation with moisture also decreases the density of the air medium although insignificantly. Emerging local air flows are spiraled by the Coriolis force. Alternating cold and hot air masses divided by fronts are sucked into the low pressure zone by the spiral. The center of the cyclone often has a so-called "storm eye," which is a small quiet area of 10–30 km in diameter. There are few clouds in the area, the wind is weak, the temperature is high, and the air pressure is very low. However, if the cyclone is above the sea, there is disorderly broken water at the bottom of the whirlpool and the waves can be as high as a 5–7-storey building. The "wall" that surrounds the "storm eye" is the most active zone of the cyclone with gale winds that may exceed a speed of 120 m/s. Thick clouds produce heavy rain, thunderstorms, and hail. The "wall" can be several dozens or hundreds of kilometers thick (Fig. 1.15).

From the zone of origin the cyclone with a diameter of 500–1,000 km and a height of 10–12 km begins to move westward, accelerates, and spirals on. It then turns to the north and moves northeastward at the final section of the path. Sometimes the cyclone reduces speed, backtracks, makes one or two circles, and then continues its movement along the previous course. It thus moves along the whole path, which can be several thousand kilometers long.

According to statistics, there are annually close to 70 cyclones which are named differently in various parts of the globe: in China and Japan they are called typhoons, in the Philippines – tropical cyclones, "willy-willy" in Australia, and hurricanes on the coast of North America.

Tropical cyclones rival earthquakes and volcanic eruptions in devastating force. They are the worst natural calamities in terms of death toll. In only 1 h such a whirlpool with a diameter of 700 km discharges energy equal to 38 thermo-nuclear medium-yield bombs. The Flora hurricane that hit the Islands of Tobago, Haiti, and Cuba in October 1963 at a speed of 70–90 m/s caused a flood on Tobago, ruined numerous settlements on Haiti, killed 5,000 people, and left 100,000 homeless.

Downpours accompany cyclones and typhoons. Precipitation of tropical cyclones looks incredible against intensive rain in temperate latitudes. For example, one hurricane poured 26 billion tons of water over Puerto Rico in 6 h. If we divide the

Fig. 1.15 Appearance of tropical cyclone

volume per unit of area, the precipitation would exceed the annual norm, for example, in Batumi (Georgia) an average of 2,700 mm (Fig. 1.16).

Hurricanes over land destroy buildings, communication and power transmission lines, damage transport communications and bridges, damage and uproot trees; over the sea they trigger immense waves 10–12 m high and more, and damage or wreck ships. Thus, for example, a typhoon near the Philippines in December 1944 hit the U.S. Third Fleet and sank three destroyers, damaged 28 warships, damaged, destroyed or washed off 147 planes on aircraft carriers and 19 hydroplanes on battleships and cruisers, and killed over 800 men.

Small-scale whirlpools often emerge in cyclone areas. In the United States where the phenomena happen most frequently they are called tornados, while in Europe – tromps. The phenomenon developing over the sea surface is often called a wind spout. It has the same name in regions of rare occurrence. Sand storms are frequent in hot deserts, for example in Sahara (Egypt). Usually cold air meets hot air at the frontline and pushes it up to the upper layers, which can also happen at the periphery of a cyclone. Sharply falling pressure in the area triggers a local whirlwind (mezzo-cyclones), which can develop into a tornado (wind spout, tromp.) Tornados are accompanied by thunderstorms, downpours, and hail. The United

Fig. 1.16 Tornado over the ocean

States is classic tornado country. For example, 1,100 devastating whirlwinds were registered in the USA in 2000.

1.2.2 Floods

Flood means a temporary flooding of land due to water rising above its ordinary level due to major precipitation, intensive snow melting, ice and snow jams on rivers, tsunami, high wind surge on the sea coast, as well as destruction of dams.

It should be noted that in the past centuries anthropogenic factors have played a growing role in the increased frequency and devastating force of floods. Forest felling (which increases the maximum surface water flow 2.5–3-fold) and irrational agriculture are the main factors (Fig. 1.17).

Long and intensive rains (monsoons) caused by tropical cyclones are the main reasons for floods in most parts of the earth. Floods on rivers in the Northern hemisphere are caused by rapid snow melting and ice jams. Foothills and mountain valleys are flooded by breakthrough glacier and rock-dammed lakes. Surge floods are frequently caused in coastal areas by strong winds or tsunamis resulting from underwater earthquakes and volcanic eruptions.

Fig. 1.17 Flood in Ukraine

Water rapidly rises and floods the adjacent territory. Flooding of residential settlements, buildings, communications, arable land, and nature reserves by quick currents results in ruinous consequences for the objects, economy, and the population. Its scope depends on the area and duration of floods and under flooding, the speed of the water flow, the density of the population and the scope of economic activities on the flooded terrain, the presence of protective dams and antiflood measures taken by rescue and emergency services.

Depending on the scope floods are classified as low (small), high, outstanding, and catastrophic. Low (small) floods cover small coastal territories and inflict inconsiderable damage. They flood less than 10% of arable land and the rhythm of life of the population remains practically unaffected. Low floods occur once in 5–10 years. High floods (recurring every 20–25 years) inflict major material and moral damage, affect 10–15% of arable land, and disrupt the normal economic and everyday way of life of the population making partial evacuation of people necessary. Outstanding floods result in mass evacuation of people and material values. The final catastrophic category inflicts major material damage, kills numerous people, and affects immense coastal areas. Such developments flood over 70% of arable land, numerous residential settlements, industrial enterprises and engineering networks, and change the way of life of populations for a long time. Catastrophic floods recur every 100–200 years.

The devastating consequences of a flood are estimated, as with other natural calamities, by direct and indirect material (financial) damage, which has a 70–30% ratio as a rule. Direct damage includes:

- Damage and destruction of residential houses, railways and roads, power transmission and communication lines, melioration systems;
- Destruction of cattle and agricultural crops;

- Destruction and damage of raw materials, fuel, food, fodder, and fertilizers;
- Expenses for temporary evacuation of the population and transportation of material values to a safe place;
- Washed-off fertile layer and soil pollution by sand, clay, or stones.

Indirect flood damage includes expenses to acquire and deliver food, clothes, medicines, construction materials and equipment, and cattle fodder to the affected areas; reduced industrial and agricultural output and economic development slowdown; deteriorating way of life of the local population; increased depreciation expenses for the maintenance of buildings and constructions, etc. A serious consequence of rarely recurring floods are major changes in river beds when the fertile layer of arable bottom lands is washed away or silted which leads to a considerable deterioration in land use and decreases crop yield.

Floods are a most devastating natural calamity which take millions of human lives and inflict damage worth billions of dollars. For example, according to world statistics, in the past 3 years floods killed over 170,000 people; over 150 million had to be temporarily evacuated, while the total damage exceeded 250 billion dollars. The growing flood damage in the United States is even more impressive, as in the beginning of the twentieth century the average annual damage comprised 100 million dollars but exceeded one billion dollars in the second half, and in some years of the past decade comprised 10 billion dollars.

1.2.3 Droughts and Fires

Droughts occur when precipitation is considerably below the normally fixed levels which results in major violations of the hydrological balance and farm land yield decrease (Fig. 1.18).

Droughts are a natural phenomenon and an integral part of the changing climate. They are a delayed process with no exact date of origin that lasts longer than the original cause. Droughts that trigger major shortages of water occur everywhere, however their characteristics differ considerably in various regions of the world. Droughts are extreme hydrological and climatic phenomena that negatively affect the social, economic, cultural, and other functions of the region during dry weather periods. Severe droughts that last for several climatic seasons running or years often trigger major economic, environmental, and social consequences. The phenomenon of drought is generally recognized as a major ecological risk.

Droughts are classified by medium. The generally recognized perspective classification divides droughts into meteorological, hydrological, agricultural, and socio-economic.

Meteorological drought occurs when annual precipitation is less than the many-year average. If a meteorological drought lasts for a long time period, it results in a hydrological drought that decreases river flow and underground water reserves.

Fig. 1.18 Hydrological drought

Agricultural drought occurs when water provision for farm land decreases to a level that negatively affects grain crops and, consequently, agricultural production in the region. The socio-economic drought is accompanied by a long period of meteorological and/or hydrological drought, which affects human life and causes famine and intensive migration.

Aerospace drought monitoring is a key factor for managing social and natural risks especially in economically weak parts of the planet.

Droughts often trigger such dangerous natural calamities as fires, which represent uncontrolled burning process resulting in a loss of human life and destruction of material values. Although man is to blame for 90% of fires, mass forest and steppe fires usually occur in dry weather (Fig. 1.19).

The main types of fires as natural calamities that cover, as a rule, large territories of several hundred and even thousand and million hectares include landscape or forest fires (creeping, crown, underground) and steppe (field) fires. For example, in 1913 forest fires in West Siberia (Russia) destroyed close to 15 million hectares of taiga. In summer of 1921 long drought and gale winds triggered fires that destroyed over 200,000 ha of precious Mari pine. In summer of 1972 peat and forest fires that

Fig. 1.19 Forest fire

broke out in the Moscow region because of a long meteorological drought affected major territories and destroyed several big peat deposits.

Forest fires are divided by blaze intensity into weak, medium, and strong, while creeping and crown fires are divided by the type of blaze into blowup and persistent. Forest creeping fires burn subsoil, topsoil and under bushes and leave tree crowns intact. Creeping fires with blaze height of 1–2 m and maximum temperature of 900°C spread at a speed of 0.3–1 m/min (for weak fires) up to 16 m/min (for strong fires). Crown forest fires usually develop from creeping fires and burn tree crowns. Blowup crown fires spread mostly from crown to crown with a top speed reaching 8–25 km/h and sometimes leave whole forest sections unaffected. Persistent crown fires burn both tree crowns and stems. The blaze travels at a speed of 5–8 km/h and affects the whole forest from topsoil to tree crowns.

Underground fires occur for natural reasons, i.e., spontaneous peat ignition or creeping or crown fires and spread along the peat layer several meters deep. They burn slowly as there is little air and travel at a speed of 0.1–0.5 m/min discharging a lot of smoke and producing burnouts. The fire may last for a long time even in winter time under snow.

Steppe (field) fires occur on open terrain with dry grass or ripe crops. They are seasonal and often occur in summer as grass (crops) ripens. They are rare in spring and absent in winter. They can spread at a speed of 20–30 km/h.

Anthropogenic causes of blazes and fires include safety rules violations, technogenic accidents and disasters, and short circuits in power transmission lines. Aerospace monitoring can provide timely information to detect the early signs of fires.

Fires destroy material values, the environment, and can have a big death toll. For example, according to the Russian Emergencies Ministry, 1,547 major fires were registered in the country in 10 months of 2006 that killed 3,522 people.

1.3 The Aftermath of a Natural Calamity

The aftermath of a natural calamity is assessed by various criteria with the main ones being the affected space, the number of casualties and scope of destruction, environmental damage (relief, biocenosis), and economic consequences for concrete countries and regions. However, there is no universal common approach to such assessments to date: in each case the level of damage and socio-economic danger of a forecasted or occurred natural calamity is assessed individually by competent agencies at the national and (or) international level.

In 1988 the Center for Research on the Epidemiology of Disasters (CRED) [9] in Brussels, Belgium, began to compile a databank and study devastating natural calamities across the globe. The databank includes only major disasters that killed or injured at least 100 people. In particular, the Center collected wide and reliable statistical information on 6,385 natural calamities in the world that took place over 35 years from 1965 to 1999.

Data analysis makes it possible to list with a major degree of reliability the following global trends related to hazards from the elements that today manifest themselves in a growing number of natural disasters (Fig. 1.20), increased social and material losses resulting from them, as well as constantly rising dependence between the protection level of the technosphere and population of the planet and the level of social and economic development of nations. According to the World Conference on Natural Disaster Reduction (Yokohama, 1994), the death toll from natural calamities in the period from 1962 to 1992 grew annually by 4.3% on average, while the number of injured increased by 8.6% and material damage by 6%.

The authors believe that the CRED statistics cited below along with Russian Emergency Ministry data are sufficient to fully illustrate modern growth trends in risks and hazards caused by various natural calamities.

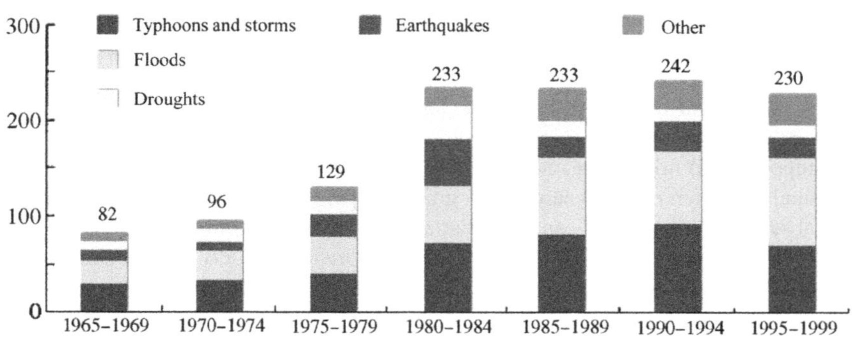

Fig. 1.20 Growth in number of natural disasters

1.3.1 Killed and Injured

In the period 1965–1999 the total death toll from the main types of natural calamities comprised (Fig. 1.21) close to four million people. In late 1980s and the first half of 1990s the number of casualties of natural calamities remained practically at the same level (52–58 thousand per year), while over 5 years - from 1995 to 1999 - it decreased to 33,000. There was no regularity in the annual breakdown of casualties by types of calamities; however the death toll from floods was evidently on the rise.

The death toll dynamic in 5-year periods shows the number of killed changed irregularly: from 25,000 to 359,000. Maximum figures were reported in 1970–1974 and 1980–1984 when droughts in Africa resulted in the death of thousands of people [2].

Surprisingly, droughts posed the main danger and took over a half of the total death toll. They are followed by cyclones (typhoons and storms) and earthquakes that accounted for 26% and 17%, respectively.

Geographically the death toll of natural calamities breaks down as follows: over a half (53%) in Africa and 37% in Asia. Droughts were the most fatal in Africa and tropical cyclones, storms, and tsunamis in Asia.

The total number of people affected by the seven main calamities in the period from 1965 to 1999 (Fig. 1.22) comprised 4.4 billion people, which testifies to a general trend of decreased protection level of the population against natural calamities. Thus, for example, the average annual number of affected people grew from 33 to 208 million, i.e., over sixfold. Floods accounted for the most growth of affected people. In 1965–1969 their share comprised 22% and this rose to 81% in 1994–1999.

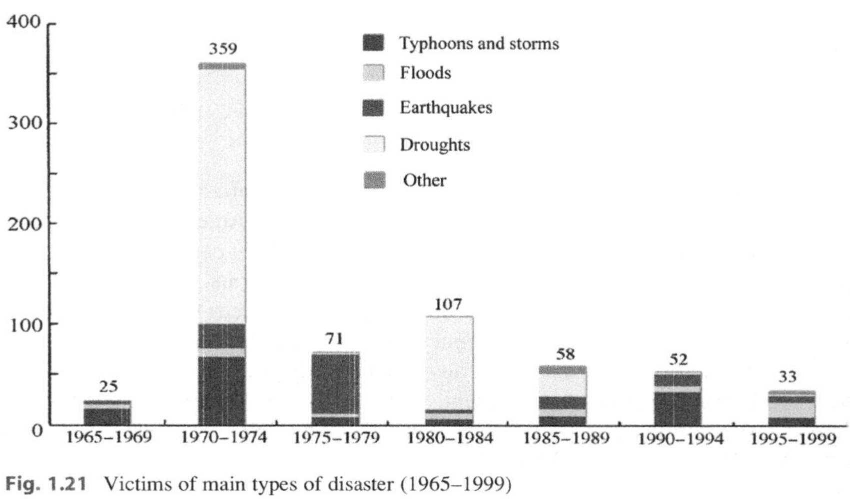

Fig. 1.21 Victims of main types of disaster (1965–1999)

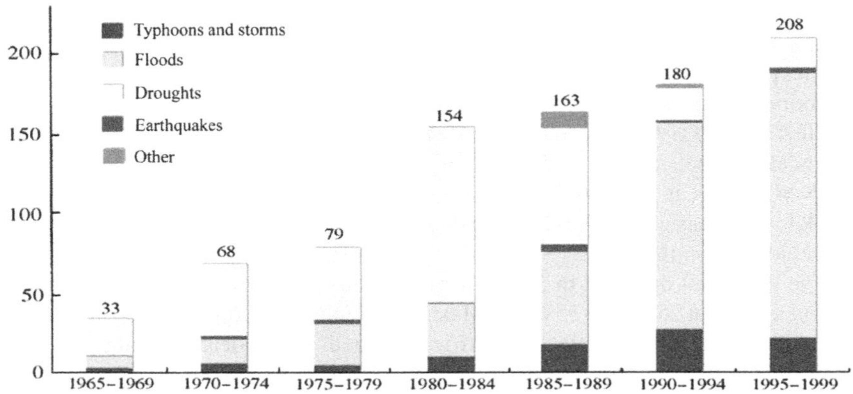

Fig. 1.22 Sufferers of main types of disaster (1965–1999)

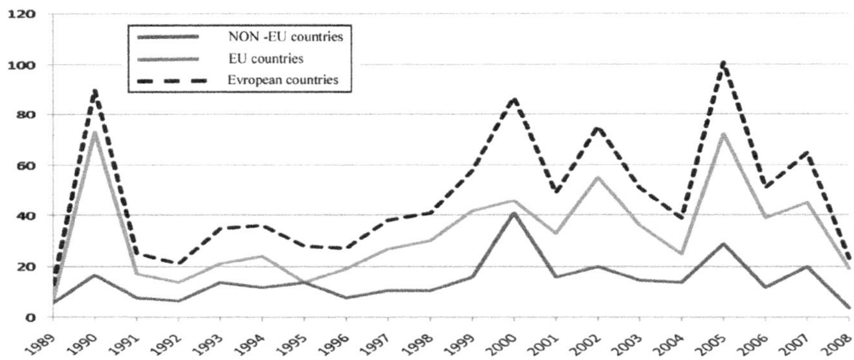

Fig. 1.23 Disasters in European countries

Asia clearly dominated all continents by the number of affected people (89%). Africa was second with a large gap (6.7%) followed by America, Europe, and Oceania that jointly accounted for 5%. In Asia close to 55% of people were affected by floods, 34% – by droughts, and 9% – by typhoons and storms. The Asian ratio of affected people to the population of the continent twice exceeds the African figure, six times the American level, and 43 times the European number.

The vulnerability of various countries to natural disasters is closely linked to their social and economic development. Economies were divided by the World Bank (WB) according to 2008 Gross National Income (GNI) per capita calculated using the WB Atlas Method. According to 1992 data, the population in low, middle, and high-income economies comprised 3,127, 1,401, and 817 million people respectively, while their aggregate GNI equaled 1,097, 3,474, and 16,920 billion US dollars correspondingly [10] (Fig. 1.23) (Table 1.3).

Table 1.3 Natural disasters and economic state of suffering countries (1965–1999)

Groups of countries according to GNI per person	Low	Medium	High
Population (million)	3,127	1,401	817
Square (10,000 km^2)	3,883	4,030	3,168
Density of population	80.5	34.3	25.8
Average year's GNP (billion US$)	1,097	3,474	16,920
Number of destructive disasters (1965–1992)	1,524	1,714	1,341
Casualties (thousands) (1965–1992)	3,166	408	33
Suffering (million) (1965–1992)	2,775	216	16
Direct economic losses, million US$ (1965–1992)	67,906	98,841	171,253
Average ratio economic losses to GNI, %	0.22	0.10	0.04

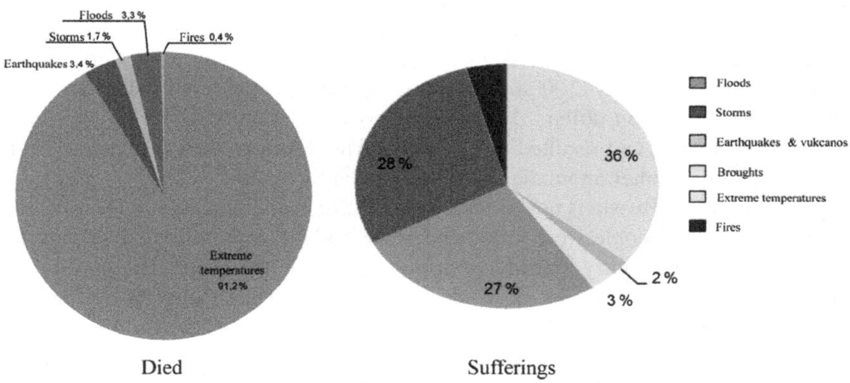

Fig. 1.24 Aftermath of natural disasters in European countries from 1989 to 2008

Vulnerability analysis of the three groups to natural calamities showed that low-income economies incurred the most social risk (death and injuries). Thus, the first group of countries (low-income) with a population comprising 58% of the total population of the planet accounted for 88% of the killed and 92% of injured and affected people by natural calamities in the period 1965–1992. The total number of casualties and affected people is 5.8 times higher in low-income economies than in middle-income countries and 45.2 times higher than in high-income economies.

Absolute economic damage figures are higher for developed nations because of the major concentration of capital and wide infrastructure. However, the ratio of direct losses to the Gross National Income shows that low-income economies sustain most damage in relative terms (5.5 times higher than in high-income economies). Thus, both economic and social damage from natural disasters is the greatest for poor countries.

In the past 20 years European nations also sustained considerable losses from climatic natural calamities (Fig. 1.24). In the period from 1989 to 2008 there were totally 953 calamities in Europe. They killed 88,671 people and affected over 29 million (the breakdown of the killed and affected people by types of disasters is

Table 1.4 Dead and suffering around the world

	2008 year	2000–2007 year's level
Number of disasters	354	397
Number of suffering countries	120	118
Casualties suffering	235,264	66,813
Casualties suffering (million)	214	231
Economic losses(US$) (billion)	190	81.8

provided in Fig. 1.22). Among the European countries Spain accounted for the highest death toll.

Scientists forecast floods and storms in Europe are likely to become more frequent and stronger in the future. The 2009–2010 developments confirm the trend. For example, the Klaus storm that hit Southern Europe in January 2009 and had a death toll of 25 lives. The earthquake in L'Aquila (Italy) in April of the same year killed nearly 300 and affected 49,000 people while damage was estimated at 2.5 billion dollars. June floods in Central Europe were the strongest in the past 50 years and resulted in 12 deaths. The developments in the winter of 2010 related to weather anomalies confirm the alarming trend.

The situation in Russia is not better. According to available data for 1965–1999, over 4.5 thousand people were killed and some 540,000 were affected by natural calamities in our country (USSR). Earthquakes pose the main danger for the Russians as only two of the disasters on Shikotan (1994) and in Neftegorsk (1995) killed close to 2,000 people. Floods, landslides, avalanches, mudslides, hurricanes and whirlwinds also caused a considerable death toll. Floods pose the main danger for Russian municipalities (746 towns affected) followed by earthquakes (103), whirlwinds (500), avalanches (5), mudslides (9), and tsunami (9). Russia registered 120 natural calamities in 1989–2008.

Statistics of the recent 2 years confirm the negative trend of increased danger from natural disasters. For example, only in 2008 (Table 1.4) they killed over 235,000 and affected 214 million people across the Globe. Although the databank registered 354 natural disasters, which is less than the average annual figure for 2000–2007 (397), the death toll exceeded the figure of the period threefold (66,813 people). That happened because of two major disasters: the Nargis cyclone (Myanmar) that killed 138,366 people and the earthquake in Sichuan province in China that took a toll of 87,476 lives. Asia continues to be the most affected continent by natural disasters, as nine out of ten countries with the highest death toll from natural calamities are located there. If China, the Philippines, the United States, and Indonesia have the highest number of natural calamities, Angola, Tajikistan, Djibouti, and Somalia are in the lead in terms of the number of casualties per 100,000 residents (Fig. 1.25).

As natural calamities are accompanied by growing number of technogenic catastrophes, in the future many countries may be unable to recoup losses from frequent emergencies with increased devastating force.

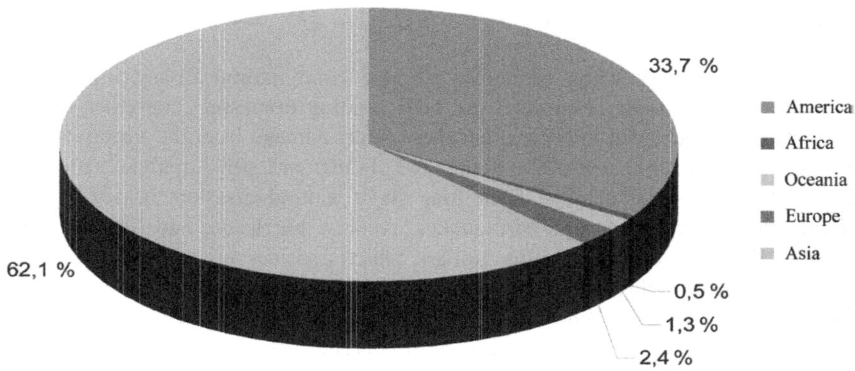

Fig. 1.25 Percentage of disasters by continent

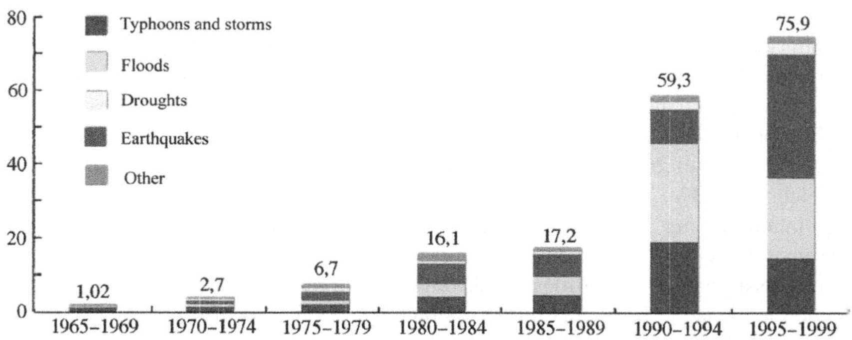

Fig. 1.26 Economic damage from natural disasters (1965–1999)

1.3.2 Economic Damage

Economic damage from natural disasters is also rapidly increasing (Fig. 1.26). In the 35-year period under review it grew 74 times (without US dollar inflation adjustment). Thus, the annual economic damage in the 1960s comprised slightly above one billion dollars, in the 1970s –4.7, in the 1980s –16.6, and in the 1990s it reached nearly 76 billion dollars. It thus totaled 895 billion dollars, of which 676 billion were accounted for in the past decade of the twentieth century (1990–1999).

It should be noted the estimates refer to seven basic types of natural disasters and do not account for the current decade of the new century. Taking into account other circumstances and events the total damage will comprise the above-mentioned figure of 1.5 trillion dollars.

Typhoons, storms, floods, and earthquakes inflicted the most damage. The calamities account for 95% of the damage in the period under review. Asia sustained nearly a half of it (46%) followed by America (26%) and Europe (23%).

In absolute figures the damage was worth 412, 234, and 210 billion dollars, respectively.

There were cases when economic damage from natural disasters in some countries exceeded the amount of the GNI sending domestic economies into a critical state. It occurred in Nicaragua where direct damage from the earthquake in Managua (1972) comprised 209% of the GNI of the poor Latin American nation. In the United States the damage from four major natural disasters in 1989–1994 (Loma Prieta and Northridge earthquakes, Andrew hurricane, and floods in the Middle West) comprised 88 billion dollars, which considerably affected the economy of the most highly industrialized nation of the world. Today many countries, such as Japan, have to spend up to 5% of their annual budget (0.8% of GDP) or 23–25 billion dollars a year to counter natural disasters. In some years the expenses in Japan upped to 8% of the annual budget. In China annual damage from natural disasters comprises an average of 3–6% of GDP. In the past decade it grew from 6.3 (1989) to 36 (1998) billion dollars [3].

Cities with maximum population and technogenic infrastructure concentration sustain the biggest material, social, and economic damage. Experts estimate the total annual social and economic damage from the 19 most dangerous processes on Russian urban territories at close to 9.7–11.7 billion rubles (in 1991 prices).

European countries sustained damage from natural disasters in 1989–2008 (Fig. 1.27) worth 269 billion dollars. Most losses were inflicted by floods (40%) and storms (33%). The significant damage figures are explained by the high population density and by the costs of the infrastructure affected by floods and storms in Europe, which is much more expensive than in Asia or Africa.

In Russia natural calamities break down as follows according to the inflicted economic damage: sheet and ravine erosion (close to 24% of losses), underflooding (14%), floods and marginal erosion (13% each), landslides and devolution (11%), and earthquakes (8%) [10].

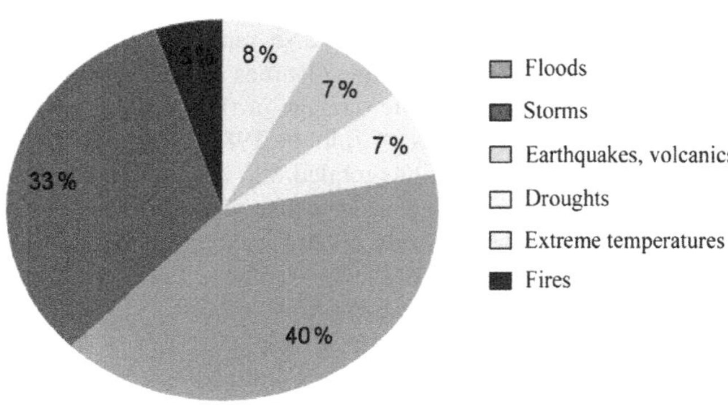

Fig. 1.27 Economic damage from natural disasters in European countries (1989–2008)

1.4 Aerospace Predictions of Geological Natural Calamities

To predict geological natural calamities means to determine their place, time, and force (magnitude). Seismologists monitor changes in various earth characteristics hoping to establish correlation between the changes and geological natural calamity occurrence. Those characteristics of the earth which regularly change before earthquakes and volcanic eruptions are called precursors, while deviations from normal parameters – anomalies [11]. According to Russian standard GOST 22.0.03-97, a precursor of a seismic event is "a sign that manifests itself as foreshocks preceding the main tremor, deformation of the surface of the earth, changes in parameters of geophysical fields, underground water content and regime, the state and characteristics of substance in the zone of potential earthquake focus."

Rich experience in precursor monitoring in various countries gives ground to suggest they really manifested themselves in several cases even at major epicenter distances. It is also suggested that precursors can emerge and disappear several times before a seismic event and can be conditionally divided into long, mid, and short-term ones [12]. Predictions are also divided into long, mid, and short-term.

Long-term predictions related to seismic zoning cause little debate. The predictions are based on monitoring the emergence of seismic standstill zones, changes in lithosphere strain and its seismic transparency. Survey of such processes can provide information about an upcoming earthquake in a period from several months to several years. For example, in mid 1960s, the research of the Pacific seismic expedition allowed designing a method of long-term seismic forecasting of strong (magnitude $M > 7.7$) Kuril and Kamchatka earthquakes at depths less than 100 km based on the concepts of "seismic holes" and "seismic cycle," which successfully justified itself in the early 1970s. The prediction method has been recognized practically in all countries of the world, including Japan and the United States [13].

Mid-term predictions provide a possibility to issue a warning weeks or months before a seismic event and are practically useful. The predictions offer a scenario of destructive processes based on current monitoring of geophysical fields, changes in the inclination of the earth's surface, regular surveys of the yield and chemical composition of water sources and deep water, oil, and gas wells. It uses formalized criteria to assess the statistical significance both of each precursor and all of them in a complex. Mostly empirical connections established between parameters of precursors and subsequent earthquakes are used to suggest the place and magnitude of the expected seismic event.

Short-term predictions are issued several hours or days before the event. They are the most sought after predictions, as they are used in decision-making related to targeting urgent measures aimed at decreasing the potential damage from an earthquake. But they are also the hardest to make. The strictest requirements are imposed regarding the reliability of short-term predictions because of their major public significance. Scientists and officials have to be especially responsible while issuing a "seismic alert." To comprehend how complicated the situation is we shall recall a sensational prediction of Chinese seismologists. In 1975 they numerously sounded alert in the area of Haichen and even evacuated the population. Several alerts were false, but did not result in major economic losses for the agrarian area. However, one evacuation was carried out 2 h before a nine-point earthquake and saved thousands of lives. A year later the scientists registered precursors of a pending quake, but refused to issue an alert for the city of Tangshan with a population of 1.3 million people and a developed mining industry. The Great Tangshan quake killed hundreds of thousands of people [14].

After the Tangshan quake and failed multi-year experiment to predict an earthquake in Parkfield (California, USA) in mid-1980s, a period of skepticism reigned in prediction research regarding the possibility to resolve the problem. Thus, generally recognized theory capable of explaining certain tectonic precursors was based on laboratory research of rock samples at very high pressure. Known as the "dilatancy theory," it was first developed in 1960s by W. Brace of the Massachusetts Institute of Technology (USA) and A.M. Nour, Stanford University. Dilatancy means increased volume of rock under deformation. Earth crust displacements increase strain in the rock and create microscopic cracks that change physical qualities of the rock. Seismic wave velocity in the lithosphere rock decreases as water flows into the cracks increasing the volume and changing electric resistance. Water spreads along the whole expanding zone increasing porous pressure in cracks and decreasing rock strength. During a quake or volcanic eruption the accumulated strain releases and drives water out of pores, while many previous rock qualities are restored [15].

To predict geological events on the basis of the behavior of the earth's crust, research models have been designed and used that are theoretically based on mechanical and physical provisions for rock destruction. The models pay different attention to the scope of the considered geological faults – cracks, their location in space, additional physical and mathematical factors that affect crack formation. The most advanced are the following models: the avalanche-type **Unsteady Crack Formation Model (UCF)** developed by experts of the Institute of the Physics of the Earth (G.A. Sobolev and others). It states that various stages of crack formation accompanied by changing formation velocity in the focus area and outside it inevitably result in changes of the physical qualities of the medium. It is reflected in varying seismic regimes, i.e., in changing numbers of weak quakes, their size and location [16]. Experiments confirmed the basic provisions of the UCF model. In particular, it was proved that changes in the elastic deformation field and seismic regime can be considered as long-term precursors of earthquakes and volcanic eruptions. I.P. Dobrovolsky [16] offered a theory to explain the nature of long-term precursors on a solidification basis that says the final stage of origination of a geological natural calamity is explained by the same avalanche-type unsteady crack formation. However, the model has so far failed to expose any reliable short-term precursors of tectonic events. American scientists developed the **Dilatancy-Diffusion Model (DD)**. It explains the emergence of precursors by water arrival in the focus area of a future seismic event after a sharp increase in tectonic strain triggers massive microcrack formation there. The model has acquired quantity assessments of late. The **Creep model**, which is based on the phenomenon of gradually accelerating motion of edges of an existing fault, is of interest as earthquakes and volcanic eruptions mostly occur in faults or close to them. Although the model is attractive, it encounters several difficulties in explaining the nature of short-term earthquake precursors. Firstly, the spread of the wide areas of such precursors, as well as major areas of their origination remain unclear. Secondly, even in the San Andreas Fault in California, where the model is most suitable, it failed to register short-term precursors before most of the quakes.

To properly comprehend seismically hazardous processes and the physics of precursor origin we shall review in detail the analysis of the nature of tectonic faults that cause earthquakes and volcanic eruptions. Tectonic faults are areas of strain concentration, canalization for fluids, aerosols and gases, nonuniform rock magnetism and electric conductivity, high electric potential emergence etc., and the characteristics change with time. The tectonic fault uses the parameters to provide information on the mantle and its state that gives grounds to expect a seismic cataclysm.

Despite very different assessments of the physical qualities of the earth's mantle and mostly of its viscosity, heat convection can be considered highly probable there and will determine the basic features of lithosphere dynamics. Therefore, spatial structure and the scope of convection are very important for geodynamics. As viscosity of the upper mantle can be considerably less than of its lower part, several researchers suggested that convection is taking place only in the upper mantle. Some scientists also supposed that convection will develop independently in the upper and lower mantle. Others said there were no long-term grounds for such suggestions and convection should cover the whole of the mantle [17].

The mantle is vertically rather heterogeneous. Therefore, the main convectional circulation will be three-dimensional and create polygonal cells. At the same time the circulation can co-exist with small-scale secondary flows in the near-surface boundary layer. Such currents may have the form of bars flowing from the external edge of the cell (where the main flow is likely going upwards) to the center [18]. Conversely, several experiments [17] with viscous liquids showed that sharply nonstationary upward flows (thermals) are typical for big time intervals. And finally, if viscosity differs considerably within the boundary level it is impossible to fully rule out the possibility of three-dimensional small-scope secondary currents that resemble the basic cell convection.

Thus, the main (large-scale) convection cells of the mantle will have a radius that exceeds its thickness (some 3,000 km). If the boundary between two adjacent cells goes under the oceanic crust, it has increased heat flow created by the upward flow from the mantle (increased heat flows are observed along oceanic ridges).

Proceeding from observed velocities of lithosphere motion we can suggest that the main convection current in the upper mantle has horizontal velocity of several centimeters a year. The boundary layer thickness can then be estimated at close to 100 km. Small-scale near-surface flows, if any, shall in that case be hundreds of kilometers in size. Figure 1.28 (right) shows quasi bidimensional bars observed in a laboratory experiment [18].

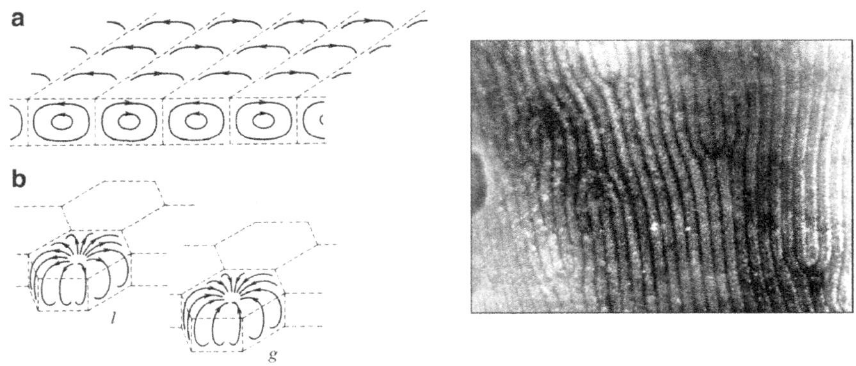

Fig. 1.28 The scheme of convectional mantle structures: (**a**) bidimensional bars (right – quasi bidimensional bars in laboratory cuvette); (**b**) hexagon cells that differ in circulation direction

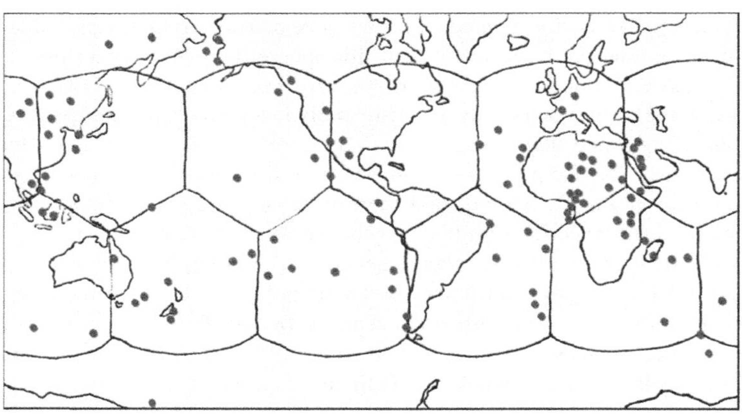

Fig. 1.29 Three-dimensional mantle cells and seismically hazardous areas of the earth

If the physical characteristics of the substance in the boundary layer are relatively homogenous, the secondary flows should be bar-like and can be responsible for the emergence of transform faults. Figure 1.29 shows the location of three-dimensional cells (from the core boundary to the earth's surface) determined for suggested homogeneity of the mantle [17]. Faults of the seismically hazardous Taman Peninsula (where the axis of the Caucasian ridge sharply turns to the west and the Black Sea) demonstrate a line of nearly parallel elevations (anticline folds) running at a distance of 20–30 km from each other. Such a small distance makes one think about the reasons behind the folds origin. Most likely it is the work of the above-mentioned near-surface convection bars. The activity of the formations testifies to a connection between the folds and geodynamics. Chains of underwater Black Sea mud volcanoes run along the folds.

Regardless of the reasons for the surface displacement (be it changing convection movement in the depths of the earth or release of locally accumulated strain as occurs during landfalls, etc.) destruction is always preceded by the emergence of the above-mentioned numerous microcracks in the rock and increased degassing flow. To better understand the initial stage of future seismic development a series of experiments was carried out in laboratories to destroy the rock under pressure, which confirmed that particles of micron and submicron size appear in the process [19]. In natural conditions the products of such numerous micro destructions move up through a subjacent system of cracks by increased degassing flow and manifest themselves as aerosol flux.

1.4.1 Earthquake Precursors and Their Detection

The conference on earthquake prediction project assessment held by the Royal Astronomical Society jointly with the unified Geophysical Association in London on November 7–8, 1996 considered the principle issue of earthquake prediction project efficiency from its, so to say, philosophical aspect. The conference was deeply pessimistic not only about the general state of the problem, but also regarding plans for the foreseeable future. Participants actually repeated the reasons voiced by Charles Francis Richter in his time to prove the impossibility of earthquake prediction [3]. However, despite the categorically negative tone of the discussion it did not and could not cite any proof that predictions are impossible in principle. All cited reasons boiled down to the difficulty and poor research of the issue. However, no lack of knowledge can prove impossibility [3].

Today there are hundreds of different earthquake precursors across the globe that can be united into two groups. The first is the most numerous and well-studied group of geophysical precursors related to the regular behavior of various geophysical fields at the stage of earthquake origination. Precursors of the group actually cover the whole prediction range – from long-term to short-term (everyday) and are divided into seismic, hydro-geo-dynamic, deformation, geochemical, atmospheric, thermal, gravitational, and electromagnetic. The second group comprises short-term precursors related to unusual behavior of biological objects before an earthquake.

Requirements in relation to observed phenomena that can be considered as geographical precursors of earthquakes and can be registered both by ground-survey instruments and aerospace monitoring means and further used to predict natural calamities are as follows:

- Clear physical meaning of prediction precursor;
- Physical substantiation of the connection of each precursor with the earthquake origination process;
- Provision of every precursor with observation data both in time (long-term lines of precursor parameters) and in space, i.e., mapping possibility;
- Availability of a formalized procedure to expose precursor anomalies based on a model of their behavior during earthquake origination;

- Possibility to obtain assessments of retrospective statistical characteristics of each precursor: probability of successful prediction (detection), probability of false alerts, prediction efficiency (information content), etc.

Seismologists currently study traditional lithosphere earthquake precursors that are united into three groups:

1. *Accompanying seismic activity.* Dislocation and the number of small earthquakes of various magnitudes can serve as a major indicator of a pending strong quake that is often preceded by a string of earth shocks. Their detection and calculation demand numerous seismographers and corresponding data-processing devices [11].
2. *Earth crust displacement.* Geophysical networks can use triangular ground grid and space satellite observation to detect large-scale earth deformations (changes of the form). A very precise photo survey with the use of a coherent source of light is necessary for this. Follow-up surveys demand a great deal of time and money and therefore are sometimes carried out with several-year breaks, which does not allow one to timely detect surface changes and dateline them. Nevertheless, such changes are a major indicator of earth crust deformations [11]. Vertical movements of the earth's surface can be measured by a precise ground leveling survey or mareographs at sea. As mareographs are installed on the ground to register sea level, they detect long-term changes of the average water level that can be interpreted as land alleviation or depression [11]. To measure the degree of inclination of the earth's surface special devices are used that are usually installed close to faults at a depth of 1–2 m and their readings show changes that occurred not long before a weak earthquake originated [11].
3. *Rock deformation.* To measure rock deformation wells are drilled to carry special devices that register the size of relative displacement of two points. Then deformation is calculated by dividing the relative displacement value by the distance between the two points. The devices are sensitive enough to measure surface deformation caused by Earth tides that originate due to gravitational attraction between the Moon and the Sun. Earth tides are earth crust motions that resemble sea tides and change the land height with an amplitude of 20 cm. Special devices are used to measure creep or slow relative motion of tectonic fault wings [11].
4. *Seismic waves.* Seismic wave velocity depends on the rock through which they travel. Changing velocity of dilatational waves – initial decrease (up to 10%) and backtracking to normal level just before the quake – is explained by changing qualities of the rock that accumulates inner strain [11].
5. *Geomagnetism.* The magnetic field of the earth can experience local changes due to rock deformation and earth crust displacement. Such changes were observed before earthquakes in most of the regions where magnetometers were installed [11] – special devices to measure small magnetic field variations.
6. *Terrestrial electricity.* Earthquakes can trigger changes in electric resistance of the rock. The necessary measurements are carried out with the help of electrodes placed into the soil at a distance of several kilometers from each

other. Experiments held by seismologists of the U.S. Geological Service detected a certain correlation of the parameter with weak earthquakes [11].

7. *Radon content in underground waters*. Radon is a radioactive gas permanently discharged from the depths of the earth into the atmosphere. Changing radon content before an earthquake was first detected by seismologists in the Soviet Union where decade-long growth of radon amount diluted in deep well waters was replaced by a sharp drop on the eve of the Tashkent earthquake in 1966 (magnitude 5.3) [11].
8. *Water level in wells and shafts and changing chemical composition of water and gases*. The level of ground waters often rises before earthquakes as occurred in Haicheng in China likely because of changing strain of the rock [11]. All geodynamic and active zones of the earth are characterized by the presence of water and gases with very variable and temporally unsteady chemical and isotope composition, which creates conditions for the incursion of water and gases with changed chemical composition into water-bearing layers.
9. *Changes in temperature regime of near-surface layers* (*underlying surface*). An invisible thin heat layer that is several centimeters thick is created close to the earth's surface by inner excitation and is a type of thermal blanket for our planet. Numerous facts have been accumulated to date that confirm changes in the temperature regime of near-surface layers during seismic activity periods [20].

The above-mentioned list covers anomalous developments in the upper lithosphere before earthquakes that are registered by sensors directly at the point of registration. The necessary information can be collected with the use of aerospace devices and transmitted to ground centers for processing. Lithosphere earthquake precursors also include mounting seismic activity and the emergence of seismic standstill zones, changing strain in lithosphere substance, changes in its seismic transparency, and other similar phenomena deep underground. Internal deformation processes in the lithosphere that cover an area with an estimated radius R = exp(M) also cause anomalies of geophysical fields. For strong M ~ 5 the quake's R is ~500 km. However, because of the heterogeneous earth crust the mentioned anomalies cannot be detected at each point of the deformation zone, but only in the so-called "sensitive areas" where during the origination period of hazardous tectonic processes underground thunderstorms are possible, as well as ignition of underground gas, excitation of the underground radio-magnetic background, failures of radio navigation systems, etc.

Precursors of upcoming seismic events can manifest themselves over a radius of hundreds and even thousands of kilometers from the supposed epicenter. Future earthquake zone estimates (**ρ**) are provided in reports [21–22], which offer approximate (but rather accurate) solution of the problem: $\rho = 10^{0.43M}$, where M is the magnitude of the predicted tremor. According to the formula the magnitude will comprise 4.65 if the zone radius is 100 km. It is easy to calculate that origination processes of an earthquake with a 7–8 magnitude will cover at least 5,000 km. It is extremely difficult to create ground sensor network that would, on one hand, cover such immense territories and, on the other, provide sufficient spatial and temporal resolution of the received information. The task can be accomplished only by engaging information obtained through remote sounding of the earth's surface and atmosphere from space.

Research has intensified of late of those anomalies in the atmosphere and lithosphere that can be potentially considered as precursors of an upcoming earthquake and registered by airborne and space-based means of remote sounding. Such anomalies include, in particular:

- Sharp changes in the concentration of the electronic component in the F_2 layer of the ionosphere, as well as the appearance of large-scale homogeneities in it;
- Ultra low frequency, super low frequency, and high frequency electromagnetic oscillations;
- Anomalous changes in the quasi-constant electric field and magnetic induction vector;
- Anomalies in the composition, concentration, flow velocity, and temperatures of the ionosphere plasma;
- Intensive airglows at frequencies corresponding to the oscillation bands of atomic oxygen and hydroxyl;
- Radon and metallized aerosol emission in the near-earth atmosphere;
- 3–5 increase in the surface temperature of the earth in the area of future earthquake focuses;
- Accumulations of clouds above active earth crust faults before earthquakes;
- Proton and high-energy electron precipitation in the upper ionosphere.

Major earthquake precursors in the lithosphere, atmosphere, and ionosphere of the earth with the advance time of their emergence before seismic events of five-point and more magnitude are provided in Table 1.5. The data collected and processed during many-year observations can be used to design and create special aerospace monitoring equipment for efficient predictions of natural calamities. However, the main task in researching any earthquake precursor and using it in empirical prediction is to determine the connection between the time of its manifestation (ΔT, time interval from the precursor to the seismic event), the force of the quake M, and the epicentral distance R. Using the connection and several monitoring stations it is possible to determine the force of the upcoming quake and epicenter coordinates. Let's consider one atmospheric precursor as an example, which is anomalous strain of the electric field of the atmosphere (***E***) that is dependent on two phenomena: increased content in near-earth air of its natural ionizer, radon $^{222}R_n$, and the emergence at the same time of a negative voluminous electric charge (*electronic reverse effect*), as well as a combination of one-way and reciprocating motions of fault flanks accompanied by mechanical and electrical energy transformations and the emergence in the fault zone of an aggregate electric moment that differs from zero.

Both phenomena emerge due to increased deformation of surface layers of the earth's crust in the zone of earthquake origination. Research exposed with an acceptable degree of reliability a common ΔT interval of 1–30 h for both phenomena related to (***E***) when the second-type anomaly initially emerges in the epicenter followed by the first-type anomaly, which are then manifested also in the ionosphere.

Table 1.5 Earthquake precursors and their manifestations

	Manifestation time			
Earthquake precursors	Months (1–12)	Days (1–30)	Hours (1–24)	Minutes (1–60)
Lithosphere				
Series of accompanying weak earthquakes	3-1	30-1		
Electrotelluric field excitation	3-1	30-5		
Increased amplitude and character of surface convulsion		30-5		
Geomagnetic field excitation		30-5		
Geomagnetic pulsations at frequencies of 0.02–0.1 Hz			4	10
Increased geomagnetic field amplitude registered by ground sensors		5	5	
Increased radon concentration	3	30-1	23-1	
Increased concentration of submicron aerosol		3-1	30-24	40
Low frequency noise in the band of 3–8 kHz (ELF/VLF)		2	6-2	
Changes in the level and chemical composition of underground waters		3-1		
Temperature changes in near-earth and surface layers		30-1		
Changes in local gravitation field	6	30-1		
Seismic activity	3	30-3		
Earth crust motion, deformation (tidal, etc.)	120-12			
Atmosphere				
Increased radon concentration	3	30-1	23-1	
Increased concentration of submicron aerosol	1.5-1	30-2	5-1	
Electric potential excitation		30-10		
Airglow, optical emissions		5-2	23-1	59-1
Electric field anomalies			30-1	
Appearance of aerosol clouds		3-1	30-1	40
Temperature anomalies in the near-earth layer		30-1		
Changes in electric conductivity of the near-earth atmosphere			30-1	
Changes in the chemical composition of atmospheric gases	3	30-1	23-1	
Changes in the thunderstorm situation		5-3		
Ionosphere				
Geomagnetic field excitation		30-10		
Changes in the lower ionosphere structure		30-10		
Increased intensity of electromagnetic field emission in upper ionosphere			2	10-20
Geomagnetic pulsations at frequencies of 0.02–0.1 Hz			2	10-30
Interaction of geomagnetic emission with plasma particles			12-2	

(continued)

Table 1.5 (continued)

Earthquake precursors	Manifestation time			
	Months (1–12)	Days (1–30)	Hours (1–24)	Minutes (1–60)
Lower ionosphere modification		5		
Deformation of lower ionosphere edge		30-10		
Increased flow of energy particles from magnetosphere to upper ionosphere		5	8-1	59-0
Increased inhomogeneity in F_2 layer		3-2		
Changes in critical frequencies and density of E and F layers of ionosphere			10	59-0
Temporary variations of the full electronic content of the ionosphere		3-1	23-2	

It will be demonstrated below that increased (25–100%) concentrations of aerosol and radon can serve as precursors of an upcoming quake. If you add seismic and acoustic oscillations, hydrogen discharges, excitations of electric field, several-fold increased low-frequency electric conductivity of the atmospheric air, decreased strain of the atmospheric electric field you can predict an earthquake 2–5 h beforehand. Two days before an earthquake the radiation background increases (related to radon) in the near-earth layer of the atmosphere and ionosphere, while several days before its aerosol clouds appear.

The appearance of cracks in earth crust rock during the first manifestations of seismic activity results in the discharge of accumulated radon and consequently in increased intensity of radon flows, which can be considered as a precursor of an upcoming earthquake, although such an increase does not always predict it. Nevertheless, the relative simplicity of radon flow registration promoted a popular idea of using it in prediction schemes. It is true that during powerful seismic events increased radon flows signaled an upcoming cataclysm, while after the earthquake the flows weakened due to intensive degassing that washed radon out (Fig. 1.30). A case was registered when radon flow intensity increase before an earthquake reached some 700 km in length along the fault line. It has also been noted that intensity is subject to changing atmospheric characteristics – temperature, pressure, humidity, etc. [23].

As tectonic processes create an uninterrupted source of gas and aerosol discharge into the atmosphere, radon concentration curves also correlate with the night-time critical frequency of the ionosphere (when it is free of the impact of the sun's emission) because aerosols change the electric conductivity of the medium and affect the state of atmospheric electricity more than radon. The impact of the mentioned components on electric characteristics of the atmosphere has been calculated [24]. The effect of enhanced aerosol discharge in the process of earth degassing seems to be an especially promising earthquake precursor also because the metallic aerosol component can variously manifest itself by increasing electric conductivity of the atmosphere, reflecting the impulses of electromagnetic waves, etc. The so-called "tectonic cloudiness" is also known, which is composed of

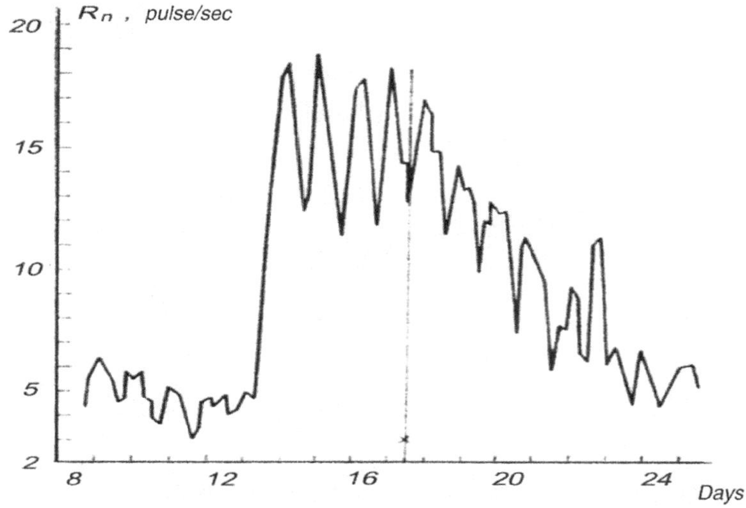

Fig. 1.30 Radon concentration in thermal mineral waters of the Tashkent artesian basin from 1956 to 1967 in the period of the Gazli earthquake on May 17, 1976 (M = 7.3)

submicron aerosols discharged as a result of active tectonic zone degassing, which manifests itself 4 days before an earthquake in two clearly distinctive maximums.

To research in detail the relation between ground and ionospheric factors in the area of active tectonic faults in the territory of Central Asia temporary changes in radon concentration were matched for half a year with daytime and night-time critical ionospheric frequency [24]. As a result it was established that curves for radon and night-time critical ionospheric frequency (Fig. 1.31) clearly correlate with each other and radon concentration changes seem to trigger changes in the ionosphere (night-time state of the ionosphere depends on the ground ionizing factor to a greater extent than in daytime because it is not affected by the sun's emission).

Figure 1.32 shows a correlation between radon decay and relative air humidity (current 5-day average) in the area of Acapulco during earthquake origination with a magnitude M = 7.4 in Copala (Mexico) on September 14, 1995. It is clearly seen that radon concentration minimums coincide with relative humidity maximums and vice versa. Naturally, the ionizing process is not the determinant one, however the modulation of atmospheric parameters related to ionization is clearly visible. It should also be noted that the origination process (in the form of radon concentration increase) lasts for several months while the earthquake occurs when the radon anomaly declines. As radon is six times heavier than air, it trails along the surface. Therefore, infrared measurements from satellites were sometimes erroneously interpreted as increased temperature of the earth's surface. Unfortunately, the unconstructive position of most seismologists in the world (except for Turkey), who erroneously claimed that radon had no qualities to serve as an earthquake

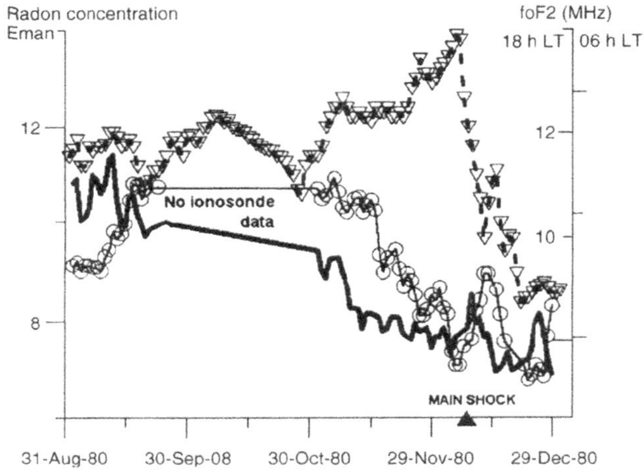

Fig. 1.31 Changes in radon concentration (*full line*) together with daytime (*triangles*) and night-time (*circles*) critical ionospheric frequency. The *arrow* on the abscissa shows the moment of earthquake initiation near Tashkent

precursor, resulted at that time in the cessation of practical radon monitoring in most countries.

Today various seismic regions of the world often register in the near-earth layer of the atmosphere anomalous changes in the electric field emerging in the first dozens of hours and in the first hours before a quake. Together with other precursors of the same time range they can be used to study processes occurring during the final stage of earthquake origination and for their short-term prediction. For this it is necessary to know the specifics of anomaly manifestations, which have not been studied fully so far.

Geologists have long been aware of the connection between cloudiness and tectonic processes in the earth's crust. Cloud monitoring data can be used to determine the network of faults in the earth's crust, as well as to observe mounting tectonic processes before an earthquake. Such monitoring is very promising, as there is a wide network of meteorological stations and cloud monitoring is also conducted with the help of numerous simultaneously operating spacecraft. However, it is important to know cloud behavior over faults in detail. Photos of cloud anomalies (Fig. 1.33) over the Haiti quake epicenter made a day before the tragic event are very informative.

"Tectonic clouds" mostly emerge due to the injection of submicron aerosols during degassing in tectonic activity zones as aerosols become the centers for water steam condensation. Consequently, the cloud shape often corresponds to that of the tectonic fault. Clouds have specific form with sharp round edges and several other features. Maximum clouds are observed 66–42 h and 30–24 h before an earthquake. Approximately 3 days before an earthquake the temperature sharply falls and reaches the maximum of its structural function however it is no precursor, but

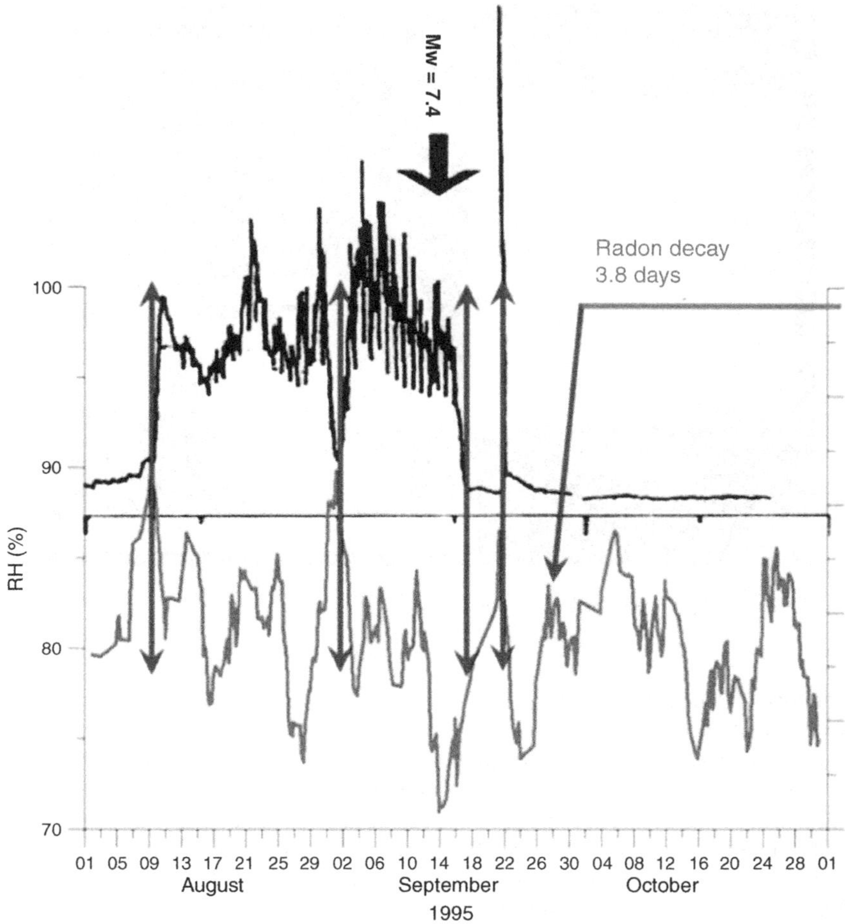

Fig. 1.32 Radon decay variations and relative air humidity

only an indication of heat discharge before an earthquake. Thunderstorm clouds as precursors appear 3–5 days before an earthquake. However, no thunderstorm has been registered during the main shock. At the same time seismic atmospheric excitation was reported 6–5 h before it.

Several researchers analyzed space images of the clouds over the island of Taiwan where organized accumulations are visible above deep faults and ocean trenches. The same effects were observed over New Zealand adjacent to faults. It is to be noted that specific thermodynamic conditions and increased humidity are necessary for tectonic cloudiness to originate. Consequently, statistically reliable predictions can be achieved only in the case of permanent and long-term

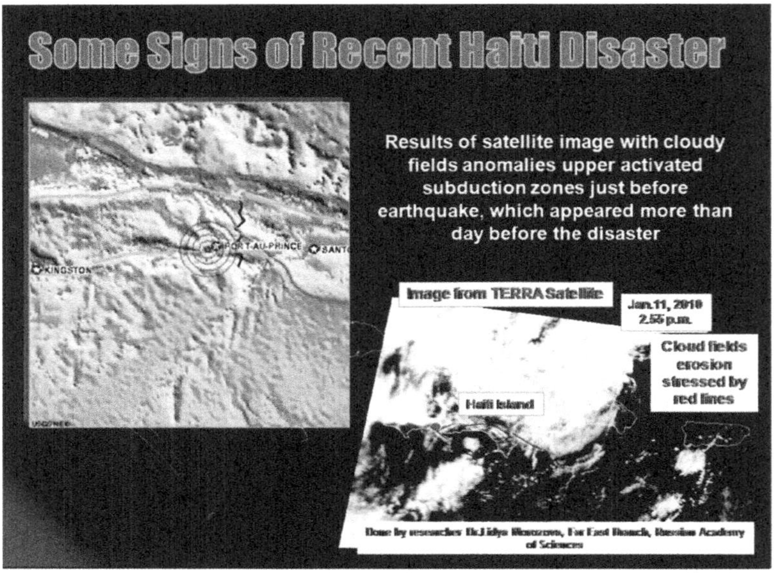

Fig. 1.33 Cloud precursors of Haiti earthquake

monitoring. The Crimea and North Caucasus are favorable research areas as they have available aerosol flow data.

For example, the cloudiness dynamic has been researched over the most active area of the Crimean seismic zone, for which extensive meteorological and seismic databanks were available. Multi-year observations established that the main earthquake areas are concentrated in the shelf zone near Yalta and coincide with the fault zones especially where they cross with coordinates of 33.5–34.5°E. and 44–45°N. 296 earthquakes were selected from the catalogue of Crimean seismic activity and meteorological data from Yalta and Ai-Petri weather stations were analyzed for a period of up to 4 days before the seismic events [25]. This was an attempt to establish a connection between earthquakes and the level of cloudiness measured from 0 to 10 points (0 – no clouds, 10 – clouds fully cover the sky) by analyzing the dynamic of cloudiness in 6-h intervals 4 days before the earthquake.

It was established that the 4-day period before the earthquake was characterized by relatively dense clouds. Those ranging from 0 to 6 points comprised only 15% (although their average annual value in the area comprised 5 according to many-year observations). Specific attention was paid to clouds ranging from 7 to 10 points. It was established that clouds intensify every day before an earthquake and two clear maximums were determined: 24–30 h and 42–66 h before the earthquake (Fig. 1.34) when cloudiness of specific configuration served as an earthquake precursor. Analysis of satellite images showed that features related to earth crust faults also manifest themselves in cloud fields.

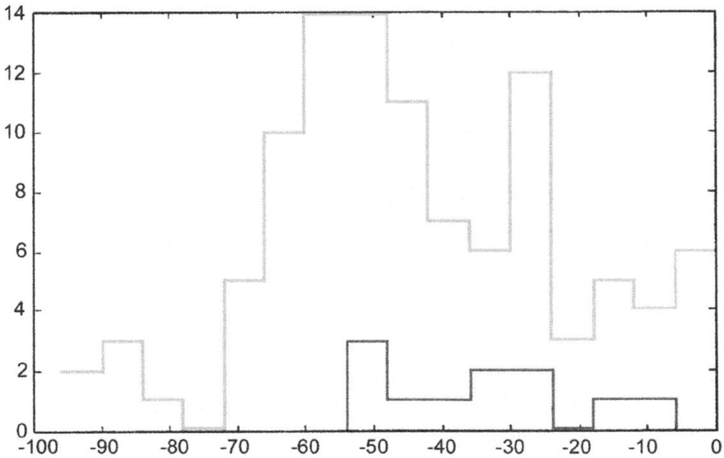

Fig. 1.34 Temporal cloudiness variations over Yalta from 1936 to 1981 [18]. The abscissa shows hours before the earthquake, the vertical axis – the number of days with 9–10-point cloudiness (*yellow*) and 7–8 points (*blue*)

This can be explained by increased intensity discharge into the atmosphere at a certain stage of seismic activities of radioactive substances, charged aerosols, and degassing [17, 26], which increases condensation nucleuses in the troposphere and influences cloudiness features, first and foremost. The report [25] showed that 3–4 days before an earthquake the appearance of 9–10 point cloudiness becomes more frequent.

In recent years connection between three processes was determined: emergence of thunder clouds, electromagnetic excitation of the ionosphere, and an upcoming earthquake. Ionosphere data from ground vertical sounding stations in Simferopol were used, as well as photos of thunder clouds during earthquakes in the area of the Crimean Peninsula, and 1957–1959 data from the ionosphere station of the Crimean astrophysical observatory in the settlement of Nauchny. Joint analysis of the data from the above-mentioned observations allowed specifying several regularities in the behavior of seismic event precursors [25]:

- Thunder clouds as earthquake precursors precede ionospheric effects and emerge 3–5 days before the main seismic shock. However, no thunderstorm has been ever registered at the moment of the main shock.
- Daytime thunderstorms promoted a weak electronic concentration increase (10–15%) in the maximum of the F_2 layer of the ionosphere.
- The ionosphere weakly reacted to earthquakes, however as ionosphere measurements were carried out practically in the earthquake epicenter, seismic and ionosphere changes were observed 5–6 h before the seismic shock.

The temperature of the underlying surface can be closely linked to the state of the cloud layer in seismically active regions. Research on the connection between the temperature and earth crust processes began because of factors exposed during several earthquakes: the drying of grass and drying-out of water wells, and intensive snow melting in winter (as occurred before the Spitak earthquake in Armenia). The statistical analysis of observation data for the relationship between earth surface temperature and seismic activity of a Crimean area for the period from 1945 to 1995 reliably determined the following:

- Differing temperatures of the underlying surface during an earthquake and in other days of the month can signify additional heat discharge 3 and 7–8 days before the earthquake.
- Although several anomalies were observed in the temperature dynamic of the underlying surface before the quake, it cannot be considered as a precursor.

The most convincing arguments in favor of heat developments in the atmosphere rather than on the surface of the earth are satellite-registered anomalous latent heat evaporation flows [27, 28] and outgoing heat emission flux [29]. The latent heat emission flux is calculated on the basis of estimated water steam content in an air column which testifies the proposed lithosphere–atmosphere–ionosphere model is operational.

Figure 1.35 shows the anomalous flux of latent heat evaporation registered a week before the earthquake with magnitude M = 7.6 in the Indian state of Kashmir on October 8, 2005. It is evident the flux is oriented along the boundary between tectonic plates.

Heat flux registration above cloudiness over the area of a future earthquake at an altitude of 12 km is even more convincing. It is clear that no heat can reach such an altitude from the earth's surface. Such a flux is possible only in the case where heat is emitted directly in the atmosphere. The infrared emission flow is registered in the so-called "transparency window" (Fig. 1.36) with a size of 10–12 μm. That part of the infrared spectrum is not blocked by clouds and escapes into outer space. Such a flux is called Outgoing Long-wave Radiation (OLR). Figure 1.37 shows the OLR dynamic before the earthquake with magnitude M = 6.7 in Japan on July 16, 2007. It can be seen that the location of heat sources changes with time, but the sources are concentrated along the active boundary between tectonic plates.

Upcoming quake indicators are not only the atmospheric anomalies as such, but their time evolution as well. In particular, several strong earthquakes displayed a certain form of daily air temperature and relative humidity changes [30, 31]. Figure 1.37 shows a typical form of changes in daily temperature for three strong earthquakes in Mexico and California.

Figure 1.38 shows simultaneous changes in daily temperatures (black curve) and humidity (blue curve) before the earthquake with magnitude M = 8.1 in Michoacan (Mexico) on September 19, 1985. It is evident at least that the daily temperature range does not exceed the limits of changes during the month, i.e., there is no anomaly from the statistical point of view. However, its form and time maximum before the earthquake coincides with those observed for other quakes (in Mexico and California), while humidity reports an absolute monthly minimum at the same time. Thus, in the given case the earthquake indicators are not the

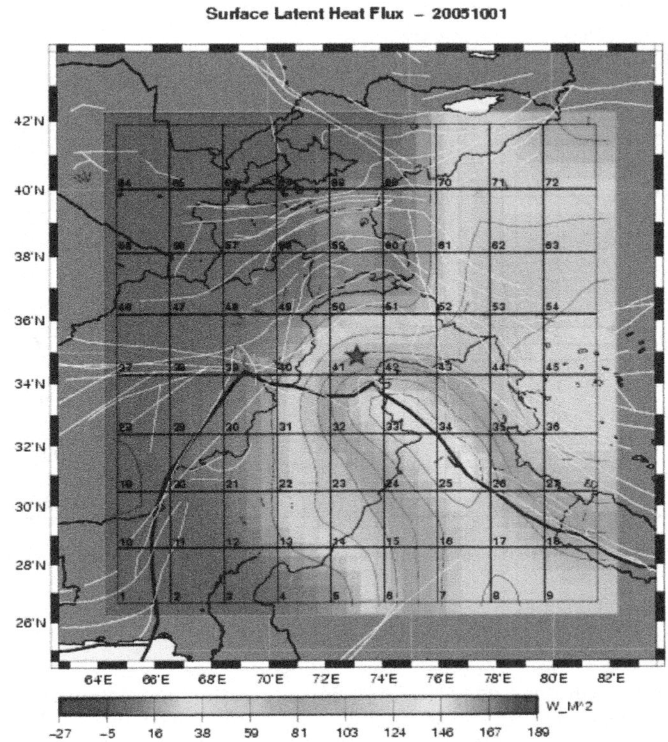

Fig. 1.35 Anomalous latent evaporation heat flux during an earthquake in Kashmir, India (8.10.2005)

changing parameters (which are not anomalous against monthly variations), but the form of changes that repeats itself in various quakes (Fig. 1.39).

While interacting with seismogenic emission at frequencies of 0.1–10 Hz the radiation belt of the earth emits charged particles [32–34] that ionize the ionosphere. Reports [35, 36] cite research results that show the presence of anomalous cloud accumulations and variations of OLR and geothermal anomalies on the earth's surface [29, 37–41] preceding strong and devastating earthquakes on the basis of data obtained from weather satellites.

The fact was confirmed in reports [42–47], which describe anomalies of the electromagnetic field of the earth over major faults determined by measurements from Intercosmos-19 and Intercosmos-24 spacecraft. Thus, the ionosphere is an important indicator of the processes taking place not only in the atmosphere and outer space, but also in the lithosphere of the earth.

Reports [45–47] provide research results that expose negative variations of electronic density with the use of the onboard ionosphere probe of the Intercosmos-19 satellite that were observed 1–2 days before the earthquake of

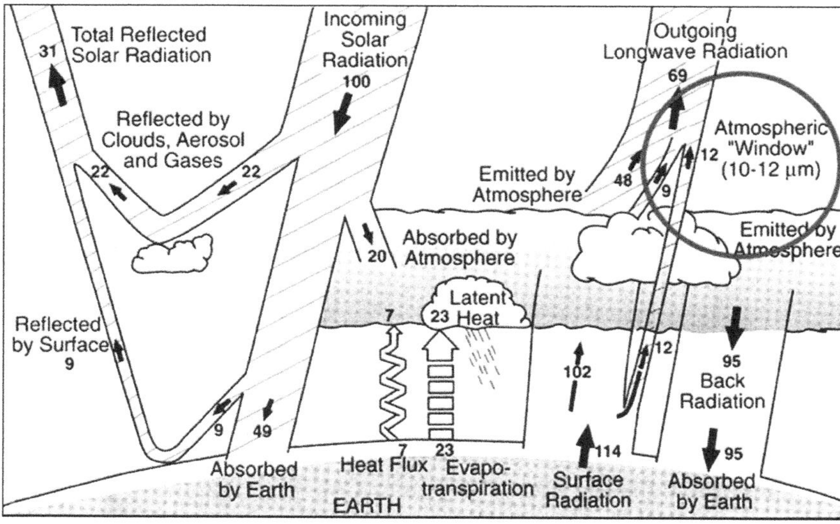

Fig. 1.36 Heat exchange in the earth–atmosphere system: transparency windows and OLR-flux

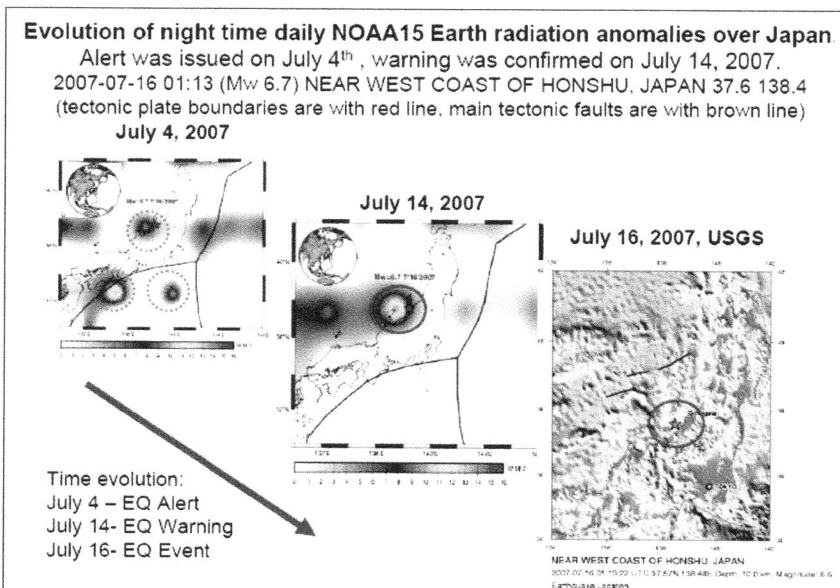

Fig. 1.37 OLR-flux dynamic during an earthquake in Japan (16.07.2007)

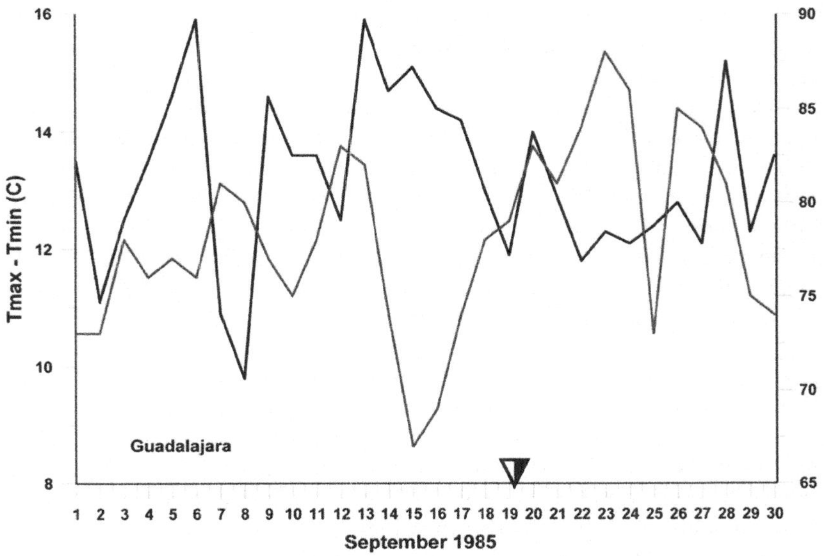

Fig. 1.38 Daily changes in temperature and humidity during an earthquake in Michoacan (10.09.1985)

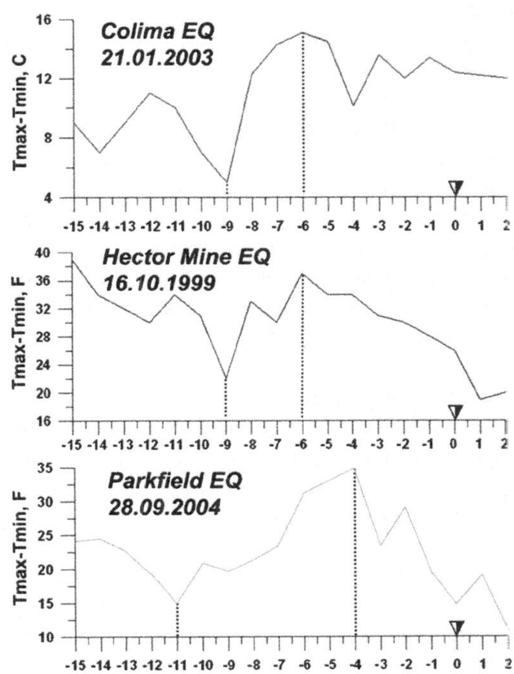

Fig. 1.39 Form of changing daily temperatures

16.07.1980 near Australia and New Guinea islands. A similar phenomenon was registered by the vertical ionosphere sounding station near Tashkent when the critical frequency f_0 of the $\boldsymbol{F_2}$ layer increased on average 1–2 days before local seismic events against controlling station data and then decreased.

There is no doubt that vertical or inclined sounding methods used by ground or space-based ionosphere probes provide invaluable information about the state of the ionospheric plasma. However, it is impossible to obtain information about the state of the upper atmosphere below or above the maximum layer of the ionosphere with such measurements [48–51]. The mass of the whole ionosphere is studied with incoherent dispersion radars. However, their deployment and operation is very costly and there are only a few units on Russian territory [52].

Today signals from artificial satellites of the earth are practically the only available instrument to study ionospheric plasma and diagnose the state of the mass of the whole ionosphere. A leap forward in the quality of research of the state and processes in the ionosphere was promoted by the startup of satellite navigation systems. Ionosphere research on the basis of signals from global navigation systems is widely practiced in the whole world. Ground monitoring stations have been deployed in Brazil [53], India [54], China [55], Norway [56], the United States [57], Canada [58], Malaysia [59], and France [60]. At present Russia is actively recovering its geophysical network and carries out research on ionospheric processes in various regions, for example in the North-West [61], in the Sakhalin region [62], and on Kamchatka [63]. Research is also underway on the spatial distribution of full electronic concentrations [64, 65] aimed at exposing anomalies of the full electronic content in the ionosphere related to seismic activity [66–74].

In the recent years excitation of the ionosphere caused by earthquakes was clearly specified. The main morphological and physical regularities of seismic energy and the influence of deep aerosol flows on the ionosphere were determined. Ground and satellite research [75–78] showed that the atmosphere and ionosphere are very sensitive to tectonic processes.

Thus, seismic hydration in the upper atmosphere creates major ion clusters of aerosol size, which radically decrease conductivity of the near-earth air layer because of their low mobility. As a result, the fall in the potential of the effective atmospheric resistance accelerates while the ionospheric potential increases above the area of earthquake origination. Because of high conductivity the ionosphere works to even the potential over the area of the anomaly. Thus, a horizontal difference of potentials emerges in the ionosphere which causes plasma drift and creates concentrations of inhomogeneities measurable by radiophysical methods, such as vertical sounding, space tomography, full electronic content measurement, etc. Depending on radon emission intensity atmospheric electricity can change with various degrees of intensity. The ionosphere can produce both large-scale anomalies and small-scale inhomogeneities in plasma density. Their interrelation is confirmed by the same time evolution of atmospheric anomalies.

It has been statistically confirmed that ionospheric anomalies occur on average 5 days before a seismic shock [79]. A local ionosphere changeability index has been designed to identify seismic ionospheric anomalies even against the background of

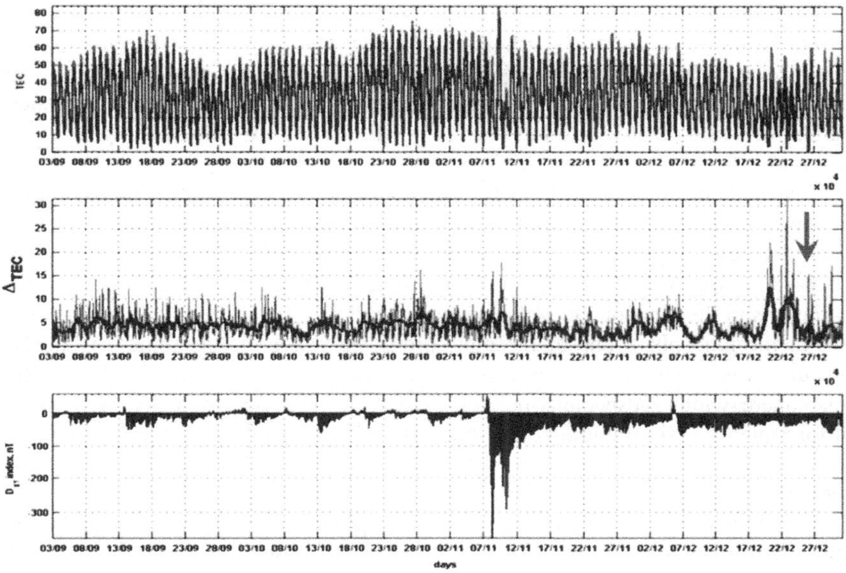

Fig. 1.40 Local ionosphere changeability index changes during the Sumatra quake (26.12.2004). Maximum (*red line*) and minimum (*blue line*) compete electron contents for GPS receivers at Sumatra region; local ionosphere changeability index in the Sumatra area (September–December 2004); global index geomagnetic inductivity in the Sumatra region

geomagnetic excitation and storms [80]. Figure 1.40 shows index indicators calculated from the data of GPS receivers close to the epicenter of the strongest earthquake on Sumatra with magnitude M = 9.0 on December 26, 2004. The figure shows that several days before the Sumatra quake the index considerably exceeded the parameters registered during the strongest magnetic storm on November 7 of the same year.

It should also be noted that density in the ***F***-layer maximum begins to grow on average several days before a seismic event and has a spatial scope of several thousand kilometers along the parallel and 1,000 km along the meridian. At the same time there is a relatively decreasing trend in the layer density 1 day before the event, which has a whirl-like structure with a scope of ~103 km localized at a distance of 500 km from the epicenter. One to two days before the event long-period (2-h time evolution) quasi wave excitations occur two times more frequently on average in the ***F***-layer variation range. Reports [22, 81–96] say the acoustic-gravitational wave may serve as a potential earthquake precursor along with it. In particular, report [81] offered a model of acoustic-gravitational wave travel from a pinpoint source on the earth to ionospheric heights. It was shown that the wave has sufficient energy to create 0.2–0.3°C heat anomalies and ionize the ***F***-layer maximum of the ionosphere.

The first reports about anomalous phenomena observed in the earth's crust several days before strong earthquakes appeared in the 1960s. However, they were practically ignored. A breakthrough occurred after the launch of the Intercosmos-19 satellite in 1979. It registered anomalous noise in a low-frequency band, in a certain belt which was narrow in latitude and extended in longitude and the center of which was later discovered to be above the epicenter of a future earthquake. The noise began to appear hours before the first shock. The phenomenon was officially recognized and registered as a discovery. Thereafter it was confirmed by data from a series of other satellites, including the Soviet-French Oreol-3 equipped with sensitive probes to register magnetic and electric field variations in the ionosphere of the earth.

For the ionosphere one of the most promising ways to determine earthquake precursors is to monitor changes in electron concentrations in the ***F***-layer of the ionosphere, which is especially reactive to tectonic processes. Statistical processing of ionospheric radio sounding data obtained by Soviet satellites of the Intercosmos series in July 1980, as well as up-to-date information from foreign spacecraft showed the following:

Very indicative variations of the critical frequency in the early hours above the area of an upcoming strong (M ~ 7.3) earthquake on New Guinea islands 2 days before it, 1 day before, and 1 day after the quake [97]. Figure 1.41 clearly shows hollowness formation in the ionosphere close to the vertical quake epicenter projection a day before the seismic shock, during the shock, and after it. In some cases a similar phenomenon occurred also in the conjugated ionosphere area in the other hemisphere (as was the case with the registration of magnetically conjugated belts of Very Low Frequency (VLF) precursors) which testifies to the transmission of the impact from the earthquake origination area along the whole force tube of the geomagnetic field.

Thus, the ionosphere and especially its ***F***-layer (100–1,000 km) are very sensitive to tectonic processes. Two days before the first shock a hollowness appears over the epicenter (or close to it) and sometimes in the other hemisphere of the earth as well. The hollowness disappears a day after the shock.

Atmospheric plasma variations in the ionosphere occur in the period of 5 days to several hours before the underground shock. Before an earthquake the conductivity of the lower ionosphere increases, but decreases during geomagnetic excitation. The effect is registered 2–6 h before the beginning. A maximum VLF-wave decay ratio is registered at an altitude of 60–90 km and is three orders less at the altitude of 1,000 km. During standstill only Extremely Low Frequency (ELF)-components are registered in the 400–500-Hz frequency band. During geomagnetic activity ELF-components are registered in the frequency band less than 0.7–1.5 kHz up to 2 kHz. Increased seismic activity results in the following phenomena: ELF-spectrum expands up to 3 kHz; VLF-components appear, as well as whistling atmospherics of abnormally high intensity $(S_{max}/S_0)^2 \sim 8$ and a wide spectrum ΔF from 100 to 10,000 Hz without a minimum at frequencies of 2–3 kHz. Magnetosphere channels appear over the seismic zone during the origination stage of an earthquake. With earthquakes of 5–5.5-point magnitude ionospheric excitation varies by amplitude

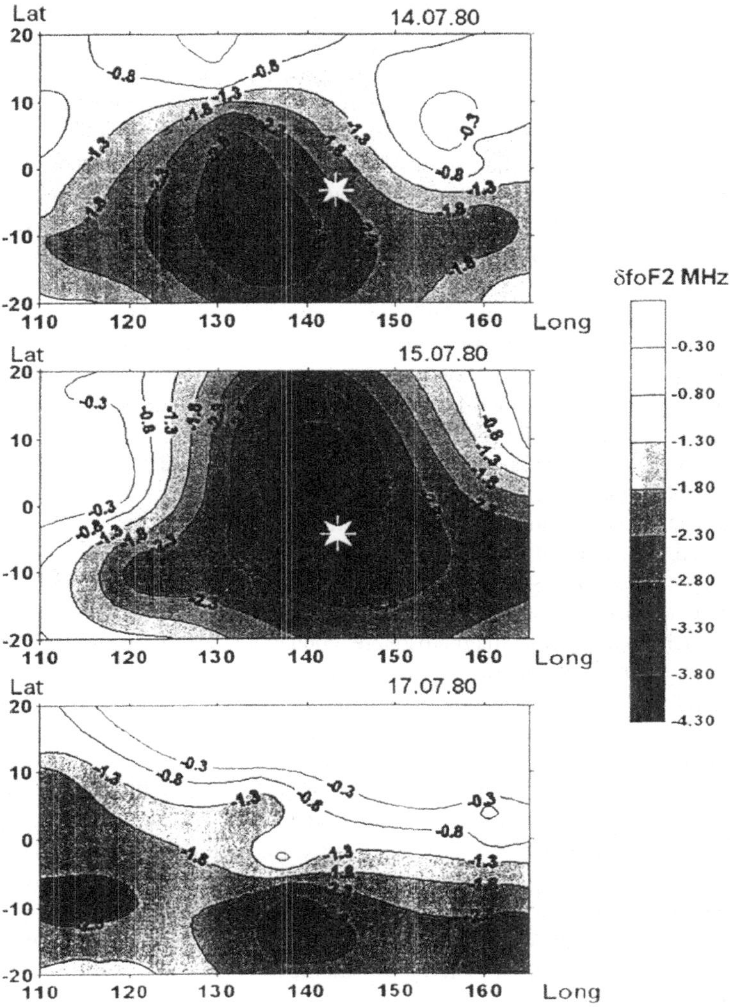

Fig. 1.41 Hollowness formation in the ionosphere before, during, and after the earthquake

from 15 to 100% and more (depending on the time of day). The size of the excitation area comprises 300 km both by latitude and longitude.

Report [98] provides a comprehensive review of research on the earth's electromagnetic field, including a description of the main mechanisms of emergence of ionospheric anomalies. The research registered electro-kinetic effects that generated signals at frequencies of 0.1–100 Hz from the earth's surface. It also listed effects caused by the advance of acoustic-gravitational waves, as well as migrating ionospheric excitations and said it was necessary to take into account these effects in the research of ELF-VLF electromagnetic anomalies related to earthquake origination.

Thus, observations in Yakutsk [99] disclosed earthquake manifestations in changing intensity of VLF-noise that emerges over earthquake areas in the limits of the Fresnel zone. It was established, for example, that 3–4 days before an earthquake the night-time intensity of VLF-noise decreases, which the authors of the report interpreted as an earthquake precursor. In Japan [100, 101] antennas in artificial wells registered anomalies of the electric field of the earth in the ELF-VLF band during earthquake origination, as well as in the advance of impulse signals in the ELF–VLF band directly before the quake.

In some cases local seismic maximums coincided with the maximums of geomagnetic storms. Thus, Russian scientists discovered a previously unknown phenomenon, i.e., the connection between seismic and geological activity in certain parts of the planet and coronal holes on the Sun, which under specific conditions can "launch" an earthquake through poorly studied mechanisms of the Sun–Earth relationship. In other words, under specific conditions earthquakes can be "born on the Sun," while their precursors can be detected long before the first shock with the help of spacecraft used to research solar-terrestrial relation.

Detailed research of changing flows of ionosphere-captured high-energy particles promoted the discovery of a new natural phenomenon related to the impact of seismic activity on the earth on the internal boundary of the radiation belt of the earth, which is a seismic magnetosphere connection [102] that opens up new prospects for earthquake prediction from space.

The epicenter of an upcoming earthquake produces electromagnetic emission over a wide frequency band which emerges (as mentioned above) due to mechanical motion of underground rock – friction, cracking, the piezo-effect, etc. However only emission in the frequency band of 0.1–10 Hz can penetrate into the earth's crust and atmosphere without loss and reach the radiation belt of the earth. After reaching the lower boundary of the radiation belt it interacts with captured electrons and protons. The interaction actively engages particles related to magnetic power lines (power line *tubes*, to be more exact) that go through the epicenter of the upcoming quake. If the oscillation frequency of the particles between the so-called "mirror" supporting points of the radiation belt arch coincides with the frequency of seismic electromagnetic emission, the interaction acquires quasi-resonant character that manifests itself in "precipitation" of particles from the radiation belt. Because of the longitudinal drift of the captured particles the precipitation circles the earth along the magnetic latitude where the epicenter of the upcoming quake is located and creates inhomogeneity that can exist for 15–20 min (until all charged particles precipitating from the earth's radiation belt *perish* in the atmosphere). Flying under the radiation belt and crossing the epicenter latitude of the upcoming earthquake a properly equipped spacecraft can register the fact of the precipitation, while a subsequent analysis of the energy and temporal distribution of the precipitating charged particles will help determine the place and time of the predicted seismic event.

1.4.2 Volcanic Eruption Precursors and Their Detection

The geochemical background of the environment is formed under the influence of natural and anthropogenic deep-earth substance flows that rival each other in intensity [75, 76]. Two major sources have to be specified – tectonically active zones with magma and mud volcanoes, seismically active areas, rift folds with hot and cold thermal springs on land and in the ocean. Researchers are especially interested in major eruptions with powerful discharge that affects the global aerosol background and atmospheric temperature. Gas and aerosol flows from such activity are less studied, while flows from mud volcanoes and seismically active zones are fully ignored.

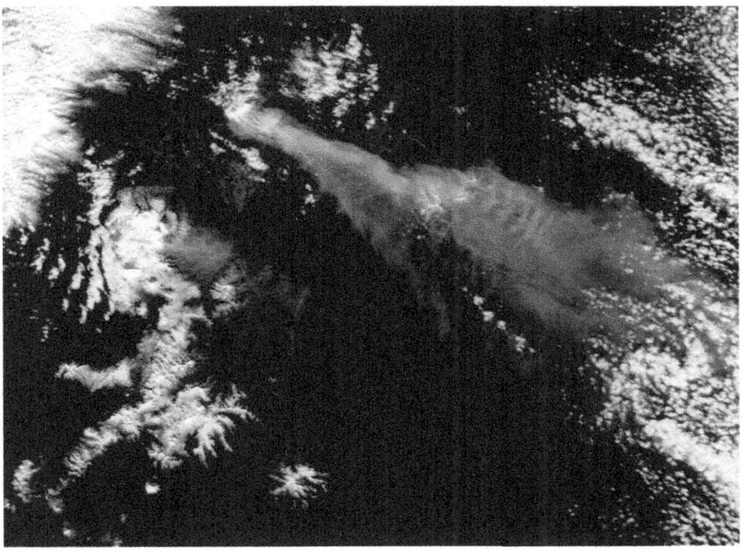

In the physical sense the task of predicting the time of a volcanic eruption is similar to the forecasting of a breakthrough in a high-pressure pipeline. The growing acoustic emission and cracking of the pipe wall before a breakthrough can be compared to the growing number of quakes during the destruction of the plug in the volcanic orifice before the eruption. In some cases the day of its destruction and the beginning of the eruption were predicted as happened with the Shiveluch volcano in 1964. The successful prediction of the Tolbachek eruption and the time of subsequent magma breakthroughs is well known. Kluchevskoy volcano eruptions on Kamchatka on March 5, 1980 and March 8, 1983 were also predicted. However, unexpected eruptions also occur and the explosion of Mt. Saint Helens volcano on May 20, 1980 was a complete surprise for American scientists. The strong summit eruption of the Alaid volcano on the Island of Atlasov (North

Kurils) that began on April 28, 1981 did not provide any significant seismic precursors either.

There are currently close to 2,000 volcanoes in the world and a fifth of them erupted in recent 150 years. Volcanic activity is caused by energy and substance transfer from the mantle to the earth's surface and proceeds in two stages: quiet fumaroles and hydrothermal activity and the eruption when gas pressure in the magma hotbed exceeds the strength of the rock. Eruptions eject volcanic gases, aerosols, and debris into the atmosphere, and exude mud and basalt lava on the surface (eruptions annually eject to the surface of the earth close to five million tons of basalt).

To estimate aerosol discharge from fumarole activity of magma volcanoes research was carried out in August-September 1981 close to Avacha volcano on Kamchatka to study the content of microelements in aerosols that precipitated on volcano crater and flanks with fresh and seasonal snow, as well as in surface water suspensions and in volcanic ashes [77]. The concentration was measured by Roentgen-fluorescent method. The discharge of elements into the atmosphere was estimated by multiplying average concentrations in ashes by the volume of eruption.

The research showed the composition of volcanic gases and aerosols is changing in time which is a sign of volcanic activity level. Uninterrupted monitoring of the composition of discharged gases and particles would help timely predict eruptions. It is the most reliable prediction indicator among others. The Avacha volcano erupted on January 13, 1991 without any registered seismic precursors. In summer snow samples were taken with 5-cm layer intervals from survey pits in flank glaciers of the volcano and precipitation accumulation picture was provided for the whole winter period [78]. The analysis of the microelement snow composition established those 1.5 months before the eruption confirms that there was a discharge of aerosols rich in microelements. The highest concentration peaks were registered for Fe, Ca, Mg, Al, Mn, Be, B, and Br, which exceeded concentrations in other snow layers by one order. Such a set of elements testified to the beginning of the destruction of the lava plug in the crater [103].

During enhanced volcanic and seismic activity on Kamchatka in 1991–1992 independent research was held to study the dynamic of aerosol flow and its element composition and to precisely measure deformation surface motion by a levels line that was 2.6 km long and located 20 km from Avacha volcano practically perpendicular to its seismic focal zone [78, 103].

Microelement concentrations in snow aerosols can be considered invariable during the whole snow accumulating season of 1990–1991. Only two periods are distinguished by sharp anomalies. The first one occurred approximately from November 20 to December 6, 1990. The microelement group reported a high peak that exceeded three standard deviations for the following elements: Fe, Al, Mg, Ca, Sr, Be, Mn. Peaks in the limits of statistical error were reported for Zn, Na, Cd, Cu. The second period of anomalous element concentrations occurred in January 1991 when Avacha volcano erupted. Specific manifestations of dissolvable and insoluble element connections call to control the concentrations of both. Active concentration dynamic was reported for the following elements: Be, B, Na, Mg, S, K, Ca, Mn, Fe, Cu, Zr both in the dissolvable and insoluble phases. Dissolvable phase dynamic was registered for Al, V, Cr, Co, Ni, Sr, Sn, Pt, Pb, Bi, U. Only primary samples containing all phases react to Si, Cd.

The results are noteworthy as most research targeted only solutions and a limited number of elements. The research determined that dissolvable phase reacts weaker than aerosols to changes in geological environment strain.

A major aerosol discharge in December 1990 was the precursor of an upcoming eruption of the Avacha volcano and occurred in parallel with a strong deformation process at the geodesic range which allows interpretation of the processes as a manifestation of one geodynamic impact. The Institute of Volcanology and the Institute of the Physics of the Earth measured surface deformations on Kamchatka by the geomagnetic method, which is held as one of the traditional earthquake prediction methods. The leveling network in the area of the Avacha-Koryak group of volcanoes and the city of Petropavlovsk-Kamchatski covers a territory of over 25,000 km. Routine measurements at such ranges are carried out twice a year, thus allowing one to monitor background geodynamic processes, but provides no mid-term prediction of earthquakes and eruptions.

Local precise level-line measurements in the Kamchatka fault zone carried out over 3 years established a connection between surface deformations and seismic processes. The line was located 15 km away from Petropavlovsk-Kamchatski and 20 km south of the Avacha-Koryak group of volcanoes towards Avacha Bay. The longitudinal line was practically perpendicular to the axis of the seismic focal zone which goes along the Kamchatka Peninsula at a distance of 100–150 km from the coast and is a junction of major northeastern and northwestern tectonic dislocations of the deep fault zone and Avacha volcanoes. The 2.8-km long level line was cut by three active faults with concentrated super intensive vertical motion reaching 8–10 cm on the 100-m basis. The line had 28 stations at a distance of 80–100 m from each other and measurements were made 1–2 times a week with a standard error of 0.11 mm.

As a result five local sections were discovered on the lines that were distinguished by considerable surface subsidence with pulsating character. The anomaly amplitude ranged from 1–2 cm to 10–13 cm in various time periods, and the anomaly width ranged from 200 to 500 m. During the whole monitoring period from November 1989 to July 1992 the five most active sections were distinguished with characteristic peak-like form and by the pulsating character of the subsidence. Intensive subsidence periods alternated with relative standstill periods after which the process resumed.

Anomalous deformations in the fault zone might be caused by rock displacement due to impregnations of lower rigidity and greater looseness. The research carried out by volcanologists on the line coincided with the study of aerosol discharge from Avacha volcano. A connection was established between the distribution of metals in aerosols and the velocity of the surface vertical motion, as well as anomalous sections corresponding to aerosol discharge before and during the Avacha volcano eruption that occurred at minimal levels of the mentioned deformations. It is noteworthy that an earthquake with Mn, Fe, Cu, Zn, Be, Na precursors preceded the eruption by 30 days.

During an eruption deformation of the earth's crust degasses volcanic rock and light gases, such as hydrogen, helium, methane, and carbon dioxide that carry into the atmosphere heavy gases – radon and thoron. Deformations create microcracks in the rock and gases carry to the atmosphere submicron-sized aerosols. In principle, tectonic aerosols have a major significance for predicting volcanic eruptions and accompanying earthquakes [75, 76]. However, it is rather difficult to estimate the total aerosol flow during the processes. It usually can be done through snow analysis after snowfalls around active volcanoes and at faults. However, the effects caused by tectonic aerosols in the near-surface layer of the atmosphere can be registered from space and act as precursors of the considered seismic manifestations.

Radon and aerosol precursors of volcanic eruptions can coincide as happens with earthquakes. Conversely, there is a correlation between geomagnetic field anomalies and increased radon concentrations in subsoil gases and waters that occur during deformation of the earth's crust in the focus area. The absence of aerosol flow measurements resulted in misinterpretation of undergoing processes and added a major share of uncertainty as aerosols can trigger major changes in the electric fields in the near-earth layer of the atmosphere before earthquakes and eruptions of mud volcanoes, in particular.

To determine the connection between radon and aerosol fields and atmosphere conductivity as precursors of volcanic eruptions research was carried out in October 1999 at the Bugaz fault in the southwestern part of the Taman Peninsula. The fault, related to the tectonic location of the Black Sea hollowness, is characterized by ongoing landslides and produces a large number of mud and gas springs (close to 100) that are unevenly distributed along the fault line. Researchers also measured atmosphere transparency, its electric conductivity, as well as subsoil radon and hydrogen concentrations in the near-earth layer. The most informative sites at the Bugaz fault were discovered after aerosol flow was measured by a microlidar that registered aerosol clouds over intensive gas discharge locations. The second research object was the mine gallery in the area of Novorossiisk where radon discharge was surveyed by film detectors and hydrogen shooting. Detectors were installed every 10 m throughout the length and breadth of the fault which allowed the creation of a radon flow intensity line for 10 days. Simultaneously the same sensor registered radon flow in the gallery near Novorossiysk to model aerosol discharge during a man-made explosion.

Work was also carried out to determine temporal statistical characteristics of electric field fluctuations and air conductivity [23] during a period of relatively weak seismic events (Fig. 1.42).

Hydrogen mapping results showed its discharge varied along the fault between the limits of 3.7–2.6 ppm. Temporal characteristics of radon content measured in the mine gallery near Novorossiisk were less informative than at the Bugaz fault. The above-mentioned seismic events were observed at minimum radon emanations

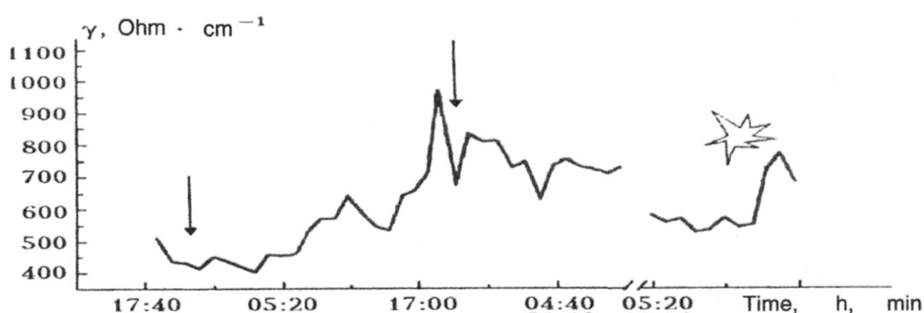

Fig. 1.42 Temporal progress of atmospheric air conductivity during a period of seismic activity and as the result of a man-made explosion (Novorossiisk area, October 19–22, 1999)

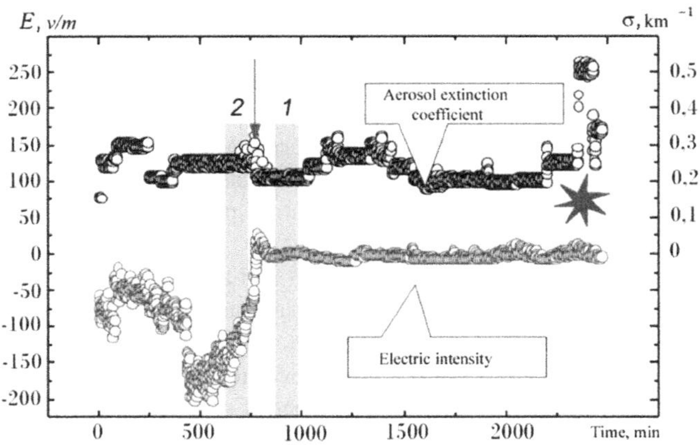

Fig. 1.43 Timing code of electric intensity E and aerosol extinction coefficient σ. The *arrow* points to the moment of a seismic event, the star points to the moment of an artificial explosion, unit 2 and 1 connected to the measurements which were made

with their subsequent sharp increase (similar data were earlier obtained at mud volcanoes in Italy).

Measurement results are provided in Fig. 1.43 which shows aerosol diffusion ratio **σ** changes (left ordinate axis) and electric field ***E*** strain (right ordinate axis) for the whole measurement period of over 40 h from October 17 to 22, 1999. The schedule shows a sharp and over twofold increase in aerosol diffusion ratio at the moment of a seismic shock (marked by an arrow) and the explosion of 500 kg of ammonite (marked by the seven-pointed star). Schedule ***E*** has bigger fluctuations specifically in the first 10 h of measurements when the electric field changed considerably and then dropped intensity practically to zero. At the moment of the man-made explosion such an intensive leap was not observed. The figure also shows considerable growth (up to 170 V per m) in electric field intensity 8–10 h before the earthquake and a sharp fall to background values directly before the seismic shock.

Research of spectral characteristics of aerosol diffusion ratio **σ** fluctuations showed they depended to a considerable extent on earth crust activity (during measurements). Figure 1.44 offers amplitude spectrums for light diffusing characteristics of the atmosphere both for "quiet" periods and seismically active times. Data analysis reported maximum values of amplitude spectrums during active periods and showed they differ 1.5–2 times. The figure also shows maximum values of amplitude spectrums for electric field ***E*** intensity fluctuations during "quiet" and "active" states of the earth's crust that differ more than 40 times. Such a significant difference shows that tectonic processes play a major role in the formation of electric field structure in a seismically active region.

It is noteworthy that seismic ionosphere manifestations comprise only a part of fundamental connections between the lithosphere, lower atmosphere, and ionosphere. Earthquake-scope electric fields can be generated also before volcano

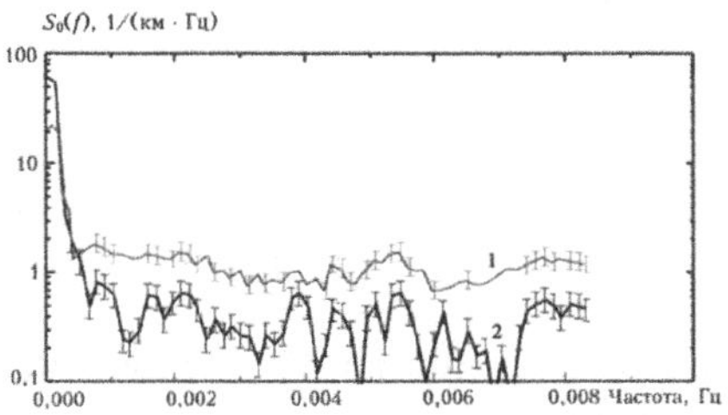

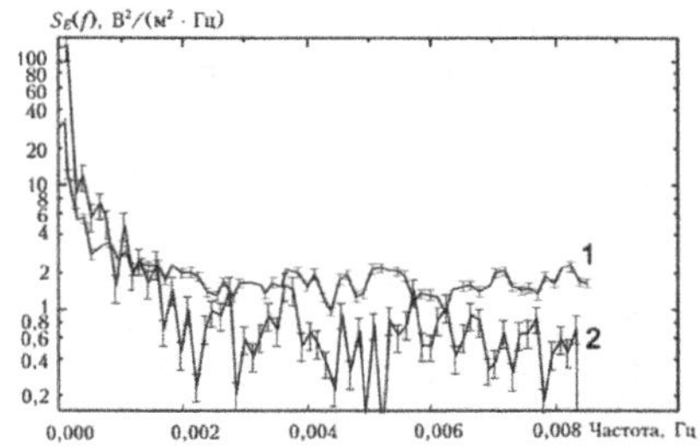

Fig. 1.44 Spectral analysis of aerosol diffusion ratio fluctuations and electric field intensity in the area of volcanic activity. Amplitude spectrums of aerosol diffusion ratio fluctuations in a quiet period (curve 1) and under seismic impact (curve 2); amplitude spectrums of electric field intensity fluctuations in a quiet period (curve 1) and before a seismic impact (curve 2)

eruptions, by the impact of strong atmospheric fronts, and other natural calamities. Creation of a specialized system for prediction of global catastrophes will help resolve numerous problems of early warning providing protection against such calamities.

1.4.3 Detecting Geological Calamity Precursors from Space

The above-mentioned precursors of seismically hazardous events – anomalous geophysical phenomena in the lithosphere, atmosphere, and ionosphere of the earth that can be registered from space along with traditional ground sensors – allow one to collect sufficient data to design prediction models and provide practical predictions of natural calamities. To register such precursors from satellites (including small satellites from 100 to 400 kg) relatively small devices can be used that have already been designed in Russia and abroad – ion-balloons, magnetometers, low and high-frequency emission recorders, elementary particle detectors, infrared radiometers, and other geophysical equipment.

Research and development units of the Russian Federal Space Agency jointly with organizations of the Russian Academy of Sciences (the Institute of the Physics of the Earth and ISMIRAN weather forecast center) confirmed the principle possibility of considerably enhanced efficiency of short-term earthquake predictions from space satellites capable of detecting anomalous physical electromagnetic and plasma phenomena in the near-earth space that can be considered as earthquake precursors (also short-term). Such phenomena include anomalous low-frequency (ULF–ELF–VLF) (ULF: Ultralow frequency) electromagnetic emission and geomagnetic pulsations at frequencies close to 1 Hz; excitation of quasi-constant electric fields and local variations of ionosphere plasma temperature; and precipitation of high-energy particles. Space equipment shall be capable of measuring ULF–ELF–VLF waves, electric field parameters, plasma density and the composition of its ion and neutral components, high-energy particle flows, atmospheric emissions in various oxygen and nitrogen lines in subsatellite areas, and hydroxyl emissions observed from the direction of the earth's limb. Onboard equipment shall thus include the following devices:

- ULF–VLF wave complex to measure characteristics of the electromagnetic emission spectrum in the frequency band of 0.1–23 Hz;
- High-frequency wave complex to measure electronic density and characteristics of the electromagnetic emission spectrum in the frequency band of 0.1–15 MHz;
- Device to measure three components of the quasi-constant electric field;
- Spectrometer of energy particles to measure energy distribution and intensity variations of electron and ion flux with energies ranging from 20 KeV to 2 MeV;
- Optical complex to measure atmospheric emission changes;
- Plasma complex to measure ion and neutral composition, variations of density and plasma drift velocity.

In 2001 an expert taskforce created by the Russian Space Agency carried out a comparative analysis of potential earthquake prediction methods with the use of spacecraft and outlined the most promising of them. Recommendations from the taskforce led to the Federal Space Program receiving a Volcano guideline called "Space System for Monitoring and Prediction of Natural and Technogenic Disasters" aimed at collecting, processing, and analyzing data about disaster precursors, development, and aftermath. The system has to comprise two groups of small spacecraft in circular orbit of 400–500 km (low-orbit group) and 900–1,000 km (high-orbit), as well as a complex of ground observatories, data receiving and processing stations, and a situation center.

The first step in creating and testing the Volcano system was the launch in 2001 together with the Meteor-3M weather satellite of the experimental COMPASS (Complex Orbital Magneto–Plasma Autonomous Small Satellite) spacecraft with onboard equipment designed by Russian, Hungarian, Greek, Ukrainian, and Polish institutes. The small satellite carried a ULF–VLF wave complex, high-frequency radio spectrometer, plasma complex to measure basic characteristics of heat plasma, optical complex comprising a TV camera and photometer, spectrometer for energy particles, and a magnetometer. The launch of the satellite was timed to the predicted earthquake on the western boundaries of the Pacific lithospheric plate. However, the failure of the equipment did not allow researchers to complete the experiment aimed at comprehending connections between the lithosphere and ionosphere. Still the obtained data were very inspiring. In 2002–2003 a similar experiment was carried out with the use of the Meteor-3M satellite. At the end of this mission space prediction was seen to have been correct for 44 out of 47 earthquakes registered by ground seismic stations in the world. To continue the program the COMPASS-2 satellite was orbited in 2006. Despite equipment failures it provided valuable information.

COMPASS-2 carried a low-frequency wave receiver SHASH-2 (Fig. 1.45) to measure electromagnetic fields in the band from several units of measure to 20,103 Hz. Its use in the ELF–VLF bands aimed at establishing connection between seismic activity and VLF–ELF phenomena, determining on the basis of two-component measurements the Umov-Poynting vector sign, observing plasmasphere boundary variations with the use of whistling characteristics, and uninterrupted monitoring of ULF–VLF wave activity in the near-earth space.

Fig. 1.45 Onboard low-frequency wave receiver SHASH-2

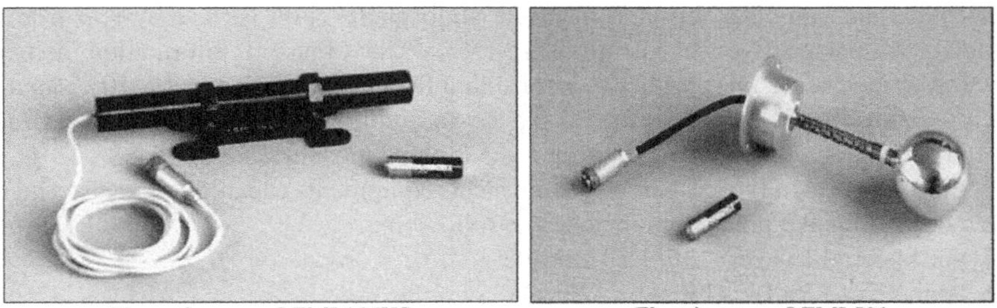

Magnetic sensor LEMI 106HS | Electric sensor LEMI 501

Fig. 1.46 Onboard sensors of the low-frequency wave receiver SHASH-2

To register magnetic and electronic components of VLF emission the receiver used magnetic LEMI 106HS and electric LEMI 501 sensors (Fig. 1.46). It also carried GID-12T equipment for precise measurements of high-altitude distribution of electronic concentration (from the bottom of the ionosphere to the altitude of the spacecraft) by transionospheric satellite sounding data (with the use of signals from GLONASS/GPS navigation systems) for assessment of the global distribution of high-altitude ionosphere, diagnosing active impacts on ionosphere plasma, as well as anthropogenic effects in the ionosphere and effects related to natural processes in the earth's atmosphere also above seismically active regions.

In addition, the satellite was equipped with a radio-frequency analyzer to measure local waves by the high-frequency radio spectrometry method and the Tatyana set of scientific devices to register the influence of various space emission components (protons, electrons, etc.) on the radiation situation in near-earth space.

Measurement means and methods planned for the Volcano system have long been tested onboard the International Space Station (ISS) in the framework of the Hurricane program. To promote the study of the impact of seismic activity of the earth on the inner radiation belt the ISS was equipped with the Vsplesk (Splash) complex to assess registration efficiency of charged particle fluctuation anomalies in near-earth space that can be related to various geophysical (also seismic) processes on earth. The experiment also registers charged particle flux by the ARINA device installed on the Russian "Resource-DK1" spacecraft, which increases the reliability of obtained information and provides a possibility to map the undergoing processes with a higher level of precision.

Unfortunately, unstable and insufficient financing did not allow researchers to complete the creation of the "Vulcan" space system although the obtained data gives grounds for optimism regarding technical aspects of proposed prediction methods and dedicated hardware implementation.

Promising Russian-designed space-based devices that can register earthquake precursors include orbital light radar (lidar) to detect aerosol concentration anomalies in the surface layer of the atmosphere and infrared Fourier spectrometers for remote sounding of the atmosphere of the earth.

Infrared Fourier spectrometer IKFS-2 designed by the Keldysh Research Center measures the outgoing emission from the atmosphere–earth surface system from Meteor-M perspective spacecraft (Fig. 1.47). The obtained information helps restore temperature lines with 1 K error, humidity and ozone content – 10%, land temperature – 1 K, water surface – 0.5 K. In addition, data can be obtained to determine CH_4, N_2O content, and trace gas in the atmosphere.

Similar foreign space-based geophysical equipment capable of registering certain earthquake precursors include the following:
Hyper Spectral Imager – "EO-1" satellite;
Poseidon Type Radar Altimeter – "Jason" satellite;
Total Ozone Mapping Spectrometer (TOMS) – "Quick TOMS" satellite;
Multi-Beam Laser Altimeter (MBLA) –"VCL" satellite;
Geosciences Laser Altimeter System (GLAS) – "Ice Sat" satellite;

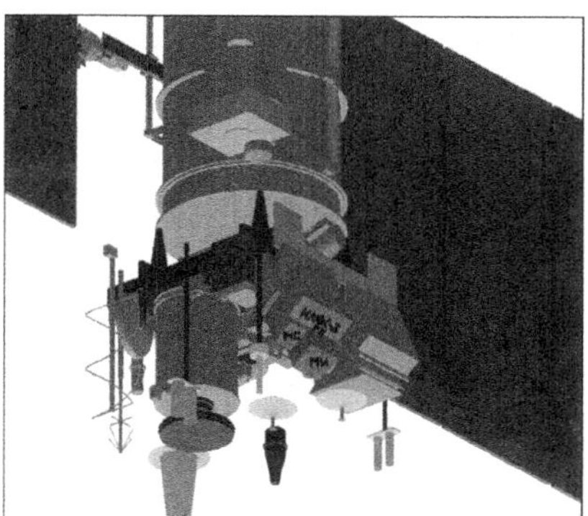

IKFS-2 Fourier spectrometer characteristics	Values
Spectral operating range	665–2000 cm^{-1} (5–15 um)
Spectral resolution	0.5 cm^{-1}
Detection limit (wave length of 13 um)	0.5×10^{-4} $w/m^2 \cdot avg \cdot cm^{-1}$
Measurement error for emission parameters in radiation temperature terms	0.5 K
Spatial resolution	35 km (at orbit altitude of 830 km)
One-point spectrum obtaining time	0.6 s
Information flow	650 Kbit/s
Mass	50 kg
Energy consumption	50 W

Fig. 1.47 IKFS-2 infrared Fourier spectrometer designed by the Keldysh Research Center

Upper atmosphere solar radiation flow meter XUV Photometer System (XPS) – "SORCE" satellite;

Devices to study the chemistry and dynamics of the earth's atmosphere, ozone, trace gas, aerosols, sulfur dioxide, nitrogen – "Aura" satellite;

Cloudiness and aerosol meter Cloud Profiling Radar (CPR) – "Cloud Sat" and "Picasso-Cena" satellites;

Atmospheric pollution meter Moderate-Resolution Imaging Spectrum-radiometer, Measurements of Pollution in the Troposphere, Clouds and the Earth's Radiant Energy System (MODIS – MOPITT – CERES) – "Terra" satellite;

SCIAMACHY (Scanning Imaging Absorption Spectrometer for Atmospheric Cartography) to measure the content of trace gas, aerosols, and clouds in the troposphere and stratosphere (up to 45 km) – "Envisat" satellite;

GLI (Global Imager), POLDER (Polarization and Directionality of the Earth's Reflectance), and ILAS (Improved Limb Atmospheric Spectrometer) to measure aerosols and ozone, carbon oxide and dioxide, nitrogen, methane, water steam – "ADEOS" satellite;

Michelson Interferometer for Passive Atmospheric Sounding (MIPAS) – "Envisat" satellite.

Fourier spectrometers to measure the content of over 20 types of trace gas at altitudes from 5 to 150 km are to be installed on perspective satellites "CrIS", "NPOESS", "IASI," and the European "METEOR" weather satellite.

Thus, the blueprinted MTSFS infrared spectrometer designed by Glavcosmos possesses similar parameters to foreign and Russian analogues, but has smaller mass and energy consumption.

As one can see, there is currently a wide list of Russian and foreign space-based equipment that is used or can be used to register precursors of seismic events and earthquakes. At upcoming work stages in the sphere of predicting natural calamities and technogenic disasters it will be necessary to select a rational set of onboard measuring equipment for the orbital aerospace monitoring group.

There is yet another aspect in the use of satellites for predicting hazardous seismic events. Multi-year observations of degassing processes in earth crust rock held by ground stations located along the fault in Italy, Dagestan and the Taman Peninsula exposed simultaneously mounting activity of the processes along a large distance which resulted in earthquakes in Turkey (1999) and Iran (2003). Although mounting activity in both cases was observed approximately 2 days before the earthquakes, it did not give grounds to reliably predict the place of the event. The strained 2,000-km long fault made it impossible to predict the exact place of strain release. To determine the epicenter it is necessary to compare the local activity level at relatively close points along the fault line. That demands an impossibly – high number of ground stations. However, it would be possible to entrust the task of monitoring small and practically edge-adjoining sections of the fault line by measuring physical values related to earth degassing to a group of satellites in orbit that crosses the fault in the desired manner. This has already been seen. For example, since 1980 increased intensity degassing flux, enhanced aerosol discharge, and anomalous intensity of electromagnetic emission have been observed over the Taman Peninsula. Several anomalies were confirmed by satellite data obtained by Intercosmos-19 [104], which documented a characteristic signal intensity increase while flying over the seismically hazardous region.

1.4.4 Problems in Practical Prediction of Seismic Activity and Their Solutions

It has been mentioned above that the problem of reliable seismic activity prediction remains unresolved although numerous examples of theoretical research have been accomplished to date to study methods of detecting earthquake precursors on the basis of analysis of seismological, geophysical, hydrogeological, geochemical, and other data that offer the basis for "seismic weather" forecasts.

Though large number of precursors is available, none of them can precisely predict the time, place, and force of an upcoming earthquake. In various seismically active regions various precursors perform differently and provide a major spread in place, time, and force of a pending quake. That happens both because of the complexity of the research object itself, which comprises the focus of a seismic event, conditions for its origination and development, and the absence of a quantity theory of earthquake origination, and the major influence of jamming factors, which are difficult to exclude from analysis. The main reason for the absence of convincing results is likely to be the unique set of precursors for each type of earthquake. Consequently, it is unclear which of them should be permanently monitored. Conversely, insufficient incoming information on various geophysical and geochemical processes in earthquake origination zones also contributes to the problem.

Earthquake and volcanic eruption predictions, like weather forecasts, have a probabilistic nature. As reports about precursors of geological natural calamities are mostly segmental and random, they make it difficult and even impossible to retrospectively assess their statistical characteristics, the probability of a successful prediction, and the average expectation time of a hazardous event after the emergence of a precursor or a group of them.

Analysis of many years' data for several geophysical (mostly seismological) precursors exposed that prediction probability by any of them does not exceed 0.5. A possible way out is the joint use of several prediction signs. It is presumed that each of them has to reflect one or another side of the multifaceted and still not very clear process of hazardous seismic event origination and does not provide sufficient information from the statistical point of view. Therefore, their complex use will enhance the reliability and efficiency of predictions. The experience of the past years justified the approach at least for mid-term predictions.

Several researchers analyzed preceding studies of precursor phenomena and anomalies in various media and attempted to design a complex systemic approach to research the lithosphere–atmosphere–ionosphere relationship [93–95]. Thus, report [22] offered the Model of lithosphere–atmosphere–ionosphere interaction (Fig. 1.48), which explained the mechanisms of anomaly emergence in the media during periods preceding seismic events. It seems to be the only attempt so far to

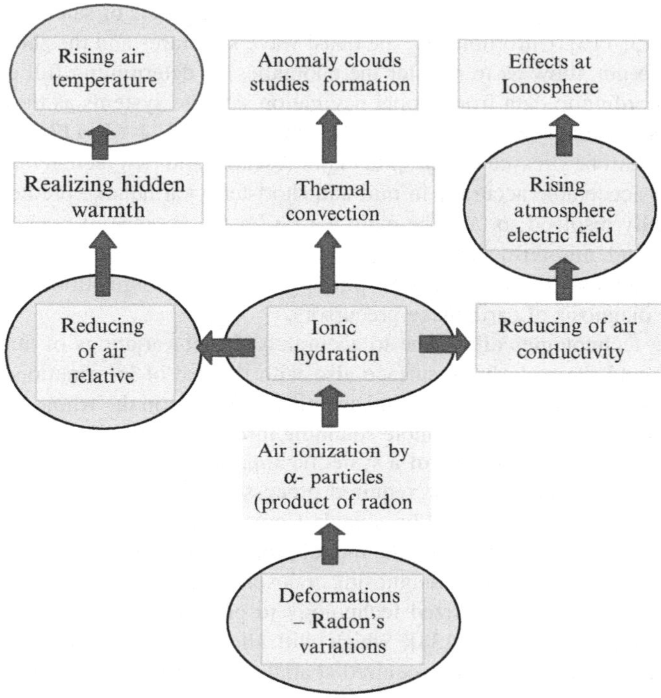

Fig. 1.48 Structural scheme of the lithosphere–atmosphere–ionosphere relationship model

systemically explain anomalous parameters before earthquakes rather than consider any medium separately from others. The model is based on research carried out in the United States, Mexico, Russia, and other countries [22, 96, 105–108] that exposed radon emanation triggered by deformation impact on the rock during an upcoming earthquake and suggested that radon discharge can be a reliable indicator of rock strain.

The physics of the process show that radioactive gas discharge into the atmosphere triggers ion hydration in the near-surface layer and creates aerosol-sized particles. As a result, air electric conductivity decreases and atmospheric electric field intensifies, on the one hand, while on the other, relative humidity falls, latent heat releases and heat convection begins in the atmosphere. The processes raise atmospheric air temperature, create anomalous cloud accumulations and both positive and negative electronic concentration variations in the maximum ***F***-layer of the ionosphere [22]. The registration of temperature and atmosphere humidity anomalies, as well as underlying surface characteristics has been properly organized at present [21–22, 29, 35–41, 100, 101, 109–114], while detection of similar variations in the ionosphere and specifically in the maximum of its ***F***-layer demand additional scientific and methodological specifications to design information processing and analysis technologies on the basis of satellite monitoring data [115, 116]. Unfortunately, the quasi wave structures and the short time of existence do not allow us to register the anomalies by determining full electronic content according to data from global navigation satellite systems as they register events of considerably bigger spatial and temporal scope [64–65, 117–120].

Thus, it shall be stressed that despite major research and rich factual material the problem of acceptable accuracy in mid and short-term earthquake predictions has not been fully resolved so far. The main reason for the absence of positive results with short and mid-term earthquake predictions is the absence of precise and confident diagnosis of the processes during earthquake origination or, in other words, the diagnosis of earthquake precursors.

Existing technologies allow one to accurately detect variations of atmosphere parameters and the underlying surface also with the use of information obtained through remote sounding of the earth. Detecting variations in the whole $\boldsymbol{F}_2$ layer of the ionosphere on the basis of remote sounding information remains an unresolved problem from the point of view of a systemic approach. Its resolution would help solve diagnosis problems for the required processes. Reliable information on the state of the ionosphere obtained by signals from satellite sounding will promote full-scale complex research to test the model proposed in report [22].

To resolve the problem of diagnosing ionosphere earthquake precursors it is necessary to design computerized technology to process ionosphere data on the radio-tomography basis [66, 138], which will allow us to expose variations of electronic concentrations of the required spatial and temporal scope in the maximum $\boldsymbol{F}_2$ layer of the ionosphere. Complex analysis of obtained data jointly with information on geodynamic lithosphere variations, as well as anomalies of other geophysical fields in seismically hazardous regions will help to determine a sufficient set of earthquake precursors to confidently diagnose earthquake-preceding processes.

Actually we need a complex experiment to confirm the principles of computerized diagnosis of ionosphere earthquake precursors within the model of the lithosphere–atmosphere–ionosphere relationship with the use of devices that operate with data from Russian and foreign space navigation systems and computerized technology of tomographic processing of information on the state of the ionosphere [93–95]. In addition to ionosphere manifestations of geotectonic activity the experiment can consider anomalous background changes in the lower atmosphere, geophysical characteristics of the lithosphere, as well as certain parameters of solar activity [91, 92]. Exposed correlation between registered anomalies (earthquake precursors) and real seismic events in the region, as well as their classification will help compile a database of the optimal number of precursors and design a complex model of earthquake precursors for the region of research.

The first step to accomplish the important scientific-applied task was the experiment held in the Far East (Sakhalin Island area) in 2007 according to the instruction of the Russian government by competent departments and organizations that aimed at diagnosing earthquake precursors. It was the first stage of a pilot project to test devices operating with data from Russian and foreign space navigation systems to promote computerized diagnosis of earthquake precursors. The Far East comprises the Kuril–Kamchatka island arch with a deep-water trench and active volcano and earthquake belts that run along it and are 1,900 km long. It is one of the most seismically active areas on earth and second only to northeastern Japan. Eighty percent of earthquakes in Russia and practically all tsunamis that hit its territory originate in the Kuril–Kamchatka area that stretches along the eastern coast of Kamchatka and the Kuril Islands on the boundary of earth crust blocs and the upper mantle of the Pacific Ocean and the Asian continent. Earthquake focus areas can be 600–700 km long. The area has 77 active volcanoes (over 10% of land volcanoes in the world). In August 2007 the Klyuchevskoy volcano powerfully erupted here (Fig. 1.49).

The pilot project and accompanying scientific experiments aimed at continuing the studies of temperature and humidity anomalies during the origination and manifestation of seismic events of various nature and intensity. At the preparation stage meteorological data obtained during strong earthquakes in Central Asia underwent a thorough statistical analysis which exposed the presence of temperature and humidity anomalies at sites of hundreds of square kilometers in size during the whole period of seismic activity [121]. Detailed analysis of the data from several strong earthquakes [30, 122] exposed the temporal dynamic of such anomalies. Thus, in particular, it was determined that daily temperature changes (difference between maximum and minimum temperatures within 24 h) offer the most indicative parameter, which in 5–7 days before an earthquake reaches its local maximum and decreases by the time of the first underground shock. It is noteworthy that the temperature maximum coincided with a relative humidity minimum. For seismic events such changes go far beyond the limits of average monthly values. However, in both cases the form of changing parameters remains invariable directly before the earthquake. Thus, the main sign of its advent is the form of changes in time (which was mentioned above) rather than absolute values of temperature and humidity variations.

Data analysis from a meteorological station network in Mexico during the origination of the Michoacan earthquake on September 19, 1985 showed the

Fig. 1.49 Klyuchevskoy volcano eruption

existence of zones with reverse in phase variations of meteorological parameters [31] and the boundary of the variation sign change corresponded to the boundary of tectonic plates and active tectonic faults. This behavior by seismic precursors corresponds to the concept provided in reports [96, 105], which use the analysis of spatial radon concentration distribution in California to detect zones of increased and decreased concentrations before the quake (which evidently corresponds to extension and compression zones). The phenomenon is likely affecting atmospheric parameter variations.

Figure 1.50a shows daily temperature range (blue line) and humidity (yellow line) variations obtained during the Far Eastern experiment in the city of Nevelsk, Sakhalin Island, in July before the August 2, 2007 quake with a magnitude of M = 6.3. It is evident that neither temperature nor humidity ranges demonstrate any anomalies and are in the limits of monthly variations, however the form of variations before the earthquake, as well as the time schedule of temperature maximum and humidity minimum manifestation several days before the quake coincides with similar manifestations in the above-mentioned seismic episodes in various parts of the planet. It is of interest that similar temperature and humidity variations in Khabarovsk (Fig. 1.50b) are practically in the antiphase to the Nevelsk parameters, which corresponds to the conclusions of report [31] on positive and negative radon emission zones.

The pilot project to diagnose precursors of geological natural calamities also analyzed cloud accumulations that emerged above the area of Nevelsk quake origination. Satellite photos exposed linear structures related to active faults and tectonic plate boundaries. The right lower corner of Fig. 1.51 shows the structure of tectonic plate boundaries near Sakhalin, while the rest of the figure is a photo from the TERRA satellite of July 30, 2007. A cut in cloudiness is clearly seen along the tectonic boundary that crosses Sakhalin (detected specifics are marked by an oval). Figure 1.51 features a photo from the AQUA satellite of July 31, 2007 which clearly shows a filamentary cloud propping practically into the epicenter of the upcoming quake (marked by oval). The size of cloud accumulations and their location testify to the scope of mounting tectonic processes and confirm the conclusions of report [31] about intensifying activity not only close to the epicenter, but also at the tectonic plate boundaries where the focus of the pending earthquake is located.

The Sakhalin experiment documented and confirmed another important factor: gas emanation into the atmosphere from tectonically active structures occurs both on land and in ocean, while corresponding short-term precursors can be successfully registered there. A proof is provided by the analysis of the OLR in the 10–13 μm band obtained from remote earth sounding radiometers [29] on satellites. Because of the transparency window the emission from the frequency band is not absorbed by cloud cover, while existing processing technology allows one to measure it above the clouds at an altitude of some 12 km. Figure 1.52 shows the advance of the ORL dynamic over Sakhalin according to night-time data from the overflying AQUA satellite. Each map provides 3-day mean-square deviation ***a*** average against the averaged monthly value of the zonal OLR index. The

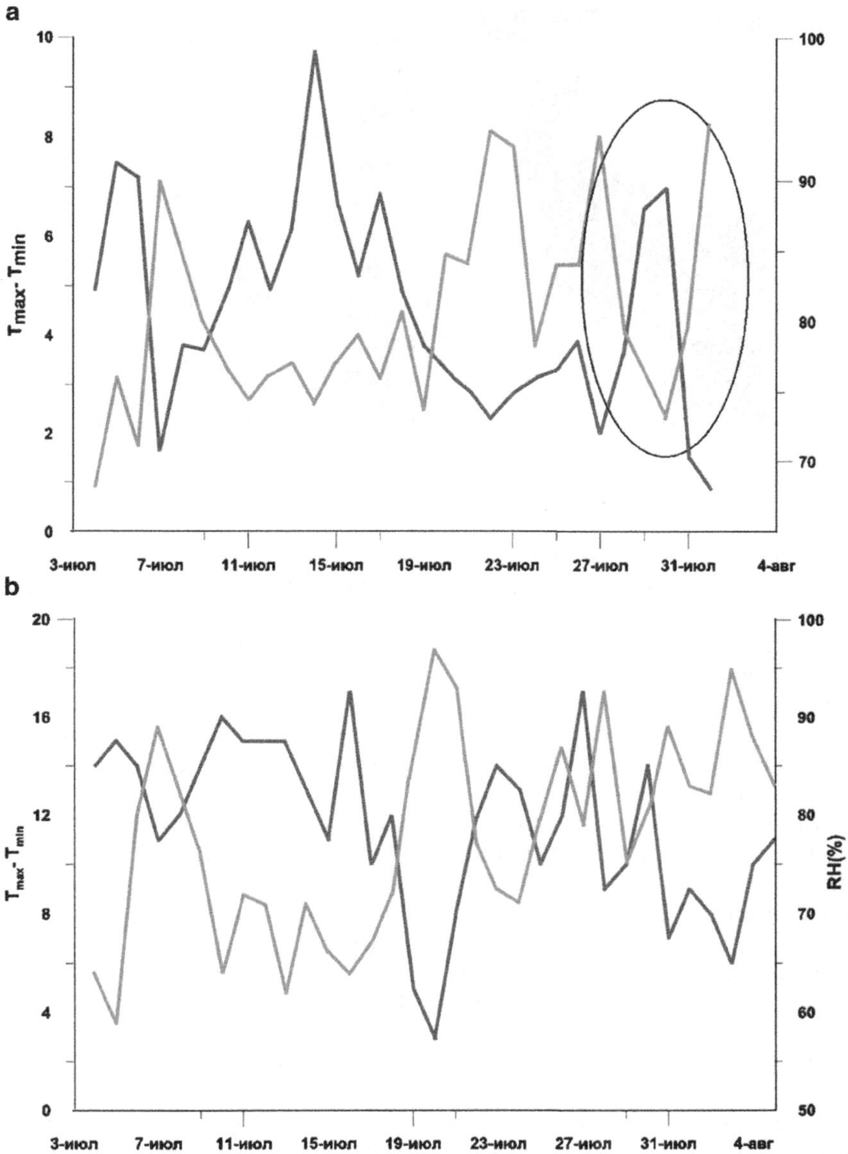

Fig. 1.50 Temperature and humidity variations during an experiment in Nevelsk (**a**) and Khabarovsk (**b**)

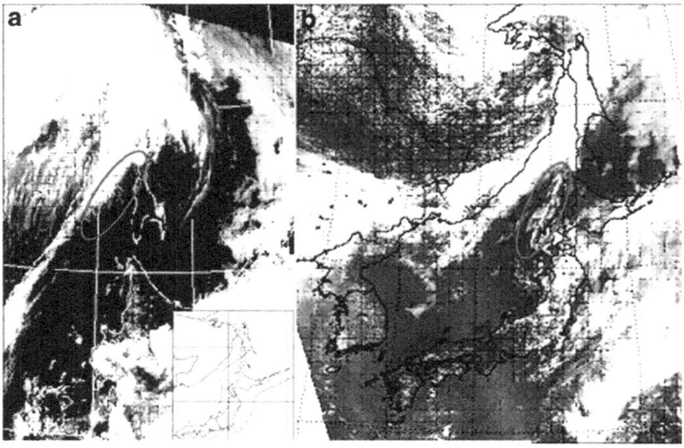

Fig. 1.51 Seismic cloud cover over Sakhalin Island

emergence of a thermal anomaly is sufficiently evident on July 26 and 29, 2007, as it travels from the beginning of the fault to the epicenter of the earthquake. On July 29 the anomaly was above the ocean.

It is clear that neither thermal waters, which are presumed to be a source of thermal anomalies before an earthquake, nor heat from the earth's crust can explain the observed phenomenon, as the anomaly is located above the water (data were verified by ocean temperature and no surface temperature increase was registered). Only latent heat discharged directly in the atmosphere because of ionization can serve as such a source.

The Sakhalin pilot project also analyzed ionospheric data. Variations of the full electronic content in July–August 2007 were calculated for four Far Eastern stations – in Petropavlovsk-Kamchatski, Yuzhno-Sakhalinsk, Khabarovsk, and Shintotsukawa (Japan). Geomagnetic excitation analysis (index $\boldsymbol{D_{st}}$, lower panel, Fig. 1.53) exposed no major geomagnetic excitations in both months.

Nevertheless, to detect possible seismic ionospheric variations they used the method to calculate a regional ionosphere changeability index [80], which detects such variations even under conditions of magnetic excitation. As Petropavlovsk-Kamchatski is located in higher latitudes and sufficiently far away from the epicenter, the index was calculated only for three stations in lower latitudes. It is seen from Fig. 1.29 that during a week before the seismic shock (from July 24 to 31, marked by an arrow) the seismic activity index increased. The second maximum after the main shock was related to successive seismic events that manifested themselves for a long time. It should be noted that index variations were not considerable, which testifies to a low radon concentration in the area. The curve of the ionosphere changeability index obtained through the analysis of signals from global navigation systems has three clear local maximums in the period preceding the seismic event: on July 24, 28, and 30.

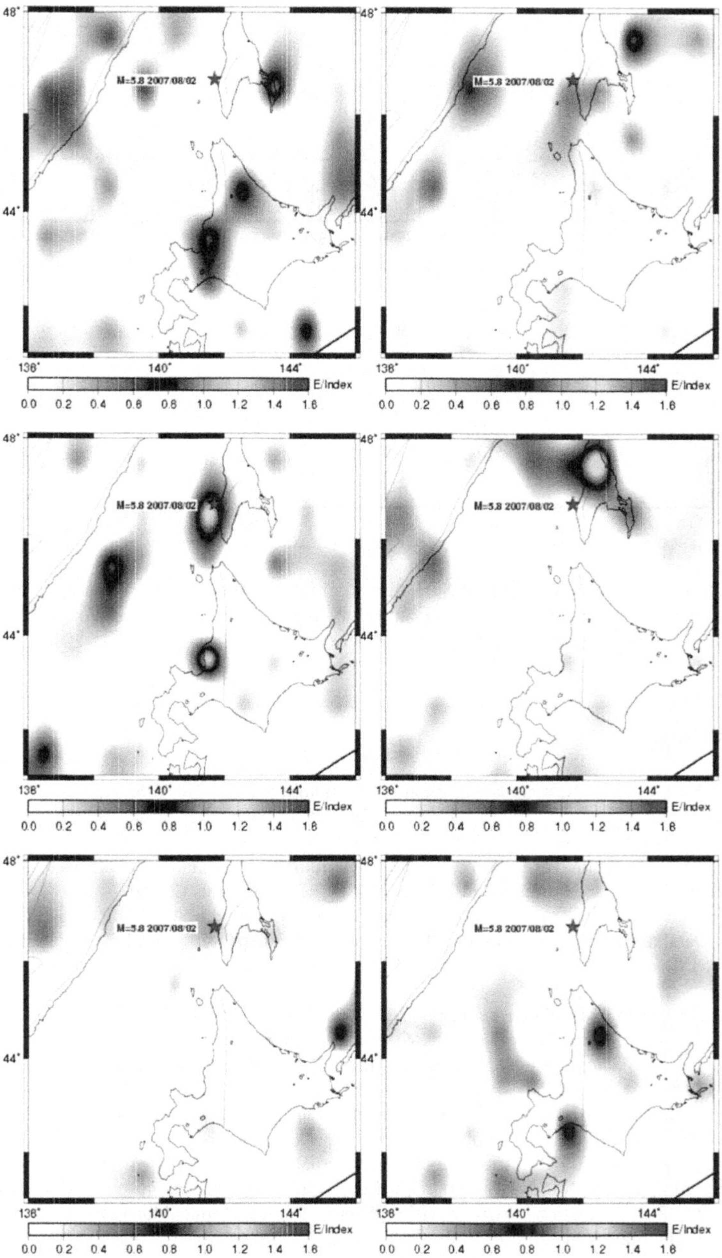

Fig. 1.52 ORL dynamic above Sakhalin Island

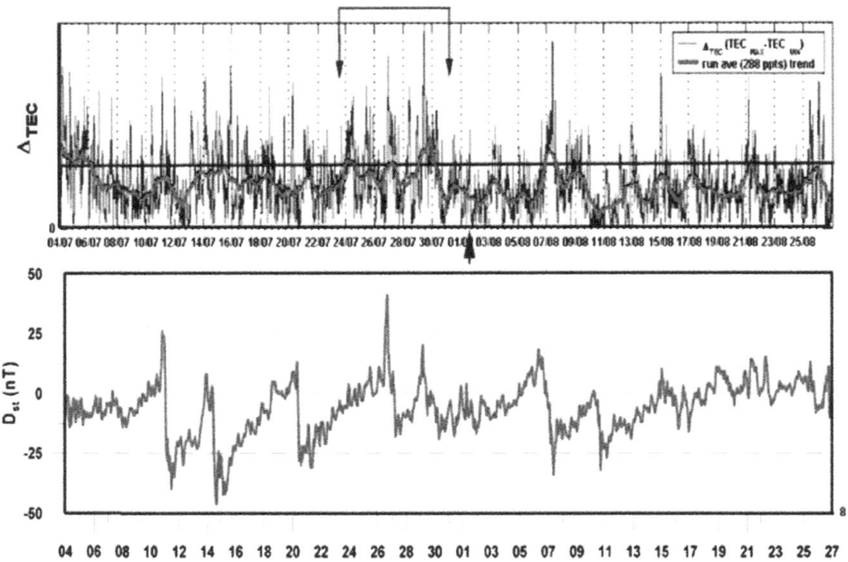

Fig. 1.53 Ionosphere behavior over Sakhalin. Regional ionosphere changeability index from GPD receiver data in the area of Sakhalin Island (July–August 2007). Global geomagnetic activity index *Dst* in the area of Sakhalin Island (July–August 2007)

References

1. UNDESA (United Nations Department of Economic and Social Affairs) Division for Sustainable Development (2002) Johannesburg Plan of Implementation. World Summit on Sustainable Development. September 2002, Johannesburg
2. Rodriguez, J., Centre for Research on the Epidemiology of Disasters, et al.: Annual Disaster Statistical Review: Numbers and Trends 2008. CRED, Brussels (2009)
3. UNEP/GRID-Europe (United Nations Environment Programme/Global Resource Information Database-Europe) Prepared for the Global Assessment Report on Disaster Risk Reduction: ISDR (International Strategy for Disaster Reduction) Secretariat. United Nations, Geneva (2009)
4. Гир Дж., Шах Х. Зыбкая твердь. – М.: Мир, 1988. – 220 с. (Russian).
5. И.А. Резанов. Великие катастрофы в истории Земли. – М.: Наука, 1980 (Russian).
6. Breining, G.: "The Deadliest Volcanoes". Super Volcano: The Ticking Time Bomb Beneath Yellowstone National Park, p.256. Voyageur Press(2007)
7. Thornton, Ian W.B.: Krakatau: The destruction and Reassembly of an Island Ecosystem. Harvard University Press(1997)
8. UNISDR/UNEP/PEDRR (United Nations International Strategy for Disaster Reduction Secretariat, United Nations Environment Programme and Partnership on Environment and Disaster Risk Reduction): Survey on Progress in Environmental Management for Disaster Risk Reduction: A Practice Area Review. Background paper for: Global Assessment Report on Disaster Risk Reduction May 2009. ISDR (International Strategy for Disaster Reduction) Secretariat. United Nations, Geneva (2009)
9. Annual Disaster Statistical Review: Numbers and Trends 2007, CRED, Brussels (2008)

10. Risk and poverty in a changing climate, Global Assessment Report on Disaster Risk Reduction (2009)
11. Апродов В.А. Зоны землетрясений. – М.: 2000. – 461 с., . (Russian).
12. Пономарев А.В., Соболев Г.А., Гитис В.Г., Чжан Чжаочэн, Ван Гусюан, Чин Синси. Комплексный анализ геофизических полей для обнаружения пространственно-временных предвестников землетрясений // Вестник ОГГГГН РАН. – 1999. – № 4(10), (Russian).
13. Федотов С.А. Выступления участников годичного общего собрания АН СССР // Вестник Академии наук СССР. – 1989. – № 8. – сс. 112–114, (Russian).
14. Моргунов В.А. Реальности прогноза землетрясений // Физика Земли. – 1999.– № 1.– сс. 79–91 (Russian).
15. Соболев Г.А. Проблема прогноза землетрясений // Природа. – 1989. – № 12 – с. 47–55, (Russian).
16. Друмя А. Землетрясения: где, когда, почему? – Кишинев, Штиинца, 1985. –195 с., (Russian).
17. Alekseev, V.A. Getling, A.V.: On the character of convective motions in the Earth's mantle. In: High-pressure Science and Technology, Sixth AIRAPT Conference. Vol. 2, pp. 321–236. Plenum Press (1979)
18. Alekseev, V.A., Getling, A.V.: Small-scale mantle convection and the origin of circular and lineal structures. In: 27th International Geological Congress. Vol. VIII, pp. 195–196. Nauka, Moscow (1984)
19. Алексеева Н.Г., Лившиц Л.Д., Алексеев В.А. Статистика мелких фракций, образующихся при разрушении горных пород // Деформирование среды при импульсном нагружении. – Киев: Наукова думка, 1985. – с. 14, (Russian).
20. Кейлис-Борок В.И. Повторный сильный толчок: прогноз возможен // Наука в России. – 1992. – № 1 – сс. 60–63, (Russian).
21. Dobrovolsky, I.P., Zubkov, S.I., Miachkin, V.I.: Estimation of the size of earthquake preparation zone. Pure Appl.Geophys. **117**, 1025–1044 (1979)
22. Добровольский И.П. Теория подготовки тектонического землетрясения. – М.: АН СССР. Ин-т физики Земли им. О.Ю. Шмидта, 1991, . (Russian)
23. Гришин А.И., Матвиенко Г.Г., Алексеев В.А., Алексеева Н.Г., Донченко В.А., Воднев С.А. Аэрозольные электрические характеристики атмосферы в сейсмоопасных районах, препринт, Институт оптики атмосферы Сибирское отделение РАН, Томск, с. 15, (Russian).
24. Alekseev, V.A., Pulinets, S.A., Legen'ka, A.D., Voitov, G.I. Radon and metallic aerosols emanation before earthquakes and its relation with variation of ionosphere parameters, Second International Aerospace Congress (IAC'97), proceed. V. 2, , pp. 274–277(1997)
25. Tectonic processes' influence on atmosphere and earthquakes prediction by cloudiness' state, Second International Aerospace Congress (IAC'97), proceed. 2, 145–147(1997)
26. Салман А.Г., Шилин Б.В. Сейсмическая активность: взгляд из космоса // Природа. – 1989. – № 12. – сс. 55–58, (Russian).
27. Dey, S., Singh, R.P.: Surface latent heat flux as an earthquake precursor. Nat. Haz. Earth Syst. Sci. **3**, 749–755 (2003)
28. Dey, S., Sarkar, S., Singh, R.P.: Anomalous changes in column water vapor after Gujarat earthquake. Adv. Space Res. **33**(3), 274–278 (2004)
29. Ouzounov, D., Liu, D., Chunli, K., Cervone, G., Kafatos, M., Taylor, P.: Outgoing long wave radiation variability from IR satellite data prior to major earthquakes. Tectonophysics **431**, 211–220 (2007)
30. Pulinets, S.A., Ouzounov, D., Karelin, A.V., Boyarchuk, K.A., Pokhmelnykh, L.A.: The physical nature of the thermal anomalies observed before strong earthquakes. Physics and Chemistry of the Earth **31**, 143–153 (2006)
31. Pulinets, S.A., Dunajecka, M.A.: Specific variations of air temperature and relative humidity around the time of michoacan earthquake M8.1 Sept. 19, 1985 as a possible indicator of interaction between tectonic plates. Tectonophysics **431**(1–4), 221–230 (2007)

32. Гальпер А.М., Дмитренко В.В., Никитина Н.В. и др. Взаимосвязь потоков высокоэнергичных заряженных частиц в радиационном поясе с сейсмичностью Земли// Космические исследования. – 1989. – Т. 27. – № 5. – сс. 789–792, (Russian).
33. Galper, A.M., Koldashov, S.V., Voronov, S.A.: High energy particle flux variations as earthquake predictors. Advance Space Research **15**, 131–134 (1995)
34. Гальпер А.М., Колдашов С.В., Мурашов А.М. Численное моделирование физических процессов на атмосферной границе радиационного пояса // Космические исследования. – 2000. – Т. 38. –№ 1. – с. 102, (Russian).
35. Pulinets, S.A., Boyarchuk, K.A.: Ionospheric Precursors of Earthquakes. Springer, Berlin (2004)
36. Haicheng Prediction of the Haicheng earthquake, Earthquake Study Delegation, Eos **58**, 236–272 (1977)
37. Ouzounov, D., Bryant, N., Logan, T., Pulinets, S., Taylor, P.: Satellite thermal IR phenomena associated with some of the major earthquakes in 1999–2003. Physics and Chemistry of the Earth **31**, 154–163 (2006)
38. Сальман А.Г., Тронин А.А. Вариации потока уходящего ИК излучения Земли в сейсмоактивных районах Средней Азии //, Известия АН СССР, Физика земли. – 1990. – № 7. – сс. 67–69, (Russian)
39. Tronin, A.A.: Thermal I.R. satellite sensor data application for earthquake research in China. Intemat. J. Rem. Sens. **21**(16), 3169–3177 (2000)
40. Tronin, A.A., Biagi, P.P., Molchanov, O.A., et al.: Temperature variations related to earthquakes from simultaneous observation at the ground stations and by satellites in Kamchatka area. Phys. and Chemistry of the Earth **29**, 501–506 (2004)
41. Горный В.И., Сальман А.Г., Тронин А.А., Шилин Б.В. Уходящее инфракрасное излучение Земли – индикатор сейсмической активности // Доклады АН СССР. – 1988. – Т. 301.– № 1. – сс. 67–69, (Russian)
42. Ларкин В.И., Мигулин В.В., Сергеева Н.Г., Сенин Б.В. Электромагнитное излучение над глубинными разломами литосферы по измерениям на спутнике // Доклады академии наук. – 1998. – Т. 360. – № 6. – сс. 814–818, (Russian).
43. Исаев Н.В., Серебрякова О.Н. Электромагнитные и плазменные эффекты сейсмической активности в ионосфере Земли // Химическая физика. – 2000. – Т. 19. – № 6. – сс. 67–70, (Russian).
44. Boskova, J., Smilauer, J., Jiricek, F., Triska, P.: Is the ion composition of the outer ionosphere related to seismic activity? J. Atmos. Terr. Phys. **55**, 1689–1695 (1993)
45. Boskova, J., Smilauer, J., Triska, P., Kudela, K.: Anomalous behavior of plasma parameters as observed by the INTERCOSMOS-24 satellite prior to the Iranian earthquake of 20 June 1990. Studia Geoph. et Geod. **38**, 213–220 (1994)
46. Ларкина В.И., Наливайко А.В., Гершензон Н.И. и др. Наблюдения на спутнике « Интеркосмос 19 » ОНЧ излучений, связанных с сейсмической активностью // Геомагнетизм и аэрономия. – 1983. – Т. 23. – № 5. – сс. 842–845, (Russian).
47. Мигулин В.В., Ларкина В.И., Молчанов О.А. и др. Обнаружение эффектов воздействия землетрясений на ОНЧ-КНЧ шумы во внешней ионосфере. – М.: ИЗМИРАН, 1982. – Препринт № 25 (390).– 28 с., (Russian)
48. Мимно Г.Р. Физика ионосферы//Успехи физических на ук. – 1957. – Т. 18 .– с. 103, (Russian).
49. Гинзбург В.Л. Распространение электромагнитных волн в плазме. – М.: Наука, 1960, (Russian).
50. Альперт Я. Л. Распространение радиоволн и ионосфера. – М.: Наука, 1960, (Russian).
51. Ратклиф Дж. А., Уикс К. Ионосфера // Физика верхней атмосферы / Пер. с англ. – М.: Мир, 1963. – сс. 339–418, (Russian).
52. Leandro, R.A.,: New Technique to TEC Regional Modeling using a Neural Network. Proceedings of the 17th International Technical Meeting of the Satellite Division of the Institute of Navigation ION GNSS, pp.366–374 (2004).

53. Jin, S., Wang, J., Zhang, H.-P., Zhu, W.-Y.: Real-time monitoring and prediction of ionospheric electron content by means of GPS. Chinese Astronomy and Astrophysics **28**, 331–337 (2004)
54. Jensen, A., Ovstedal, O., Grinde, G.: Development of a Regional Ionosphere Model for Norway. Proceedings of the 20th International Technical Meeting of the Satellite Division of the Institute of Navigation ION GNSS, pp.2880–2889, (2007)
55. Komjathy, A., Sparks, L., Mannucci, A., Coster, A.: The Ionospheric Impact of the October 2003 Storm Event on WAAS. Proceedings of the 17th International Technical Meeting of the Satellite Division of the Institute of Navigation ION GNSS, 1298–1307, (2004)
56. El-Gizawy, M., Skone, S.: A Canadian. Ionospheric Warning and Alert System. Proceedings of the 15th International Technical Meeting of the Satellite Division of the Institute of Navigation ION GPS, pp.1345–1352, (2002)
57. Zain, A., Ho, Y., Abdullah, M.: Enabling GPS Technology on Equatorial Ionosphere Monitoring During Geomagnetic Storm of July 15–17, 2000. Proceedings of the 15th International Technical Meeting of the Satellite Division of the Institute of Navigation ION GPS, pp.1341– 1344, (2002)
58. Asgari, J., Harmel, A.: Least Square Spectral Analysis of the GPS Derived Ionosphere Data. Proceedings of the 2005 National Technical Meeting of the Institute of Navigation, pp.716–724, (2005)
59. Regar, S., Sunda, S., Sivaraman, M., Bandyopadhyay, K.: Statistics of Occurrence of Ionospheric Scintillation and TEC Depletions Over Indian Region and its Effect on Satellite Navigation. Proceedings of the 19th International Technical Meeting of the Satellite Division of the Institute of Navigation ION GNSS, pp.1336–1340, (2006)
60. UNISDR (United Nations International Strategy for Disaster Reduction Secretariat): Linking Disaster Risk Reduction and Poverty Reduction: Good Practices and Lessons Learned. United Nations, New York and Geneva (2008). A publication of the Global Network of NGOs for Disaster Risk Reduction
61. Markkanen, M., Lehtinen, M., Nygren, T., Partilla, J., Hene-lius, P., Vilenius, E., Tereshchenko, E.D., Khudukon, B.Z.: Bayesian approach to satellite radio tomography with applications in the scandinavian sector. Ann. Geophysicae **13**, 1277–1287 (1995)
62. Романов А.А., Трусов С.В., Романов А.А., Крючков В.Г. Томографическая реконструкция вертикального распределения электронной концентрации ионосферы в дальневосточном регионе России // Труды научно-технической конференции « Актуальные проблемы ракетно-космического приборостроения и информационных технологий», посвященной 50-летию первого искусственного спутника Земли, 19–21 июня 2007 г. – М.: Физматлит, 2008. – сс. 278–284, (Russian).
63. Левин В.Е., Гордеев Е.И., Бахтиаров В.Ф., Касахара М. Предварительные результаты GPS мониторинга на Камчатке и Командорских островах // Вулканология и сейсмология. – 2002. – № 1. – сс. 3–11, (Russian).
64. Урличич Ю.М., Трусов С.В., Романов А.А., Новиков А.В., Романов А.А., Аверин С.В. Моделирование пространственно-временного поля ионосферной задержки сигнала спутниковых навигационных систем ГЛОНАСС/GPS в среде виртуальной реальности Avango // Мехатроника, автоматизация, управление. – 2007. – № 5. – сс. 29–31, (Russian).
65. Новиков А.В., Романов А.А., Романов А.А. Исследование пространственного распределения ПЭС над территорией России по данным радио-просвечивания атмосферы сигналами ГЛОНАСС/GPS // Современные проблемы дистанционного зондирования Земли из космоса: Физические основы, методы и технологии мониторинга окружающей среды, потенциально опасных явлений и объектов. – М.: ООО « Азбука-2000», 2008. – Вып. 5. – Т. I. – сс. 550–553, (Russian).
66. Трусов С.В., Романов А.А., Романов А.А., Новиков А.А. Современные подходы к созданию программного комплекса сбора и обработки данных спутникового мониторинга ионосферы и ионосферных предвестников землетрясений // Труды научно-технической конференции « Актуальные проблемы ракетно-космического приборостроения и информационных технологий», посвященной

50-летию первого искусственного спутника Земли, 19–21 июня 2007 г. – М.: ФИЗМАТЛИТ, 2008. – сс. 284–291, (Russian).
67. Смирнов В.М. Зондирование ионосферы методом радио-просвечивания по трассе « навигационный спутник-Земля» / В.М. Смирнов, Е.В. Смирнова // Инженерная экология-2003: материалы межд. сим., Москва, 1–2 декабря 2003 г. – М.: Изд-во ИРЭ РАН, 2003. – сс. 66–70, (Russian).
68. Марчук В.Н., Смирнов В.М. Определение электронного содержания ионосферы Земли по данным дальномерных и фазовых измерений // Электронный журнал « И сследовано в России». – 2001. – № 127. – сс. 1465–1475, (Russian) http://zhurnal.ape.relarn.ru
69. Смирнов В.М. Вариации ионосферы в период землетрясений по данным навигационных систем // Электронный журнал « Исследовано в России». – 2001. № 153. – сс. 1759–1767, (Russian) http://zhurnal.ape.relarn.ru
70. Гохберг М.Б., Кустов А.В., Липеровский В.А., Липеровская Р.Х., Харин Е.П., Шалимов С.Л. О возмущениях F-области ионосферы перед сильными землетрясениями // Известия АН СССР. Физика Земли. – 1988. – № 4. – сс. 12–20, (Russian).
71. Смирнов В.М. Интерпретация ионосферных возмущений в период слабых землетрясений // Электронный журнал « Исследовано в России». – 2003. – № 012. – сс. 121–129, (Russian) http://zhurnal.ape.relarn.ru
72. Афраймович Э.Л., Косогоров Е.А., Лесюта О.С., Ушаков И.И. Спектр перемещающихся ионосферных возмущений по данным глобальной сети GPS // Известия высших учебных заведений. Радиофизика. – 2001. – Т. 44. – № 10. – сс. 828–839, (Russian)
73. Афраймович Э.Л., Астафьева Э.И. Морфология и динамика изолированных ионосферных неоднородностей по данным глобальной сети GPS // Солнечно-земная физика. – 2003. – Т. 116. – № 3. – сс. 50–60, (Russian).
74. Takayuki Y., Naoki F., Akinori S.A. Study of the Ionospheric Effect on GBAS (Ground-Based Augmentation System) Using the Nation-Wide GPS Network Data in Japan// Proceedings of the 2004 National Technical Meeting of the Institute of Navigation, pp.502–511, 2004.
75. Алексеев В.А., Алексеева Н.Г. К вопросу о поступлении тяжелых металлов при дегазации Земли. Ядерно-физизические методы анализа в контроле окружающей среды // Труды Второго всесоюзного совещания – Л.: Гидромет, 1985. – сс. 173–179 (Russian).
76. Alekseev, V.A., Alekseeva, N.G.: Investigation of metal transfer in the biosphere during gaseous emission in zones of tectonic activity using methods of nuclear physics. Nucl. Geophys. **6**(1), 99–110 (1992)
77. Алексеев В.А., Алексеева Н.Г. Поступление микроэлементов при фумарольной деятельности вулкана Авачинский и их распределение в свежевыпавшем и сезонном снеге и природных водах // Вулканология и сейсмология. – 1989. – № 3. – сс. 30–37, (Russian).
78. Алексеев В.А., Алексеева Н.Г., Муравьев Я.Д. и др. Аэрозольный предвестник извержения Авачинского вулкана в январе 1991 г.// ДАН. – 1995. – Т. 345. – № 5. – сс. 667–670 (Russian).
79. Liu, J.Y., Chuo, Y.J., Shan, S.J., Tsai, Y.B., Chen, Y.I., Pulinets, S.A., Yu, S.B.: Pre-earthquake ionospheric anomalies registered by continuous GPS TEC measurements. Annales Geophysicae **22**(5), 1585–1593 (2004)
80. Pulinets, S.A., Kotsarenko, A.N., Ciraolo, L., Pulinets, I.A.: Special case of ionospheric day-to-day variability associated with earthquake preparation. Adv. Space Res. **39**(5), 970–977 (2007)
81. Ахметов Р.Р., Куницын В.Е. Численный метод решения задачи распространения акустико-гравитационных волн в атмосфере до ионосферных высот // Вестник Московского Университета. – 2003. Сер. 3. – № 3. – сс. 38–42, (Russian).
82. Гохберг М.Б., Моргунов В.А., Похотелов О.А. Сейсмо-электромагнитные явления. – М.: Наука, 1988, (Russian).

83. Гохберг М.Б., Пилипенко В.А., Похотелов О.А., Партасарати С. Акустическое возмущение от подземного ядерного взрыва как источник электростатической турбулентности в магнитосфере // доклад АН СССР. – 1990. – № 313(3). – сс. 568–574,. (Russian)
84. Гохберг М.Б., Сараев А.К., Пертель М.И. Исследования ионосферных возмущений, вызванных карьерными взрывами в диапазоне сверхнизких частот // Краткосрочный прогноз катастрофических землетрясений с помощью радиофизических наземно-космических методов. – М.: ОИФЗ РАН, 1999. – сс. 153–162, (Russian).
85. Андреева Е.С., Гохберг М.Б., Куницын В.Е., Терещенко Е.Д., Худукон Б.З., Шалимов С.Л. Радио-томографическая регистрация возмущений ионосферы от наземных взрывов // Космические исследования –2001. – Т. 39. – № 1. – сс. 13–17, (Russian).
86. Перцев Н.Н., Шалимов С.Л. Генерация атмосферных гравитационных волн в сейсмически активном регионе и их влияние на ионосферу // Геомагнетизм и аэрономия. – 1996. – № 36. – сс. 111–118, (Russian).
87. Pulinets, S.A., Legen'ka, A.D., Gaivoronskaya, T.V., Depuev, V.Kh.: Main phenomenological features of ionospheric precursors of strong earthquakes. Journal of Atmospheric and Solar-Terrestrial Physics **65**, 1337–1347 (2003)
88. Pulinets, S.A., Liu, J.Y.: Ionospheric variability unrelated to solar and geomagnetic activity. Adv. Space Res. **34**, 1926–1933 (2004)
89. Афраймович Э. Л., Астафьева Э. И., Гохберг М.Б., Лапшин В.М., Пермякова В.Е., Стеблов Г.М., Шалимов С.Л. Поиски предвестников землетрясений в вариациях полного электронного содержания в ионосфере по данным GPS во время землетрясения Hector Mine 16 октября 1999 г. // Материалы III международного совещания « Солнечно-земные связи и электромагнитные предвестники землетрясений». – Пос. Паратунка, 2004, (Russian).
90. Афраймович Э.Л., Перевалова Н.П. GPS-мониторинг атмосферы Земли. – Иркутск: ГУ НЦ РВХ ВСНЦ СО РАМН, 2006. – 480 с., (Russian)
91. Пулинец С.А., Романов А.А., Урличич Ю.М., Романов А.А., Дода Л.Н., Узунов Д. Первые результаты пилотного проекта по комплексному диагностированию предвестников землетрясений на Сахалине // Геомагнетизм и Аэрономия. – 2009. – т. 49. – № 1. – сс. 123–132, (Russian).
92. Романов А.А., Урличич Ю.М., Пулинец С.А., Романов А.А. Пилотный проект по комплексному диагностированию предвестников землетрясений на Сахалине: программа и предварительные результаты эксперимента в 2007 г. // Вопросы электромеханики. – 2008. – т. 105. – сс. 84–93, (Russian).
93. Романов А.А., Романов А.А., Трусов С.В., Кантаков Г.А. Проведение комплексных исследований сейсмо-гео-гелио-атмо-литосферных связей и выявление сигнатур предвестников землетрясений (The Complex Research of the Earthquake's Precursors Signatures Revealing) // Труды международного семинара по теме « Превентивные геотехнические меры по уменьшению природных и техногенных бедствий » (25–27 июля 2007 г., Россия), (Russian) .
94. Романов А.А., Романов А.А., Трусов С.В. Комплексное диагностирование предвестников землетрясений на основе гелио-геофизических данных дистанционного зондирования // Труды IV международной конференции « Солнечно-земные связи и предвестники землетрясений » 14.08.2007–18.08.2007, (Russian) .
95. Romanov, A.A., Romanov, A.A., Urlichich, Y.M.: The pilot-project on the composite diagnosis of the earthquake's precursors relying on helio-geophysical remote sensing. Asia-Pacific Space Outlook **#3**(14), 27–30 (2007)
96. Уткин В.И. Радон и проблема тектонических землетрясений // Соросовский образовательный журнал. – 2000. – Т. 6. – № 12. – сс. 64–70, (Russian).

97. Ораевский В.Н., Пулинец С.А., Ружин Ю.Я. Ионосферные предвестники землетрясений, препринт № 1(1129), ИЗМИРАН. – М., 2000. – 18 с, (Russian).
98. Johnston, M.J.S.: Review of electric and magnetic fields accompanying seismic And volcanic activity. Surveys in Geophysics **18**, 441–475 (1997)
99. Дружин Г.И., Шапаев В.И. Роль мировой грозовой активности в формировании амплитуды регулярного шумового фона // Геомагнетизм и аэрономия. – 1988. – т. 28. – № 1. – сс. 81–86 (Russian).
100. Molchanov, O.A., Hayakawa, M.: Sub-ionospheric VLF signal perturbations possibly related with earthquakes. J. Geophys. Res. **103**, 17489–17504 (1998)
101. Fujinawa, Y. Takahashi, k,: Electromagnetic radiations associated with major earthquakes, Physics of the Earth and Planetary Interiors, **105**, p.249 (1998)
102. Гальпер А.М. Землетрясения: прогноз из космоса? // Наука в России. – 1994. – вып. 1. – с. 39 (Russian)
103. Алексеев В.А., Алексеева Н.Г. Связь аэрозольных предвестников землетрясений и извержений вулканов с деформациями земной коры//Дегазация Земли: геодинамика, геофлюиды, нефть, газ. – М.: 2002. – сс. 79–81, (Russian).
104. Alekseev, V.A., Larkina, V.F., Oraevsky, V.N., Yu.Ruzhin, Yu.Ya.: Tectonic anomaly and fault of the earth crust discovered by satellite monitoring. Second International Aerospace Congress IAC'97. Proceed. 2, 301–306(1997)
105. Уткин В.И. Газовое дыхание Земли // Соросовский образовательный журнал. – 1997. – № 1. – сс. 57–64, (Russian).
106. Belyaev, A.A.: Specific features of radon earthquake precursors. Geochemistry International **39**(12), 1245–1250 (2001)
107. Segovia, N., Pulinets, S.A., Leyva, A., Mena, M., Monnin, M., Camacho, M.E., Poncianod, M.G., Fernandez, V.: Ground radon exhalation, an electrostatic contribution for upper atmospheric layers processes. Radiat. Meas. **40**, 670–672 (2005)
108. Урличич Ю.М., Романов А.А., Романов А.А., Трусов С.В. Современные подходы к созданию автоматизированной системы регистрации ионосферных предвестников землетрясений по спутниковым данным // Космонавтика и ракетостроение. – 2006. – № 1. – сс. 167–172, (Russian).
109. Schmugge, TJ., Kustas, WP., Ritchie,JC., Jackson, TJ., Al Rango: Remote sensing in hydrology. Advances in Water Resources **25** 1367–1385 (2002)
110. Biagl, P.F., Eremin, A., Cozzi, E., Khatkevich, Y.M., Gordeev, E.I.,:Hydrogeochemical precursors in Kamchatka (Russia) related to the strongest earthquakes in 1988–1997, Natural Hazards. 263–276 (2000).
111. Varshal, G.M., Zamokina, N.S., Timakova, E.P., Bakaldina, N.A., Dzeheglova, M.L., Zinger, A.A.: Halogenated and sulfide ions as earthquake precursors and their detection by use of ion-selective electrodes. In: Sobolev, G.A., Varshal, G.M. (eds.) Hydrogeochemical Earthquake Precursors, pp. 236–249. Nauka, Moscow (1985)
112. Wakita, H.: Earthquake prediction and geochemical studies in China. China Geophysics **1**, 443–457 (1978)
113. King, C.Y., Evans, W.C., Presser, T., Husk, R.: Anomalous chemical changes in well waters and possible relation to earthquakes. Geophysics Research Letters **8**, 425–428 (1981)
114. Wakita, H.: Earthquake Prediction Techniques, p. 175. University of Tokyo Press, Tokyo (1982)
115. Трусов С.В., Романов А.А., Романов А.А., Новиков А.А. Современные подходы к созданию программного комплекса сбора и обработки данных спутникового мониторинга ионосферы и ионосферных предвестников землетрясений//Сборник тезисов научно-технической конференции « Актуальные проблемы ракетно-космического приборостроения и информационных технологий», посвященной 50-летию первого искусственного спутника Земли, 19–21 июня 2007 г. – М.: Физматлит, 2008. – сс. 284–290, (Russian).

116. Андреева Е.С., Куницын В.Е., Терещенко Е.Д. Фазоразностная радиотомография ионосферы // Геомагнетизм и аэрономия. – 1992. – 32. – № 1. – сс. 104–110, (Russian)
117. Mapping and predicting the earth's ionosphere using the Global Positioning System. PhD dissertation of Stefan Schaer, Bern University, c. 205 (1999)
118. Plotkin, V.V.: GPS detection of ionospheric perturbation before the 13 February 2001, El salvador earthquake. Nat. Hazards and Earth Syst. Sci. **3**, 249–253 (2003)
119. Новиков А.В., Романов А.А., Романов А.А. Комбинированный алгоритм расчета пространственного распределения ПЭС по данным радиопросвечивания атмосферы сигналами ГЛОНАСС/GPS и моделирование // Мехатроника, автоматизация, управление. – 2007. – № 5.– сс. 24–27, (Russian).
120. Новиков А.В., Романов А.А., Романов А.А. Вычисление поля ионосферных поправок для сигналов ГЛОНАСС/GPS по данным сети контрольно-корректирующих станций// Труды научно-технической конференции « Актуальные проблемы ракетно-космического приборостроения и информационных технологий», посвященной 50-летию первого искусственного спутника Земли, 19–21 июня 2007 г. – М.: ФИЗМАТЛИТ, 2008. – сс. 49–56, (Russian)
121. Милькис М.Р. Метеорологические предвестники сильных землетрясений//Изв. АН СССР. Физика Земли. – 1986. – № 3. – сс. 36–47, (Russian)
122. Pulinets, S.A., Ouzounov, D., Ciraolo, L., Singh, R., Cervone, G., Leyva, A., Dunajecka, M., Karelin, A.V., Boyarchuk, K.A., Kotsarenko, A.: Thermal, atmospheric and ionospheric anomalies around the time of the colima M7.8 earthquake of 21 January 2003. Annales Geophysicae **24**, 835–849 (2006)
123. UNDESA (United Nations Department of Economic and Social Affairs): World Economic and Social Survey 2008: Overcoming Economic Insecurity. United Nations, New York (2008)
124. Tang, X.: Anomalous Meteorology. A General History of Earthquake Study in China, pp. 49–84. Science Press, Beijing (1988)

Further Reading

1. Меньшиков, В.А., Перминов, А.Н., Ю.М. Урличич Глобальные проблемы человечества и космоса – М.: НИИ КС имени А.А. Максимова, 2010. (Russian) – 570c
2. Меньшиков и др, В.А. "Прогнозирование землетрясений, тайфунов и цунами с помощью космических ускорительных средств геологического зондирования и мониторинга." –\ Геоинформатика. – 1997. – № 3
3. Меньшиков, В.А. "Перспективы развития космических систем получения и обработки мониторинговой информации для управления процессами ликвидации экстримальных ситуаций на земле.". \"двойные технологии". – 2001 – № 1
4. Alekseev V. A., Radikov A. V., Rembeza A.I. "The forecast of earthquakes on the basis of the analysis of their harbingers." by book "Space for Security and Prosperity of the Peoples.", editors Contant J.-M., Menshikov V., IAA,- 2010, p. 277–294
5. Ulivieri C., Martino L., Jahjah M. "Space monitoring and natural disasters case study: post tsunami Banda Aceh district." by book "Space for Security and Prosperity of the Peoples.", editors Contant J.-M., Menshikov V., IAA,- 2010, p. 108–128
6. Alekseev V. A., Danialov M. G., Malgin A. S., Ryazhskaya O. G., Saavedra O., Tronin A. A. "Precursors of the Turkey earthquakes of 1999, Iran earthquake 2003, Italian 2009, all observed in Daghestan, in central Italy and in space." by book "Space for Security and Prosperity of the Peoples.", editors Contant J.-M., Menshikov V., IAA,- 2010, p. 358–36

Technogenic Emergencies: Can They Be Reliably Predicted?

2

The current level of science, technology, and production places individual countries and mankind in general at risk. Modern society becomes more and more concerned with technogenic safety. Having concentrated in a locational sense tremendous amounts of different kinds of energy, hazardous substances and materials, present-day industry has become a constant source of significant risks associated with mass loss of life and environmental problems during emergencies such as incidents, accidents, and disasters. Although mankind cannot develop without new technologies, their increasing use does not reduce potential risks, but to the contrary often creates new threats and risks. The use of increasingly sophisticated and energy-consuming equipment inevitably results in incidents if not handled properly.

It was believed earlier that technogenic emergencies were contingent and therefore were blamed on incidents, unfavorable circumstances, human error, etc. Now a somewhat different approach toward dangers related to human activities is evolving. It has been recognized that many of them are quite predictable and can be modeled and prevented to the fullest extent possible using appropriate preventive measures. The existing methodology and practice of ensuring reliability and safety of sophisticated man–machine systems can often solve these tasks successfully.

According to estimates, about 10–15% of technogenic disasters have measurable physical harbingers that can be used for forecasting such events and preventing their adverse consequences. The majority of such harbingers can be detected using ground-based or aerial means. However, in a number of cases involving extended engineering systems (such as trunk oil and gas pipelines, railroads, sea fleets, mining industries, hydroelectric stations, nuclear power plants, etc.), these tasks can be solved only by space monitoring systems. Moreover, it is often impossible to ensure compliance with safety requirements and subsequent safe operation without space monitoring data that are used when designing complex engineering systems or in the location of seismically and hydrologically safe places for their construction, and exercising construction supervision.

V.A. Menshikov et al., *Global Aerospace Monitoring and Disaster Management*,
DOI 10.1007/978-3-7091-0810-9_2,

Each technogenic disaster is unique in its own way. However, practically all of them are associated with failures of complex organizational and engineering systems, i.e., systems that involve during all cycles of their operation not only equipment but also people, whose faultless actions are critical to the reliability and efficiency of its work. According to Lee Davis, the author of "Man-Made Catastrophes," the so-called "human factor" that is to be blamed for up to 90% of technogenic emergencies boils down almost entirely to three circumstances: "stupidity, neglect, and greed." This may prompt a more positive conclusion: the above-mentioned reasons can be dealt with using preventive measures at all stages of an engineering system's life cycle, as cutting-edge industries do in developed countries. The authors of another work, "Man-Made Disaster," Barry Turner and Nick Pidgeon, take a broader look at the problem: although a technogenic disaster can happen virtually everywhere and there is no "absolute tool" that can avert it, there are a number of factors that can help "delay" it and minimize its consequences. The authors believe that these are mainly a high level of education and a pro-active civil position of local society. The more responsible and professional the residents of a certain country are in performing their obligations to society and the better society oversees them, the lower the probability of a technogenic disaster. The training of personnel in governmental agencies and private companies to act in emergencies is also vital.

Nevertheless, we have what we have: despite achievements in the development of automated and remote control technologies, and technical personnel training, the number of technogenic emergencies grows every year, challenging natural disasters in terms of destructive effects.

2.1 Technogenic Disasters in the Modern World: Classification, Facts, Consequences, and Tendencies

For decades, the well-known and influential Center for Research on the Epidemiology of Disasters (CRED) has been keeping a database of natural and technogenic disasters on the planet. According to the center's criteria, an emergency is a disaster if at least one of the following four events has occurred: more than 10 people have died, more than 100 people have been injured, and local authorities have declared a state of emergency or requested international assistance. A technogenic disaster is a type of disaster that is caused by a failure of engineering systems resulting in an accident at an industrial facility or on transport, which as a rule is accompanied by mass casualties among the population and an environmental disaster, and poses an immediate threat to public and national security. It is considered an emergency with all the ensuing social, legal, and environmental consequences.

One of the classification options for technogenic disasters (Fig. 2.1) divides them into three main types: "industrial" (radiation and chemical contamination, explosions, destruction caused by other factors), "transport" (air, sea, and railroad accidents, etc.), and "others" (all other disasters). By origin, technogenic disasters can be caused by: unintentional, erroneous actions of the maintenance personnel whose incorrect

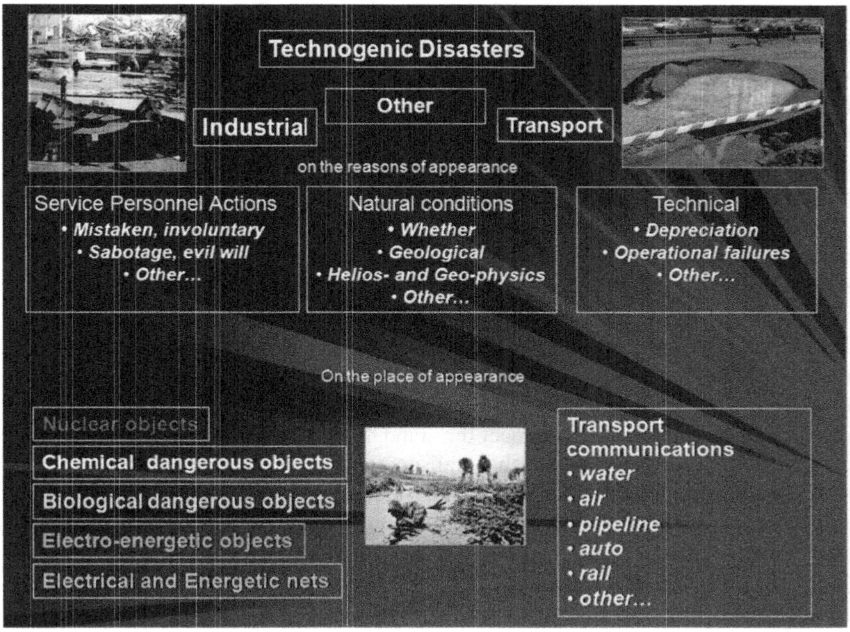

Fig. 2.1 One of the classifications of technogenic disasters

actions due to lack of attention or poor training provoked an accident (for example, the Chernobyl nuclear catastrophe), intentional actions of the maintenance personnel with evil intent or as sabotage; tectonic, natural, or weather conditions; wear and tear of equipment, or other unforeseen and undesirable consequences of standard operation of complex organizational and engineering systems (the accident at the chemical plant Union Carbide in Bhopal, India).

By the place of occurrence, accidents are divided into those at nuclear facilities of engineering and research centers and nuclear power plants (with destruction of production premises and radioactive contamination of the surrounding area); at hazardous chemical facilities (with release, spill, or leak of toxic substances); at research institutes (enterprises) that design, make, process, and transport bacterial substances and preparations, or other biological substances (with release of deadly biological strains into the environment); in waterways (with heavy casualties or pollution of port waters, coastal territories, inland water bodies); on pipelines (with mass release of the substances being transported, resulting in explosions, fires or environmental pollution); air crashes (with loss of life, which require search and rescue operations; train collisions or derailment (subway trains) resulting in casualties, considerable damage to the railway tracks or installations in the surrounding settlements); at major power plants, systems and in power mains; at waste disposal plants; and hydrodynamic accidents (dam breaks); etc.

Between 1901 and 2007, 1,125 technogenic disasters occurred in the world, affecting about 4.5 million people, of whom around 49,000 died. The total damage was estimated at $225 billion. Such emergencies occurred most frequently in Asia (651 occurrences) and much rarer in Europe (199) and North and South Americas (177). Some estimates indicate that hundreds of thousands of technogenic incidents occur in the world annually, killing hundreds of people, and causing substantial damage that exceeds tens of thousands of US dollars.

According to the UN, technogenic disasters rank third by the number of casualties after natural calamities caused by weather or geological factors. CRED statistics show that the number of technogenic disasters has been growing steadily since the late 1970s. Experts believe that the main reason for this growth is the increasingly sophisticated engineering systems used by people along with a dramatic increase in the number of functions they perform (automation), while the level of knowledge and training of those who design such systems and those who make and operate them fails to meet the standards of work at all stages of their life cycle. CRED statistics indicate that transport accidents, especially at sea and on rivers, have become particularly frequent. Since Europe and North America have a reliable and safe transport and industrial infrastructure, the biggest casualties in such accidents are sustained by poor nations in Asia and Africa. Between 1994 and 2003, the level of casualties as a result of transport accidents in industrialized countries was less than one death per one million people, while it exceeded three in all the other states.

According to the consulting firm Risk Management Solutions, the number of big technogenic disasters has increasingly exceeded the number of natural ones in the past several decades, even though the latter account for 80% of the damage caused. In 2003–2006, the number of technogenic disasters exceeded the number of natural cataclysms several times, with the death rate from them growing steadily. Following below are some of the data regarding the main types of technogenic disasters and their consequences that have occurred in the contemporary history of mankind.

Technogenic emergencies are believed to be caused most frequently by explosions and fires that often occur at industrial facilities, enterprises that mine, store, or process flammable, combustible, and explosive substances; in the coalmines, underground workings, on trains, motor and water transport, pipelines used for the transportation of liquid and gaseous fire- and explosion-hazardous products; as well as in dwelling, social, and cultural buildings and installations. The main causes of technogenic fires are operating practice breaches and fire safety violations, technical defects leading to ignition, human negligence, and (or) malicious intent.

Here are some of the examples of destructive technogenic fires that occurred in peacetime last century: December 30, 1903, Iroquois Theatre, Chicago, USA, (more than 600 people died); October 14, 1913, coalmine in Mid Glamorgan, Wales, Great Britain (439 people died); November 28, 1942, a nightclub in Boston, USA (about 500 people died); September 2, 1949, Chongqing slums, China (more than 1,700 people died); August 20, 1978, a movie theater in Abadan, Iran (more than 400 people died).

Peacetime explosions occur as a result of human mistakes and miscalculations, high concentrations of flammable gases and explosive dust in the air, operating practice breaches during transportation of flammable and explosive materials, violations of the rules of handling (production, storage, transportation, and disposal) of explosive substances and ammunition, as well as terrorist acts. A number of big explosions have occurred in contemporary peacetime history, claiming hundreds of lives and leaving thousands of people injured. These include (number of deaths in parentheses) unprecedented explosions at the Courrières mine in France on March 10, 1906 (1,060), at coalmines in Manchuria, China, on February 12, 1931 (3,000) and at an arms depot in Lianzhou on April 25, 1942 (1,549); at a mine in Johanngeorgendstadt, East Germany, on November 29, 1949 (3,700); in a convoy with dynamite in Cali, Colombia, on August 17, 1956 (1,100); and in the Salang tunnel, Afghanistan, on November 2, 1982 (1,000–3,000).

The majority of experts believe that big plane crashes (Table 2.1), which resulted in loss of life and considerable damage, were caused by engine malfunctions, aircraft element breakdown during the flight, control, power supply, communication, or piloting system failures, a shortage of fuel, crew and passenger life support disruption, piloting mistakes, bad weather, collisions with objects in midair, terrorist acts, and destruction by combat weapons.

Railroad accidents (derailment, collisions, hitting obstacles at railroad crossings, fires and explosions on the train) are caused by track defects, rolling stock, signaling, centralization and blocking equipment malfunction, railroad line and train overload, traffic operator's mistakes, or engine driver's inadvertence or negligence. Following are some of the big peacetime train crashes (the number of deaths in parentheses) that occurred on May 22, 1915 in Gretna, Scotland, (227); on December 12, 1917 near the town of Modane, France, (800); on July 9, 1918 in Tennessee, USA, (101); on March 2, 1944 near the town of Salerno, Italy, (521); on April 3, 1955 near the city of Guadalajara, Mexico, (300); on June 6, 1981 in Bihar, India, (500); and on January 13, 1985 in Ethiopia (392).

Statistics for peacetime casualties on water transport are just as strikingly gloomy. According to the International Association of Classification Societies, 1,226 ships and 8,777 people were lost in 1990–1996 as a result of sea accidents alone (Table 2.2). On June 15, 1904, the pleasure boat General Slocum was completely destroyed by fire on the East River, New York, USA (1,030 people died); on April 14–15, 1912, one of the world's biggest passenger liners, Titanic, sank after a collision with an iceberg in the North Atlantic (1,503); on September 28, 1912, the Japanese steamship Kichemaru sank off Japan (1,000); on May 29, 1914, the British steamer Empress of Ireland sank on the St. Lawrence River after a collision with a Norwegian coal freighter (1,012); on February 26, 1916, the French cruiser Provence sank in the Mediterranean (3,100); on December 6, 1917, the French Mont Blanc carrying munitions and the Belgian steamship Imo collided (causing a big explosion) and sank in Halifax Harbor, Canada, (1,600); on March 18, 1921, the Hong Kong sank in the South China Sea (1,000); on April 14, 1944, the Fort Stikine carrying munitions exploded in Bombay Harbor, India (1,300); on

Table 2.1 Some of the significant air crashes (1953–1998)

Date	Description	Casualties
May 3, 1953	The world's first jetliner Comet-1 G-ALYV crashed during a violent storm, Calcutta, India	43
March 16, 1969	DC-9 plane crashed after takeoff in Maracaibo, Venezuela	155
July 30, 1971	A Boeing-727 passenger liner collided with an F-86 combat plane over Morioka, Japan	162
October 13, 1972	Il-62 passenger plane belonging to East German airlines crashed near Moscow when making a landing approach in bad weather	176
January 22, 1973	A Boeing-707 passenger liner crashed after fire when landing at Cano, Nigeria	176
March 3, 1974	A Turkish DC-10 plane crashed near Paris due to cargo bay depressurization	346
August 3, 1976	A Boeing-707 passenger plane flew into a mountain near Agadir, Morocco	188
March 27, 1977	Two Boeing-727 planes collided on the runway at Tenerife, Canary Islands	582
May 25, 1979	A DC-10 passenger plane crashed during takeoff at O'Hare International Airport in Chicago, USA	275
August 19, 1980	A TriStar plane belonging to Saudi Airlines was destroyed by fire after an emergency landing in Al-Riyadh, Saudi Arabia	301
September 1, 1983	A South Korean Boeing-747 passenger liner was downed after violating Soviet airspace	269
August 12, 1985	A Japanese Boeing-747 passenger liner flew into Mount Ogura, Japan	520
July 3, 1988	An Iranian A-300 passenger plane was downed over the Persian Gulf by a U.S. warship	290
September 3, 1989	A Cuban Il-62 airliner crashed near Havana, Cuba	214
July 11, 1991	A Nigerian DC-8 plane crashed when landing in Jeddah, Saudi Arabia	261
November 12, 1996	An ascending Saudi Boeing-747 collided with a landing Il-76 plane belonging to Kazakh Airlines near New Delhi's airport, India	372
September 26, 1997	An Indonesian A-300-B4 airliner crashed in the mountains not far from Medan, Sumatra, Indonesia	234
September 2, 1998	A Swiss McDonnell-Douglas plane crashed near Halifax, Canada	229

November 12, 1948, a Chinese army evacuation ship sank off Manchuria (6,000); on September 26, 1954, the Japanese ferry Toya Maru sank in the Tsugaru Strait, Japan, (1,172); on December 20, 1987, the Philippine ferry Dona Paz and the tanker Victor collided in the Tablas Strait, the Philippines (3,000); on September 28, 1994, the Estonian passenger ferry Estonia sank in the Baltic Sea (912).

Table 2.2 World sea accident statistics (1990–1996)

Years	Number of ships lost	Number of people killed
1990	149	807
1991	174	1,389
1992	175	986
1993	178	915
1994	181	1,478
1995	183	1,312
1996	186	1,890
Total	1,226	8,777

There are hundreds of big nuclear energy and chemical enterprises in the world and the accumulated nuclear and chemical stocks are enough to destroy all living beings on earth several times. The most dangerous are emergencies at radiation and chemically hazardous facilities, as evidenced by the world's biggest chemical and nuclear accidents (close to disaster in magnitude) that occurred in Bhopal, India, in 1983 and at Chernobyl, USSR, in 1986.

According to estimates, about seven million artificial toxic substances have been produced in the world since the beginning of mankind, and 60,000–70,000 of that amount are in close contact with people or close to their residential areas, thus creating the risk of chemical accidents and disasters.

A chemical accident (disaster) is a violation of production processes at chemical facilities accompanied by damage to and (or) destruction of pipelines, tanks, storage facilities, or transport means, which result in a release of chemically hazardous substances into the atmosphere or biosphere, endangering biocenosis and the lives and health of people. In addition, under certain circumstances some nontoxic substances emerging during a chemical accident (explosion, fire) can produce chemically hazardous substances. In the event of a chemical accident, the surface layer of the atmosphere, sources of water, soil, and food get polluted. Chemical, pulp-and-paper and processing factories, mineral fertilizer enterprises, ferrous and nonferrous metallurgy plants have large stocks of chemically hazardous substances. The biggest number of chemical accidents occurs at enterprises that make or store chlorine, ammonia, mineral fertilizers, herbicides, and products of organic and petroleum organic synthesis.

Bad chemical accidents and disasters occur in the world with enviable regularity. In 1976, an accident at an Italian plant in Seveso led to dioxin pollution of over 18 km^2. More than 1,000 people were affected, and many animals died. The clean-up operation took 18 months. 2 years later, 28 tons of sodium cyanide leaked into the local river as a result of an accident at a chemical plant in Suzhou. This would have been enough to poison tens of millions of people, but, luckily, only 3,000 were affected.

One of the most tragic chemical disasters of the twenteith century was an explosion at a Union Carbide plant in Bhopal, India, on December 2, 1984, which killed more than 4,000 and poisoned over 40,000 people. A cloud of 43 tons of methylisocyanate (it is 2–3 times more toxic than phosgene) released from the plant by the explosion covered an area of tens of square kilometers, causing mass deaths.

In 1988, an extremely toxic rocket fuel, unsymmetrical dimethylhydrazine (heptyl), spilt during a railroad accident in Yaroslavl, USSR. About 3,000 people happened to be in the hazardous zone. A year later, about 7,000 tons of liquid ammonia spilt at a plant in Jonava, Lithuania, forming a lake of about 10,000 square meters. No one was affected only due to favorable meteorological conditions. Finally, 32 tanker cars with liquid chlorine derailed in Mexico in 1991. Specialists say that 300 tons of the chemical escaped into the atmosphere. As a result, about 500 people were affected, and 17 of them died at the scene of the accident. More than 1,000 people were evacuated from the nearest town.

Nowadays, practically every industry and field of science uses radioactive substances and sources of ionizing irradiation (nuclear materials) in increasingly growing amounts. Nuclear power generation is developing particularly fast. While opening up big energy and other prospects, nuclear facilities and technologies create big risks for people and the environment. Nuclear materials have to be stored, transported, and processed. All these operations create additional risks of radioactive contamination of the environment and harmful effects for people, plants, and animals. Import and disposal of spent nuclear fuel is another acute problem.

A radiation accident is an emergency resulting in a loss of control of radioactive material due to a natural disaster, equipment failure, unskilled actions of personnel, or other factors, which leads or may lead to radioactive poisoning of people or radioactive contamination of the environment. Radiation accidents occur at radio-active hazardous facilities or during transportation of cargoes containing sources of ionizing irradiation. Although big radiation accidents occur relatively rarely, their emotional impact on the population can hardly be overestimated.

Several nuclear accidents (disasters) have happened in the history of mankind since the start of the nuclear era: 1957 – an explosion at a nuclear waste storage facility leading to extensive radioactive contamination of an area of 15,000 km^2 and evacuation of the local population (Chelyabinsk region, USSR); the burning of graphite rods at the nuclear power plant in Windscale, England, where 500 km^2 were affected by radioactive contamination; 1979 – meltdown of the reactor core at the Middletown nuclear power plant (Three Mile Island, Pennsylvania, USA), releasing large amounts of radioactive contamination; 1981 – a spill of 400,000 l of radioactive coolant at the Sequoyah one plant in Tennessee, USA; 1986 – the world's biggest disaster at the Chernobyl nuclear power plant in Ukraine, USSR, where an uncontrolled acceleration of the reactor and the ensuing reactor core explosion released several million cubic meters of radioactive gases, which spread all over the European part of the continent. More than 20,000 km^2 were contaminated with radiation. All people were evacuated in the radius of 30 km from the scene of the accident (including forced relocation of 100,000 people from the town of Pripyat). In addition to environmental damage, nuclear accidents also cause tremendous social, economic, moral, ideological, and political impact on society.

Accidents at waterworks facilities are relatively rare. However, their consequences, especially under unfavorable circumstances, can be compared to those of meteorological disasters. For example, when a dam or a hydroelectric complex falls apart, this creates a real threat of inundation in lowland areas similar to that caused by violent cyclones. The height and the speed of the resulting wave depend on the extent of the damage to the hydroelectric facility and the height of its bottom and top barriers. The break wave moves at a speed of 3–25 km/h in plain areas, and up to 100 km/h in mountainous areas. As a result, considerable parts of territory get flooded within 15–30 min after a burst and end up under water 0.5–10 m deep, which can stay from several hours to several days.

Hydroelectric plant dam bursts, including those under construction, happen rarely. Only three such incidents have been registered in the world over the past 10 years: on May 27, 2004, the dam of the Dalongtan hydroelectric plant on the Qingjiang River broke within the boundaries of the city of Enshi (as a result, a minibus with 14 people was engulfed); on February 11, 2005, water rose dramatically due to heavy rains, breaking the dam in the province of Balochistan in the southwest of Pakistan and causing a flood that washed away several villages (more than 50 people died and 500 disappeared), and on October 5, 2007, a typhoon-caused flood broke the dam at the Cua Dat hydroelectric plant under construction on the Chu River in the Vietnamese province of Thanh Hoá.

Environmental disasters occupy a special place among technogenic disasters as they have extensive harmful effects on the environment and biosphere. In addition to the causes mentioned above, environmental incidents, accidents, and disasters are caused by gross safety violations, negligence on the part of the administration and personnel at potentially hazardous industries, political and administrative ambitions, greed, economy drives, and withholding of information about technogenic emergencies. On the basis of this definition, anthropogenic environmental disasters may include explosions, fires, and transport accidents if they release polluting, toxic, or radioactive substances into the atmosphere and biosphere in exorbitant amounts. In addition to those cited above, the most vivid examples of technogenic environmental disasters include an oil blowout at an oil platform in Santa Barbara

Bay (California, USA) on January 28, 1969, when about a million liters of oil spilt into the sea over 11 days; a release of a poisonous dioxin cloud into the air as a result of an accident at a chemical plant in Seveso, Italy, on July 10, 1976, which necessitated the evacuation of the local town for 18 months; and an accidental atmospheric release of anthrax spores at the Institute of Microbiology and Virology in Sverdlovsk (USSR), which contaminated an area over a radius of 3 km and killed several people.

Hundreds of emergencies occur in Russia nowadays. The cost of ensuing cleanup operations accounts for more than 15% of the national domestic product. If this tendency persists, very soon the national economy will not be able to carry out postcrisis rehabilitation. Since two-thirds of such emergencies have a technogenic origin and are caused by anthropogenic activities, studying their nature, developing a theory of their occurrence, and working out measures to mitigate their consequences has become a pressing scientific and practical task in Russia.

2.2 Man-Made Disasters and Emergencies in Russia: Causes and Consequences

In order of decreasing number and frequency the man-made disasters and emergencies in Russia are distributed as follows [1]: disasters at industrial facilities, road traffic accidents, motor vehicle breakdowns, emergencies in residential and social function buildings, at nuclear facilities, and chemical disasters. Their main causes are:

- engineering systems' failure resulting from inadequate workmanship and violations in operating procedures (for instance, many modern potentially dangerous facilities are designed so that the major accident probability at the facilities is rather high, being estimated at 0.0001 and higher; in real life, however, accidents at these facilities occur much more often);
- mistakes of engineering systems' operators. The statistics show that more than 60% of accidents result from erroneous actions of the operating personnel;
- concentration of various production facilities in industrial zones without due study of their interaction and mutual influence;
- high power supply level of the engineering systems;
- external deleterious impacts on power, transportation, and other facilities.

Also, there is a class of so-called "geo-technical systems" whose failure results from a synergic interaction of an engineering system and the environment.

The variety and complexity of subject–object interactions leading to emergencies in Russia are shown in Fig. 2.2. In the course of a further analysis of Russia's man-made disasters and emergencies we shall abide by this scheme in order to systematically evaluate the risks resulting from inadequate functioning of the country's facilities in power supply, transport, and industrial infrastructure.

Looking back, according to unofficial information of the Russian Ministry of Emergencies and the Federal Service for Ecological, Technological and Nuclear

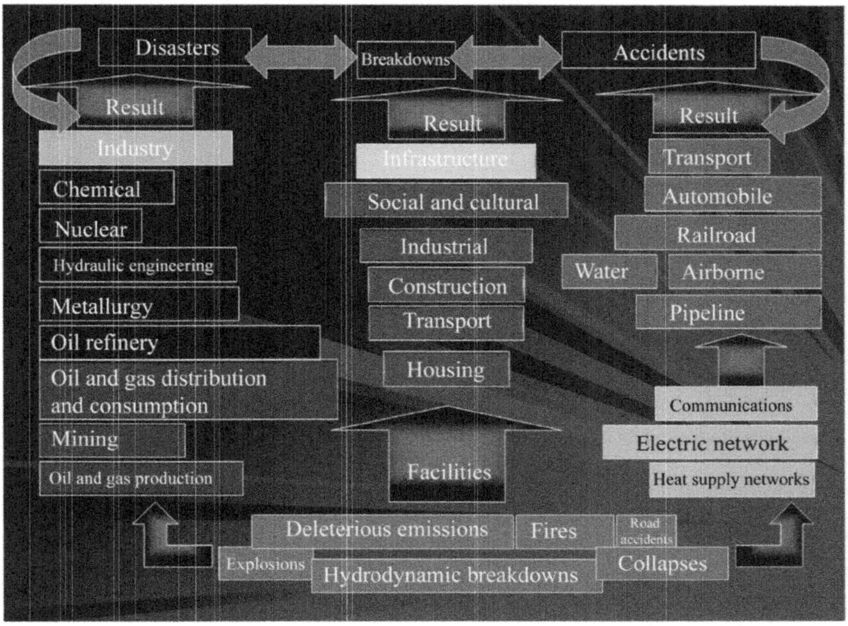

Fig. 2.2 Subject–object man-made disasters in Russia

Monitoring (ROSTEKHNADZOR) [2], in 2006 the bulk of Russia's man-made disasters were related to fires in buildings and structures (back then, in 10 months alone 1,547 major fires occurred with a death toll of 3,522). Statistically, during this period, every 5 min a fire broke out in the country, every hour one man died in a fire and 20 sustained burns and injuries; 44 major traffic accidents occurred with more than 10,000 casualties; there were 38 air crashes in which 296 people died, 15 railway wrecks with five dead, and 18 accidents on water (11,019 dead). In 11 months of 2008, 162 accidents occurred at the country's industrial facilities: 34 on oil and gas extraction sites and main pipelines, 13 at petroleum and chemical plants and oil refineries, 12 in coal production, 10 in the chemical industry, 7 in the mining industry, 4 each in metallurgy and botanical raw material processing and storage. The total casualties in those accidents reached 414.

Overall, in recent years the man-made hazard in Russia is presented by such emergencies as fires and explosions at dangerous production facilities and residential buildings, road accidents, accidents at main pipelines, communal utility and power supply facilities, and emissions of chemically dangerous compounds. Because of the general wear and exhaustion of designed service life of many technological systems of thermoelectric plants and boiler stations, and incomplete preventive and maintenance measures on account of underfunding, the number of accidents at them, including fire, is also growing.

Table 2.3 Man-made disasters and emergencies in Russia

Branch of Russia's industry	Number of accidents by year		
	2003	2004	2005
Metallurgy	4	3	No data
Mining and nonmining industry, underground construction facilities	14	14	No data
Coal production	30	35	No data
Mining industry	44	40	No data
Chemical industry	No data	11	27
Nuclear industry	No data	No data	No data
Oil and gas production facilities	No data	20	16
Main pipeline facilities	No data	48	45
• pipelines	No data	29	19
• gas supply lines	No data	19	19
• oil pipelines	No data	-	3
• oil products pipelines	No data	-	4
Hydraulic facilities	-	1	-

Given below [2] are summarized statistical evaluations of man-made disasters and emergencies in Russia's main industries during 2001–2004 (Table 2.3). According to "ROSTECHNADZOR," in the past 5 years 62 accidents of this type took place. Resulting material damage exceeded US$ 3 billion. The principal causes were physical wear of the main production equipment and structures, inadequate supervision by the facilities' monitoring services of the technical condition of the equipment, and violation of operating rules and safety regulations.

In the past 30 years the number of accidents in the metallurgical and chemical-recovery coke industry has grown by 2.5, the number of casualties by six, and the economic damage by 11 times (Table 2.3). In 2007 alone, two accidents occurred there with 17 casualties. The economic damage amounted around US$ 10 million. The principal causes were violation of service instructions in the metallurgical processes (33.4%) and technical faults of the equipment.

In recent years, Russia's underground and open cast mining industry witnessed a buildup of production and an increase in the penetration and extraction depth. This increased the danger stemming from the deteriorated mining and geological conditions. In 2006, the material damage from accidents in this area alone exceeded

US$ 40 million. Two thirds of the accidents that took place in underground works (four fires and one explosion in mines) were caused by "the human factor," i.e., inadequate safety engineering, lack of production control, violation of technological and labor regulations, and careless or unauthorized activities of operators. (Table 2.4)

The statistics show that mining production in Russia proceeds with an unjustifiable risk so that each million tons of coal costs one miner (Table 2.5). The industry's experts admit that around 80% of cases of the general injury rate and 60% of lethal cases are the result of violations in the technological process, in discipline and safety engineering. The damaging factors at Russian mines include collapse, equipment failure, transport malfunction, power outage, geodynamic phenomena, and methane explosions (the public is informed about the gas hazard of the Russian mines because of the heavy casualty effect of each such explosion as compared to other types of accidents). Each year, around 30 methane explosions occur at Russia's mines with tens of casualties (100 dead in 2003, 148 in 2004).

Table 2.4 Accidents and lethality rate at Russia's metallurgical plants from 1998 to 2007

Year	Amount of production, million tones	Number of accidents	Number of lethal cases
1998	132.1	5	26
1999	148.2	5	36
2000	177.8	4	33
2001	158.7	6	27
2002	168.6	4	23
2003	176.1	4	24
2004	181.3	3	22
2005	188.6	2	21
2007	184.0	2	17
Per years	168.4	3,9	25,4

Table 2.5 Accidents and lethality rate at Russia's mining industry in 1997–2007

Years	Coal production, million tones	Number of accidents	Lethality
1997	244.4	56	242
1998	232.4	54	139
2009	249.1	39	104
2000	254.2	34	115
2001	266.4	34	107
2002	234.2	27	83
2003	270.3	30	109
2004	287.5	35	148
...	...	...	...
2007			219

Despite the fact that year 2008 witnessed relatively low lethality rate in the mining industry when considering the past decade, two major disasters occurred in 2009 at mines of the Kemerovo region with 150 casualties. Thus, the accident rate and number of lethal injuries in Russia's mining industry kept growing each year of this century. Man-made lethal accidents exceeded 50% of the total losses (Fig. 2.3).

Russia today has over 3,000 chemically dangerous facilities. One hundred and forty-six towns with population of more than 100,000 people are located in chemically dangerous areas. There are more than 9,000 plants of chemical, petrochemical, and oil refining industries, 7,700 of which are organizations using hazardous industrial facilities. From 1992 to 1996 more than 250 accidents occurred at such facilities, during which time more than 800 people sustained injuries and 69 died. In 2005, 27 accidents occurred at the facilities of chemical, petrochemical, and oil refining industries, in which 41 people died. Seventy-three percent of casualties occurred in pouring and discharging operations, during container cleaning and repair.

The principal causes of disasters at the chemical facilities were inadequate condition of engineering systems, buildings and structures resulting from ageing of production assets. Other contributing factors were low rate of modernization of the industry, incorrect organization and violation of operating procedures, inefficient production management, deliberate disconnection of protection systems, defiance of production discipline, and careless or unauthorized actions of the operating personnel. As a result, a fourth of all accidents were caused by beyond-limit use of obsolete and worn equipment, inoperability of test instruments, and "the human factor."

Fig. 2.3 Fire at one of the chemical plants

The situation in Russia's oil and gas production and refining industries is also rather perilous. For instance, in addition to explosions and fires at the oil and gas production facilities, falls of drilling derricks and drill string-ups during deep drilling and repair of boreholes (Table 2.6), each of the last 2 years witnessed oil and gas blowouts. In 2008 alone, two such blowouts occurred at the Anastasyevsko-Troitskoye deposit in Krasnodar region and one at the Fyodorovskoye deposit in Khanty-Mansi autonomous territory. The preconditions for such disasters also exist at other deposits in West Siberia. The open (uncontrolled) oil and gas blowouts reduce the bedding energy and cause irreversible loss of much of the resources. In addition, Russia still doesn't utilize the associated gas but burns it (30% of the extracted amount). This is not only a direct loss to the national economy, but also a permanent threat to environment.

The main problem in providing safety at oil and gas production and refining facilities is the slow rate of upgrading of obsolete equipment and that with exhausted service life. The main causes of accidents and injuries in operating such facilities are typical of Russia and result from the low professionalism and work discipline of the personnel using it.

By virtue of its geographical position and the specific economic development, Russia has the world's longest main pipeline system. The state-owned network of oil and gas pipelines together with oil and gas production and transportation infrastructure covers 35% of the country's territory where 60% of its population live. The main pipelines are the most capital-intensive facilities of the oil production industry. The average length of oil transportation lines in Russia exceeds 2,000 km (800 km in the U.S.). By year 2006 more than 40% of Russia's main pipelines exceeded the accepted service life of 33 years. Over 400 pump stations and 900 tanks operate at the oil pipelines of Russia. The gas pipelines use, respectively, over 250 compressor stations and 21 underground gas storage facilities with a total capacity of more than 80 billion cubic meters.

Special classes of danger are pipeline transit sections installed on natural and man-made obstacles such as rivers, reservoirs, railways, and motor roads. The gas pipelines have more than 1,175 underwater transits with a total length of 3,500 km Over 40% of tunnels have been laid more than 25 years ago. As of 2005, the total length of the linear part of the main pipelines was more than 231,000 km, of which 161,100 km are main gas pipelines, 49,000 km are main oil pipelines, 19,500 km are main product pipelines, and 1,400 km are ammonia pipelines.

Table 2.6 Causes of disasters in Russia's oil and gas production industry

Types of disasters	2004	2005
Open fountains and injections	6	5
Explosions and fires at facilities	7	5
Fall of derricks and disintegration of their components	1	2
Fall of drill string-ups during deep drilling and underground repair of boreholes	2	1
Other	4	3
Total	20	16

The annual number of emergencies on Russian high-risk pipelines reaches 5,000–6,000. In terms of modern ecological requirements and safety rules their further operation becomes unpredictable and uncontrollable because of the landslide pattern in accident occurrence, which is now 0.4 per 1,000 km (120 accidents per year). According to the studies carried out by JSC GAZPROM, TRANSNEFT Joint Stock Company, RAO Russian Oil and Gas Production Company NEFTEGAZSTROY, the current condition of oil and gas transportation facilities located in West Siberia is unpredictable which is the cause of the growing rate of accidents on the field and main pipelines that lack due engineering forecast (along with ineffective anticorrosion measures) and ecological monitoring of the environment in areas where oil and gas production facilities are located.

Accidents of Russia's gas pipelines occur rather often. Every year, around ten of them occur on major gas pipelines alone. In most cases (the 2007–2008 statistics are shown in Table 2.7) such accidents do not lead to casualties, though many of them (in 50% of cases) are followed, as a rule, by large-scale fires. Table 2.8 shows data for the main causes of such disasters, among which the following must be pointed out.

1. Formation, in some portions of the main pipelines, of erosion patches 3–10 m deep and up to several hundred meters long. Dangerous sags and deformations occur, which dramatically increase the risk of a disaster. As a result, 15% (over 20,000 km) of main pipelines are used with a reduced pressure in order to avoid a rupture.
2. Intense development of stress-corrosion processes on large diameter pipes associated with degradation of the insulation film coating of main pipelines many of which had been built 20 and more years ago. So, whereas in 1991–1996 the share of accidents for this reason made up around a fourth of the total balance of the JSC GAZPROM accident rate, in the period from 1998 to 2003 it made up a third and, in 2005, more than half of all accidents.
3. Insufficient overhaul and inadequate fitting of main transport pipelines with tele-mechanics and automation systems.
4. Commissioning gas distribution stations currently not included in the unified gas supply system.
5. Considerable growth of instances of unauthorized tapping of oil, oil products, and gas pipelines for theft of the transported products.
6. Accidents through inadequate workmanship in the course of construction and installation during the construction boom (1970–1980s) on the main pipeline facilities by construction organizations that did not have the required equipment.

As a result, annually Russia sustains losses from accidents, gas leaks, and oil spills equal to 6–8% of the total market value of the annual production of those products less the cost of recovery of the environment.

A positive point is some reduction, in the past 2 years, of the accident rate on main and field pipelines. This became possible due to allocation of the required funds for repairing the lines and diagnosing their condition. However, as before, minor accidents occur because of mechanical disturbances, theft of transported

Table 2.7 Emergencies at Russia's main gas pipelines during 2007 to 2008

Emergency date	Brief description of emergency	The afflicted
3 April 2007	A rupture and fire at main gas pipeline Urengoi-Center-2 in Khanty-Mansi autonomous territory	No afflicted personnel
7 April 2007	A repeat rupture with ignition of a pipe at Urengoi-Center-2 main gas pipeline. The emergency occurred in preparation for commissioning of the pipeline's main branch	No afflicted personnel
3 June 2007	Rupture of a pipe in the Ukhta-Torzhok main gas pipeline and a fire in a forest in Tutayevsky district of Yaroslavl region	No afflicted personnel
3 July 2007	Rupture of a portion of the Yamburg-Zapadnaya Granitsa pipeline in Arsk district of Tatarstan, gas explosion and a subsequent fire. The high temperature at the site of the fire caused a depressurization of two more pipelines, Yamburg-Yelets-1 and Yamburg-Yelets-2 with a subsequent torch burning of methane (torch height 20–30 m, burning area around 100 square meters)	No afflicted personnel
26 July 2007	Explosion and fire on a main gas pipeline in Vsevolzhsky district of Leningrad region in Severnaya TETs Petersburg-Lavriki zone. The accident was accompanied by a heavy blowout of flame and smoke which took the form of a thunder storm, which caused panic among the local population. At the emergency site the forest and peat bog caught fire over an area of around two hectares	No afflicted personnel
28 September 2007	Rupture of main pipeline without ignition in Krasny Yar settlement of Sverdlovsk region	One person injured
1 December 2007	Rupture of the main pipeline Srednyaya Aziya – Tsntr (Middle Asia – Center) with subsequent ignition of remotely located villages in Alekseyevsky district of Volgograd region	No afflicted personnel
13 January 2008	An explosion and fire (height of flame column several meters) on main gas pipeline in Tosninsky district of Leningrad region. As the fire was localized, around 0.5 ha of surrounding terrain had burnt out	No afflicted personnel
16 February 2008	Rupture and ignition of a main gas pipeline Yamburg-Yelets-1 in Oktyabrsky district of Khanty-Mansi autonomous territory	No afflicted personnel
17 February 2008	Rupture and fire at main gas pipeline in Novgorod region between Valdai and St. Petersburg. Two private houses, located 100–120 m from the rupture site, burned down	No afflicted personnel
22 April 2008	Rupture of main gas pipeline Yamburg-Tula-2 in Nizhny Novgorod region 3 km off the Pilna compressor station	No afflicted personnel
24 July 2008	A rupture of a main gas pipeline with subsequent ignition in Volgograd region 300 m off the railway. Five districts of Volgograd region were left without gas	No afflicted personnel
24 July 2008	Two disasters with ignition at gas pipelines running across Khanty-Mansi autonomous territory: on main gas pipeline Yamburg-Yelets-1 and main gas pipeline Punga-Ukhta-Gryazev	No afflicted personnel
25 July 2008	A rupture with ignition occurred on the 627 th km of the gas pipeline Yamburg-Yelets-2 150 km north of the town of Beloyarsk of Khanty-Mansi autonomous territory	No afflicted personnel

(continued)

Table 2.7 (continued)

Emergency date	Brief description of emergency	The afflicted
14 August 2008	A rupture of a main gas pipeline with subsequent ignition of Nekrasovsky district of Yaroslavl region	No afflicted personnel
5 October 2008	A rupture of a main gas pipeline in Shchelkovsky district of Moscow region. No gas ignition occurred	No afflicted personnel
24 December 2008	A rupture and fire on a branch of the main gas pipeline Gryazovets-Petersburg in Volkhovsky district of Leningrad region	No afflicted personnel

Table 2.8 Main causes of accidents on main pipelines in Russia

Troubles	Gas pipelines	Oil pipelines	Products pipelines	Total
External mechanical impacts, including:	3	12	5	20
–tapping;	-	8	1	9
–construction systems	3	4	4	11
Corrosive destructions	14	-	-	14
Faults in construction and installation	3	2	-	5
Personnel errors	1	-	1	2
Manufacturing fault	2	2	-	4
Total	23	16	6	45

products, faults in construction and installation, violation of design regulations, and corrosion of pipes, fittings, and control appliances. As usual, the main pipelines remain vulnerable to terrorist acts, the probability of which is still rather high.

The coverage of emergencies on Russia's hydraulic facilities that occurred in the past decade (Table 2.9) should start with the 17 August 2009 disaster on the country's largest Sayano-Shushenskaya hydroelectric plant (Fig. 2.4) on the Yenisei River in Siberia. The disaster in the form of an inrush of water into the plant's engine room without collapse of the dam and flooding of residential areas was stopped. But it brought about the death of 75 personnel and fully destroyed three power sections of the plant. This heavily impacted the power supply of Siberia and neighboring regions. The recovery, estimated at more than one billion Euros, will take, according to experts, no less than 4 years.

Prior to the August 2009 disaster, only one accident in the last 30 years of operation had occurred at the Sayano-Shushenskaya plant. It was in spring of 1985. Back then, during powerful flooding the water from Yenisei rapidly washed out the concrete and metal framework of the dam thus leading to destruction of 80% of the

Table 2.9 Accidents at Russia's hydraulic facilities from 2001 to 2009

Accident date	Brief description of accident	Consequences
10 October 2001	Accident and fire at substation of Irkutsk power plant following an earthquake at Baikal	No consequences
11 March 2004	Accident at GES-10 power plant on the Vuoks River in Svetogorsk, Leningrad region, which threatened to flood a town of around 15,000 people	No consequences
6 February 2006	While lifting a transformer at Bureiskaya GES plant in Talakan, Amur region, a bridge crane broke down. Upon falling, it crushed the plant' water supply line	No consequences
19 August 2006	At Bureiskaya power plant in Talakan, Amur region, through a mistake by the plant's personnel's a hydro-unit block transformer was not useable for over a month	No consequences
13 June 2007	Fire at Zhigulyovskaya power plant in Samara region	No consequences
12 September 2007	Fire at a block transformer of Novosibirsk power plant	No consequences
8 October 2007	Damage to 500-kV power transmission lines of Bureiskaya plant caused by a rain cyclone and strong winds	Fan-type power outage of facilities in Khabarovsk
27 February 2008	Ignition on the roof of Rybinsk power plant in Yaroslavl region	No consequences

water discharge channel. Other contributing factors of the destruction were previous inadequate research, development, and installation work.

The damage from all accidents on Russian hydroelectric plants in 2001–2009 (less the Sayano-Shushenskaya disaster) exceeded US$ 80 million.

Investigating the accident rate in Russia's nuclear industry, which incorporates many enterprises for extraction, enrichment, and processing of uranium, as well as nuclear power plants, defense organizations, the authors seek, for some reasons, not to supply detailed statistics. Most experts believe that the principal source of man-made threats is the nuclear industry, functioning now for over half a century, with its danger of ecological disasters (like the one in Chernobyl) originating from radioactive emissions during blowouts.

Among the faults and problems in the work of ROSATOM (Russia's Nuclear Energy State Corporation) the most conspicuous are the slack discipline and eroded professional skills of the nuclear heat production industry personnel. Many violations and accidents, detected each year by ROSTECHNADZOR, are caused by human factor. At the same time, the main equipment continues to wear whereas its modernization is thwarted by financial constraints. The problem of processing and handling radioactive waste is not being addressed. In the course of operation, Russian nuclear plants emit into the atmosphere a large amount of such dangerous isotopes as Ts-137,

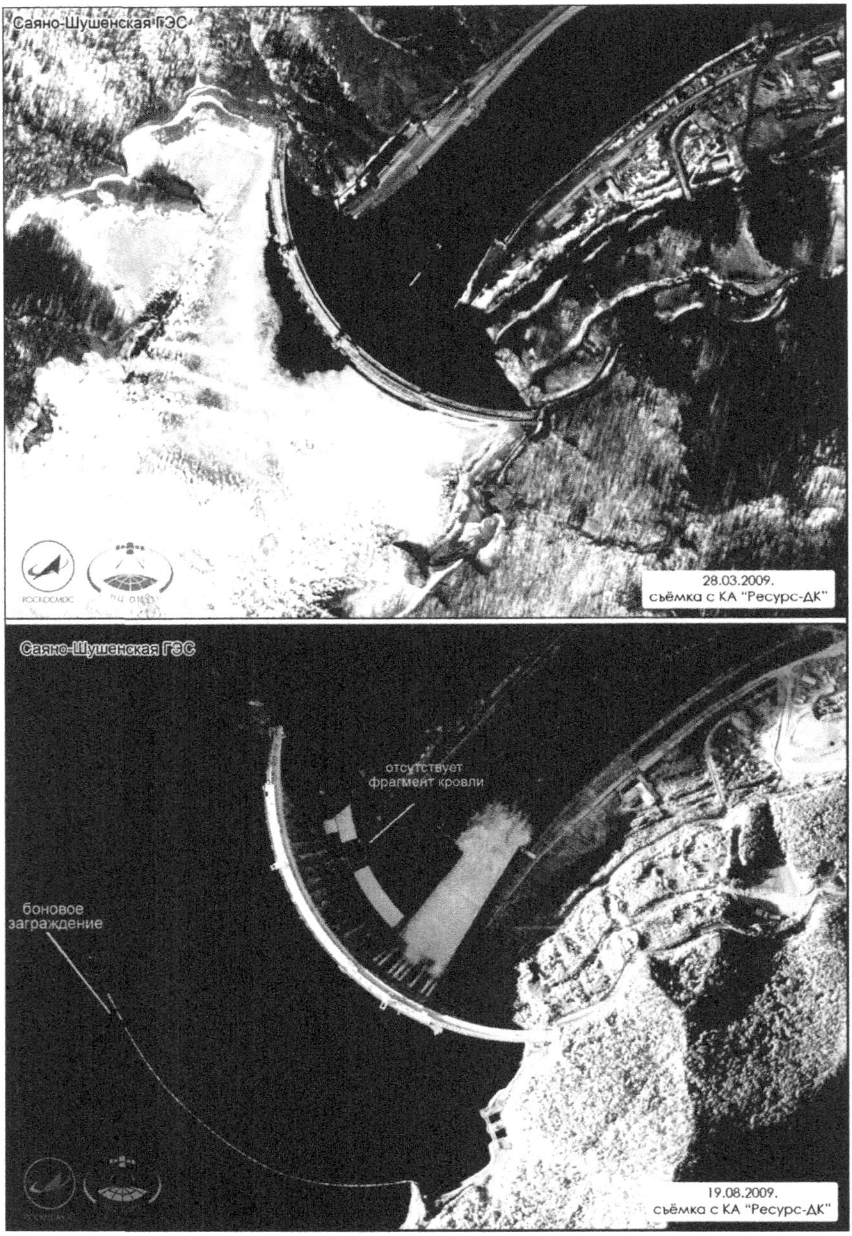

Fig. 2.4 Space images of Sayano-Shushenskaya hydroelectric power plant

I-131, Co-60, and others. Tens of thousands of cubic meters of radioactive water used for cooling the reactors are discharged into neighboring reservoirs. In 2007, at ten Russian nuclear plants 47 malfunctions were registered. Those had been caused by mistakes in design and manufacture of equipment, by neglect of safety regulations, and unprofessional actions of the plants' personnel.

On the basis of the accident statistics for Russia's industries given above, the following conclusions can be drawn.

1. Nine to sixteen accidents of various gravity occur every year
2. The main figures for accidents and casualties per one million tons of "products obtained" are, respectively, as follows:
   - in metallurgy –0.02 and 0.14;
   - in the ore mining industry –0.01 and 0.02;
   - in the coal production industry –0.12 and 0.52;
   - in the mineral resource industry –0.01 and 0.07;
   - in the petrochemical industry –0.004 and 0.005;
   - in the oil/gas production industry based on various estimates –0.03/0.04 and 0.04/0.05;
   - on pipeline transport (for 1,000 km) –0.19 and 0.02
3. Annually, on average, three emergencies occur for each Russian hydroelectric plant, and 4.7 for each nuclear plant.

Considering that 60–70% of accidents are caused by "human factor," in the future their number could be reduced at least by half through active introduction of methods and systems, including those in space that ensure reliability and safety of complex equipment. However, the forecasts of experts of the Russian Ministry of Emergencies unfortunately indicate that in the coming years the number of man-made disasters in the country will grow in geometric progression because of infrastructure ageing and the typically Russian carelessness in construction of potentially dangerous facilities.

The accident rate in the chemical, petrochemical, and oil refining industries may be expected to lessen somewhat because of the tightened supervision by the state of safety at industrial facilities and through their modernization and renovation. The probability of major fires at oil terminals and oil refineries still remains high. Equipping of ecologically dangerous facilities with accident preventive systems at some chemical production plants proceeds slowly, and that may lead to new major disasters in the form of emissions of harmful compounds into the atmosphere and damage to people.

There is still danger that water storage dams may burst and contaminate reservoirs with harmful compounds. To eliminate this danger it is necessary to establish an effective system for monitoring the condition of hydraulic facilities in industry and in the water utility service, to enhance safety, to timely repair and maintain the structures and equipment.

The decisive factor still affecting railroad safety is the wear of the rolling stock and upper structures of the road. The positive factor, however, is introduction on some roads of automatic systems for monitoring dangerous freight traffic.

Unfortunately, the road accident rate doesn't fall in Russia. The gravity of the situation (the number of dead per 100 victims) is still inadmissibly high, 14, whereas in Western countries this figure doesn't exceed five. To a great extent this situation is determined by the poor state of the road infrastructure and the soaring number of motor vehicles belonging to individuals with slack discipline on the road. The problem can be resolved with development of telematic systems and establishment of traffic corridors using space monitoring systems.

The observed trend of negative impacts of man-made disasters in Russia testifies both to the general state of security at the country's industrial facilities and to the influence of other objective factors such as physico-geographic and climatic conditions. In a cold period of the year, such factors essentially increase the load on Russia's power supply systems, contributing to harsh operating conditions, temperature and humidity difference. The periodic freezing and thawing cause soil displacement and other structural disturbance.

Global buildup of temperature is known to be fraught with an array of threats. In the twenteith century the annual mean yearly temperature had grown by 0.6°. In subpolar regions it was still higher, reaching 3–6°. In Russia, vast territories north of Baikal are located in permafrost regions where the thickness of permanently frozen soil varies from half a meter in the south to hundreds of meters in the north. The southern border of multiyear frozen rocks has moved hundreds of kilometers to the north. This applies in the first place to the European part, West Siberia, Yakutia, and the Arctic coast. The permafrost thawing rate under such circumstances reaches 20 cm per year, which impacts geo-engineering structures built on permafrost (main oil and gas pipelines, power transmission lines, railroads, elements of infrastructure, etc.) The thawing of "eternal" ice may cause a massive deformation and destruction of buildings. So, in the last 10 years the number of buildings that sustained different destructive impacts due to the foundation sagging alone increased as compared to the previous decade, by 42% in Norilsk and by 61% in Yakutsk. Their condition must be permanently monitored. Considerable efforts are needed to maintain them in a state of repair.

The growing number and power of typhoons impose a threat for the Far East, the Kurile Islands in the first place. Geomagnetic storms increase the number of mistakes made by operators at all levels. Also, they enhance the risk of failures in sophisticated electronic and electro-mechanic systems. Such risks increase the probability of man-made disasters.

Many major geo-engineering systems, such as main oil and gas pipelines, power generation facilities, motor roads, railways, power transmission lines, factories, towns, and many more, have been built in these sensitive territories. Each year they sustain heavy climatic and geophysical impacts. Under such circumstances it is logical to expect a growth in the number of accidents on those facilities caused, among other things, by chronically insufficient maintenance, improper preparation for heating in winter, untimely and inadequate repair of equipment operating in severe climatic conditions, and localization of discharges of explosive and toxic compounds into the atmosphere. Other problems include poor supervision of fire safety of residential buildings and entertainment centers.

2.3 Harbingers of Technogenic Emergencies and Ways to Forecast Them

The harbingers of technogenic emergencies are physical phenomena that provide objective evidence of potential hazardous events and processes at industrial facilities and in complex organizational and engineering systems, which can be instrumentally detected and documented for preventive purposes. Since the operation of complex engineering systems requires in the majority of cases different forms of human participation, when considering the signs of approaching disasters in this work we should exclude those that are related to human activities (lack of experience, negligence, mercenary motives or malicious intent that cause a technogenic emergency must be prevented and excluded using legal, juridical, and organizational mechanisms). So the harbingers of technogenic disasters in their entirety can be divided into four groups (Fig. 2.5). Using each alone or in combination makes it possible to forecast an accident in order to take necessary preventive measures.

The first group of harbingers includes failures, breakdowns, and defects caused by both declining reliability of complex man–machine engineering systems and their components. We know from the theory of reliability that there can be gradual failures caused by wear and tear and/or end of service life, which manifest themselves either as abnormal deviations in the work of systems and their breakdown, or as the end of operation due to the depletion of fuel, energy, or other renewable resources.

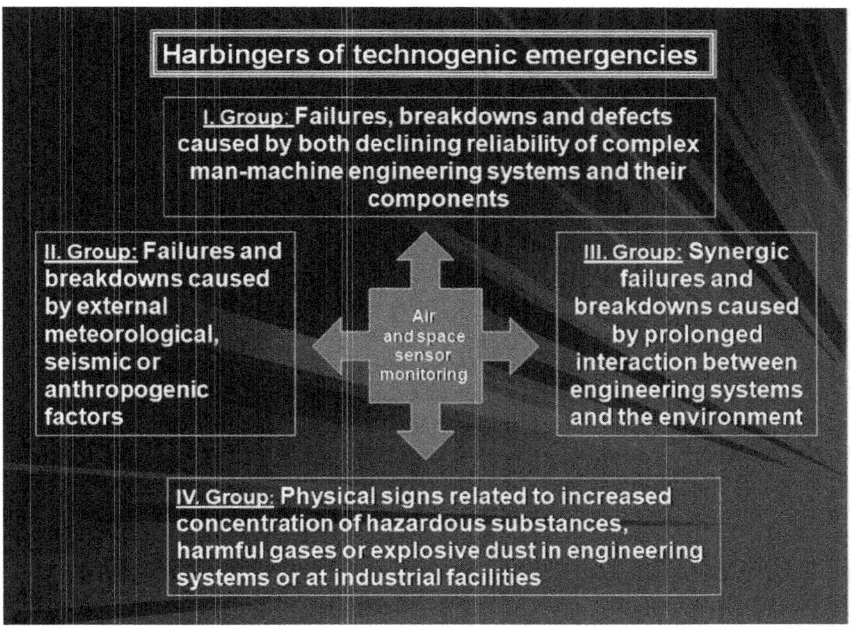

Fig. 2.5 Harbingers of technogenic emergencies

The experience of creating and operating complex engineering systems has allowed mankind to develop and introduce a method for monitoring their reliability and safety. For example, Russian industries have a system of organizational, methodological, and technical regulatory documents (national, industry, and departmental standards, guidelines, and instructions) which enables them to handle effectively increasingly sophisticated and energy-consuming equipment, and ensure its reliable and safe operation.

The second group of harbingers includes failures and breakdowns caused by external meteorological, seismic, or anthropogenic factors. Such technical anomalies occur when external loads exceed the rated ones accepted when designing equipment. Instrumental registration of such anomalies (including accumulation of relevant statistics) along with disaster early warning can help forecast technogenic accidents. It is also important that such monitoring will make it possible to correct (increase) the design parameters of systems.

The third group of harbingers includes so-called synergetic failures and breakdowns caused by long interaction between an engineering system and the environment. For example, many engineering systems, buildings, and installations are operated in seismic or meteorologically aggressive (dangerous) areas. Engineering systems and the environment influence each other in the process of operation and this can lead to an accident or a disaster. For instance, trunk pipelines, railroads, power transmission lines, compressor stations, storage facilities, and urban and industrial infrastructure are located in tropical or permafrost regions that cover big portions of the globe.

Special interest is attracted to the growing number of failures in complex mechanical systems involved in the use of subsoil or other natural resources caused by extreme resonance vibrations due to anthropogenic geodynamic impacts. This causes technogenic earthquakes, landslides, and floods. The harbingers of such failures are geodynamic movements caused by technogenic factors: growing horizontal tectonic stresses, vertical and horizontal displacement of the earth's crust, and other geophysical processes provoked by human activities. They can be detected instrumentally both using contact-type geophysical instruments and remotely.

The fourth group of harbingers includes geophysical signs related to increased concentration of hazardous substances, harmful gases, or explosive dust in engineering systems or at industrial facilities. Such signs can be registered both instrumentally in situ (sensors, gas analyzers) and remotely using mass spectrometers (multi- or hyperspectral ones) and laser tools aboard aerospace monitoring systems.

The use of aerospace monitoring systems (including those for monitoring and troubleshooting purposes) helps detect operational harbingers of technogenic emergencies listed in Groups II and III. The harbingers of Groups I and IV can be effectively detected using ground-based sensors. Data supplied by them through the "space control contour" can be automatically accumulated, processed, and promptly transmitted to the relevant bodies for making decisions regarding safety.

Possible ways to register the harbingers of technogenic accidents in the coal and oil and gas industries are listed below. Accident statistics in Russia's coal industry indicate that, as a rule, accidents are preceded by a number of deviations from the standard production process. For example, it can be growing concentration of methane or carbon dioxide. This necessitates the use of automatic control systems in the return airways of a coal mine or its sections. Modern telecommunication systems can effectively provide such control (and determine microconcentrations of hazardous gases), immediately alerting the personnel on the surface and automatically de-energizing certain equipment, if needed. Telematic systems can already control, without the presence of people, practically all aspects of safe operation of the shearer in the coal mine.

In order to rule out accidents in the Russian oil and gas sector, it is necessary to develop and improve a comprehensive monitoring and diagnostics system for ground-based facilities, especially trunk pipelines, including aerospace methods and means to supplement remote earth sensing information and management systems that already exist in the industry. Russia's Energy Strategy and the Federal Program "Highly Reliable Pipeline Systems" provided for annual comprehensive examination of 20,000–25,000 km of operating trunk pipelines, and for the creation of 100–150 incident prevention and localization units on the existing pipelines. Since major repairs and upgrading costs $120–150 million per 100 km of a large-diameter pipeline, and comprehensive examination and diagnostics requires another $200,000–400,000, it would be advisable to use instrumental control systems more widely, employing modern information and space technologies.

Yamburggazdobycha has positive experience of environmental monitoring at gas production facilities. The company has created a pilot industrial environmental monitoring zone in order to try out and arrange for the use of its main technical means, including:

- Automated systems to monitor hazardous emissions at compressor stations;
- Automated posts to monitor the condition of the air in residential and industrial areas;
- Mobile environmental laboratories to detect the contamination of soil, water (snow), and air in different parts of the Yamburg gas condensate field;
- Systems to monitor the quality of water at water intakes and after cleaning;
- Engineering and geological monitoring equipment to watch the condition of soil under production facilities in permafrost areas; and
- Sets of laboratory and analytical equipment for detailed air, water (snow), and soil analysis.

The above-mentioned means are tied together by an integrated monitoring data transmission network that uses the Yamal satellite system.

For the purpose of monitoring the technical condition of the gas transportation network, the company uses a telemetric system to watch the working parameters of field equipment, pipelines, pumping and compressor stations, storage facilities, etc., an aviation system to respond to emergencies, and a space system to discover new oil and gas fields, and also to monitor the construction and operation of gas pipelines.

Gazprom has lately created and been actively testing a pipeline aerospace monitoring system on the basis of existing and future remote earth sensing spacecraft, as well as unmanned aerial vehicles (Figs. 2.6–2.7).

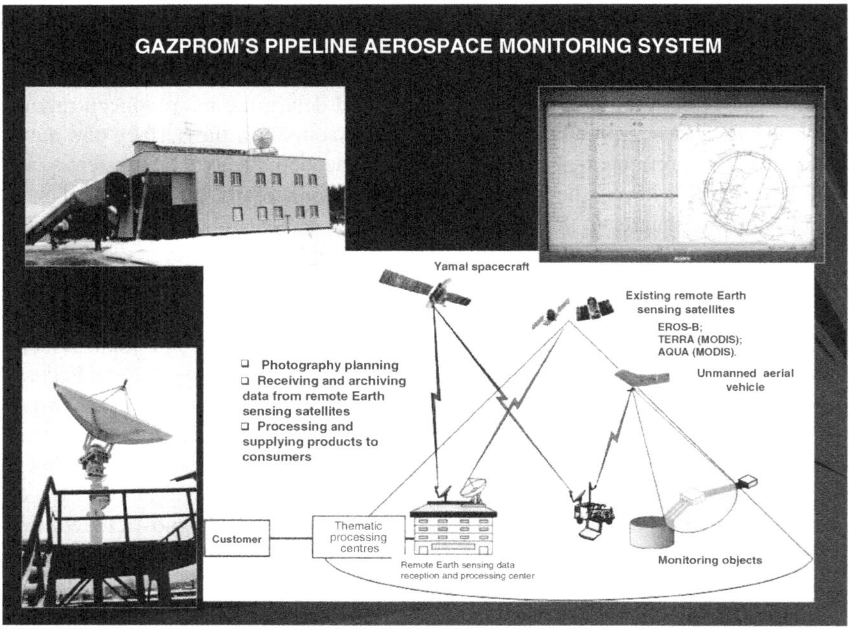

Fig. 2.6 GAZPROM's pipeline aerospace monitoring system (nowadays)

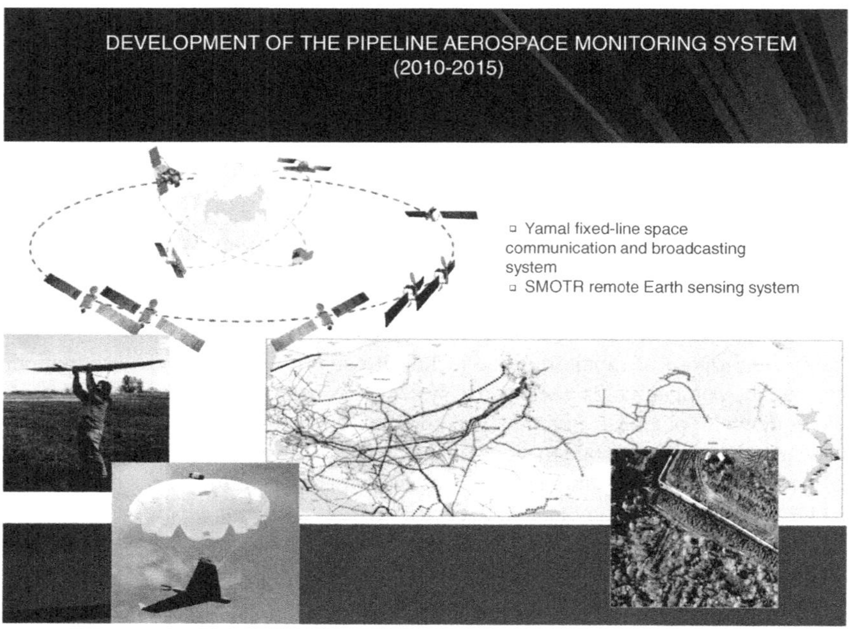

Fig. 2.7 Development of the Pipeline Aerospace Monitoring System (2010–2015)

The system is designed, *inter alia*, to detect sections of trunk gas pipelines laid at a smaller depth than necessary, monitor the condition of gas pipelines, engineering systems and other infrastructure facilities (dykes, surfaced sections, water crossings, transport routes, etc.), watch restricted areas around trunk pipelines and gas distribution stations for unauthorized presence of people and automobiles during the day and at night, illegal construction or logging operations, and oversee construction works and relocation of construction machinery and equipment.

2.4 Geodynamics, Seismic Location, and Prospects for Forecasting Technogenic Emergencies

The transformation of the theory of mobility of the earth's crust should probably be linked to A. Wegener's hypothesis of continental drift which appeared in the early twentieth century and became a theory of plate tectonics in the second half of the last century [3]. The onset of satellite technology in geodetic surveying made it possible to determine the numerical values of these displacements experimentally. Large-scale experimental studies at geodynamic testing ranges exposed intensive local anomalies in vertical and horizontal displacements of the earth's crust linked to crust fractures [4], which have a big amplitude (up to 50–70 mm/year) and are brief (0.1–1 year), localized in space (0.1–1 km), pulsing and alternating. The use of differential GPS technologies for periodic (discrete) and continuous monitoring of trend and vibrational displacements and deformations helped identify a new class of geodynamic movements in fractured areas with a period of 30–60 s and 40–60 min, and confirmed movements occurring at intervals up to 1 year and more [5]. Places with geodynamic movements tend to gravitate toward active tectonic structures and adjacent rocks.

Current geodynamic phenomena include trend and periodic, cyclic movements of rocks in the upper layer of the earth's crust, each of which creates its own specific patterns for the development of accidents (sagging, collapse, destruction of headers and water supply lines, etc.). But trend and vibrational geodynamic displacements affect technogenic objects differently. Trend displacements place an extra stress on the structure of facilities due to slow dislocations and deformations in rock mass. Therefore, facilities located in an area of active tectonic structures with strongly marked trend movements begin to fall apart when the stress within them reaches maximum levels. However, such structures occur rarely, even though trend movements can be initiated by technogenic activities as well. The results of geodynamic studies conducted in recent years and illustrating this method are contained in Table 2.10.

During cyclic geodynamic movements, the process develops in two ways: cyclic deformations and displacements affect a facility directly or in a combined manner, the latter being a result, among other things, of changes in the property of the rock interacting with the target object. In the event of cyclic movements, there can be several reasons for disintegration of technogenic facilities: during direct impact, a facility can be destroyed either by excessive stress and deformation, or by internal

Table 2.10 Some results of geodynamic movement studies for technogenic facilities

Target object	Type of monitoring	Maximum displacement, mm		Maximum deformation, $1 \cdot 10^{-3}$	
		Horizontal	Vertical	Horizontal	Vertical
Surgut, oil pipeline	Continuous	47	108	1.17	2.69
Surgut, sewage system	Continuous	57	92	1.03	1.46
Sarana, radio relay tower	Periodic	48	28	—	—
Kamensk-Uralsky, dolines	Continuous	8	22	0.08	0.37
Yemanzhilinsk, Bukhara-Urals gas pipeline	Continuous	38	63	0.18	0.17
Yasny, Kiyembayevsky open-cut mine	Periodic	335	113	0.06	0.04
Zheleznogorsk-Ilimsky, Korzhunovsky open-cut mine	Periodic	629	600	1.20	0.29
Beloyarsk NPP, Unit No. 4	Continuous	5	5	0.10	0.10
Khromtau, Donskoi ore dressing factory	Continuous	15	31	0.09	0.22

fatigue. During combined impact, destruction can occur because of excessive stress caused by three factors: a change in the property of the rock mass, the combined impact of cyclic movements and thixotropy, or internal fatigue in a facility caused by the same factors.

Cyclic movements of the earth's crust occur more often and can have a variety of effects ranging from the direct impact of cyclic deformations on a facility to that produced by a change in the property of the rock mass in the areas of crustal fractures due to alternating stresses. The pattern of direct impact of cyclic movements on engineering systems is quite simple: if the amplitude of alternating deformations exceeds the maximum stability levels designed for structural parts of a facility, this will cause structural failures resulting in an accident. If deformation is below the maximum permissible level, an accident may or may not occur depending on fatigue in the structural parts of a facility.

Of the short-lasting geodynamic vibrations, the most dangerous are those that last around 1 min or approximately 1 h and create annually 500 and 9,000 stresses, respectively. The period of time during which facilities are destroyed by such vibrations depends on the amplitude of alternating deformations as against their maximum levels. For example, corrosion increases cyclic stresses on the metal parts of pipelines tens and even hundreds of times.

A rather commonly occurring phenomenon characterizing alternating cyclic stresses is the changing property of the rock mass in the areas of crustal fractures induced by such stresses. However, its external signs are not vivid and it is hard to determine their impact on technogenic objects. This phenomenon was for the first time described in the works of the Research and Technical Firm "Geofizprognoz" after studying crustal fracture zones using the profile shooting method [6]. It was determined in particular that cyclic stresses in crustal fracture zones create areas with substantial deviations in strength and deformation properties of the rock mass. It was assumed that the rock mass in a crustal fracture zone is in thixotropic state. This helped solve numerous practical tasks such as determining the nature and patterns of a number of technogenic accidents and disasters (including an accident in St. Petersburg's subway system).

Thixotropy is known to occur in a number of soils and rocks during earthquakes. Alternating cyclic stresses cause them (while having considerable bearing power in static condition) to soften and lose a great deal of their strength, which often results in the tilting and collapse of dwelling houses and other engineering systems. However, this phenomenon occurs only briefly during an earthquake and soil regains its strength after the end of the tremors. Thixotropy in crustal fracture zones, which in many cases consist of rock with a more deformed structure, occurs implicitly for long periods of time and possibly even continuously in historical terms. This phenomenon can be described more precisely by the term "quasi thixotropy." It is believed that the quasi-thixotropic condition of rock in a crustal fracture zone is caused by alternating cyclic displacements, when induced stresses change their strength and deformation properties. Depending on the design features of installations that interact with the rock in a crustal fracture zone, different patterns and scenarios of emergencies can evolve.

For example, the north-western wall of the main open-cut mine at the Korshunovsky ore dressing factory (Urals) has been in critical condition for more than 35 years in a section crossed by a 500-meter latitudinal fracture. Big landslides with the slope angle of 22° have been occurring in the mine since 1975, even though its underlying sedimentary rocks were estimated to remain stable at bigger angles of 28–30°. The movement of surveying posts placed along the perimeter of the mine has been monitored annually to expose their cyclic displacement. The transformation of the rock mass is considered to be a transition to thixotropic state. For example, the mine has a water gallery that draws off the local river from the production area. An examination of the gallery showed that the monolithic reinforced concrete lining in the crustal fracture zone is divided by ring fractures into 7–10-meter sections through which underground water seeps into the gallery, sometimes even spurting under pressure. However, the fractures have practically never occurred in the construction joints in the lining. Therefore, an emergency at the Korshunovsky open-cut mine is developing mainly due to active geodynamic processes in the crustal fracture zone, which requires constant monitoring in order to prevent irreversible destruction.

Intensified karstification is one of the factors that provoke technogenic emergencies. Russian scientists encountered this phenomenon when studying the causes of increased karstification in the residential area of the city of Kamensk-Uralsk and in one of the sections of the Bukhara-Urals gas pipeline near the city of Yemanzhelinsk (see Table 2.9). Dolines appeared to have been caused by alternating stresses in the rocks and their transition into quasi-thixotropic state that had intensified erosion. A series of sinkholes in the crustal fracture zone that crosses the gas pipeline route had resulted in the exposure of its third section, creating a serious technogenic danger.

The above-mentioned examples illustrating the risk of technogenic accidents and disasters suggest the following schematic interconnection between them and contemporary geodynamic processes (Fig. 2.8). For example, a big group of technogenic accidents is caused by both direct and indirect impact of geodynamic processes on technogenic facilities.

Geodynamic phenomena in the earth's crust occur naturally due to tectonic processes, or are caused by anthropogenic activities. The geodynamic crustal patterns that manifest themselves through crustal deformations are the same. The difference is in scale and physics. For example, some studies indicate that the area of geodynamic processes caused by the operation of a big mining enterprise can have a radius of several dozen kilometers [7].

Slow deformations, as a rule, are not accompanied by seismic effects, and devastating consequences are limited either to the foundation of a facility or the facility itself if it is of an underground (semi-buried) type. However, despite the absence of seismic effects, the danger of such deformations is characterized by relatively high rates of changes in the earth's crust that can reach $3 \cdot 10^{-3}$ m a year even in seismically quiet regions of the Urals. This happened in 1961–1962 in two districts of Sverdlovsk region where 0.4-km-wide-open fissures had developed [8].

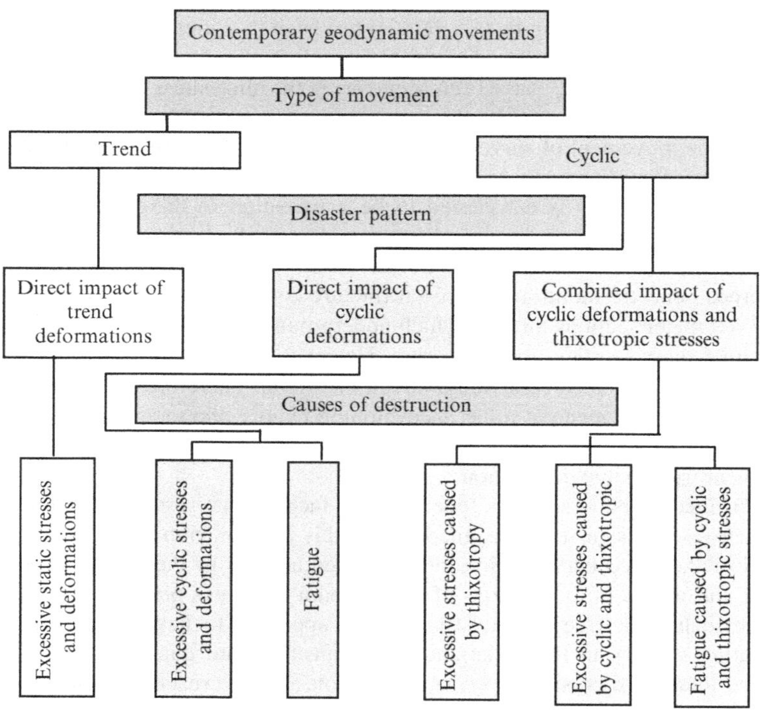

Fig. 2.8 Interdependency between technogenic accidents and geodynamic processes

The observable spread of areas of slow deformations and crustal movements poses a real threat to vital and environmentally vulnerable technical facilities. Such zones occurring in places of crustal fractures are likely to cross extended technical facilities such as trunk pipelines, railroads, power transmission lines, etc., as well as protective dams on water reservoirs, and mining sites.

Theoretical studies of large-scale technogenic effects on a given section of the lithosphere are based on classical solutions of flow mechanics. Experimental studies of deformation caused by man-made effects of mining operations are conducted by way of monitoring survey markers installed at the mining sites. Given the size of such sites, such monitoring became effectively possible only after the use of GPS geodetic devices for research purposes.

It is common knowledge that the Urals is the biggest and the oldest mining region in Russia and the risk of technogenic accidents caused by intensive mining operations is quite real there. For example, the three biggest open-cut mines in the Urals which produce iron ore (Kachkanar), asbestos (Asbest), and coal (Korkino) have already exceeded the permissible level of technogenic impact beyond which induced geodynamic processes begin. A similar situation can be seen in the areas of underground production of salt (Solimkamsk), bauxites (Severouralsk and Satka), and iron ore (Kushva, Nizhny Tagil). Specialists name at least seven mining areas that are potentially dangerous in terms of induced geodynamic processes. Such processes, including earthquakes with a magnitude of $M = 4–6$, have already occurred in some of them.

The Mining Institute of the Urals Branch of the Russian Academy of Sciences has extensive instrumental monitoring data concerning the movement of rocks at the mining enterprises in the Urals over more than 30 years. Satellite surveying technologies open up utterly new opportunities in this field. The institute has created the Urals Center for Geomechanical Studies of the Nature of Technogenic Accidents in the Mining Areas, which has GPS surveying equipment that can monitor crustal deformations caused by the technogenic impact of mining operations. The Center's laboratory studies of plate models and analytical calculations based on the theory of shells indicate that natural horizontal tectonic stresses in the crust (the first invariant of which is estimated at 30.8 MPa on average [9]) create a certain critical load that can cause a deformation. This form of buckling distinguished in the theory of shells causes immediate deformations and seismic effects. Technogenic loads in this case act as a trigger that determines the time and the place of such occurrences.

One of the characteristic zones of large-scale technogenic impacts is the asbestos production area (Fig. 2.9) in the Urals. The depth of open-cut mines from which more than five billion tons of rock have been taken and moved mainly to refuse dumps has reached 300 m. In addition to the open-cut mines and refuse dumps belonging to Uralasbest Plant, there are also other technogenic factors such as the Kurmanovsky chipping quarry and the Malysheva emerald open-cut mine, as well as the Reftinsky and Beloyarsky water reservoirs, each with a capacity of about 1 km^3. The risk of geodynamic processes is exacerbated by the presence of the Beloyarsky nuclear power plant in the area. Because of the direct risk of

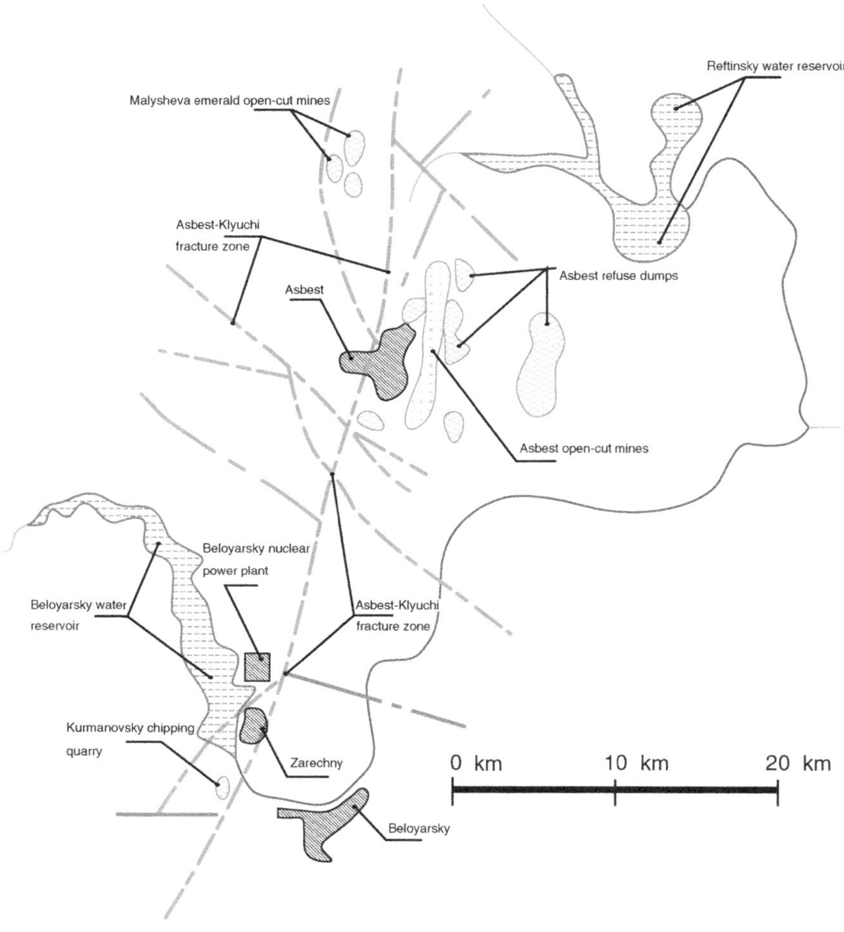

Fig. 2.9 Area of large-scale technogenic impact on the earth's crust (Asbest, Urals)

anthropogenic seismic processes in this part of Russia, scientists from the Urals Center for Geomechanical Studies have made some forecasts.

Theoretical studies using the geomechanical model of the area with a given technogenic impact level (Fig. 2.10) show that vertical crustal displacements have obvious areas of oppositely directed deformations – areas where the crust is rising due to the technogenic impact of mining operations and areas where the crust is sagging due to the loads created by refuse dumps and water reservoirs. The uplift and subsidence rates are almost identical and reach 1 m.

The results above were obtained on the basis of theoretical solutions in which the model environment (the structure of the rock mass) was assumed to be homogeneous. In a real hierarchic environment of a block, deformations are discrete.

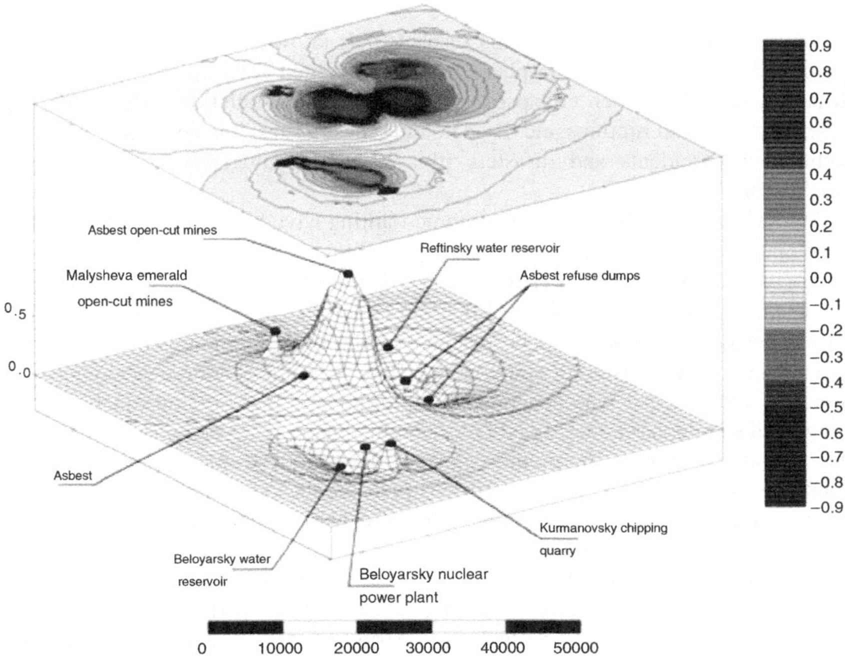

Fig. 2.10 Geo-mechanical model of large-scale technogenic impact on the earth's crust (Asbest, Urals)

This requires serious corrections to be made in projections. Nevertheless, the obtained results showed that the level of vertical displacements of the crust due to technogenic factors is comparable with the levels registered in regions of major earthquakes, and the closeness of technogenic loads opposite in sign creates areas with a high vertical displacement gradient, tremendously increasing the risk of geodynamic seismic effects in such zones.

So, stresses and deformations in large sections of the earth's crust change considerably due to long-term mining operations, which can cause geodynamic processes and disastrous consequences. Therefore, the study of deformation patterns in the upper layers of the lithosphere caused by long-term mining operations becomes particularly relevant. The forecasting of technogenic disasters, their dynamics and prevention requires geodetic monitoring using satellite surveying and aerospace surveillance technologies.

The interconnection between contemporary geodynamic processes and the yet-to-be-discovered mechanisms of technogenic disasters in the mining industry provides a theoretical basis for determining a strategy for fundamental studies related to forecasting anthropogenic emergencies. In the future, this can help solve more effectively a number of applied tasks aimed at preventing such emergencies and reducing their negative effects.

Scientific results achieved in the field of seismic location and spectral seismic surveying can be of interest in this respect. It is known that so-called "areas of tectonic faulting" in the lithosphere are often blamed for different technogenic disasters, such as sudden collapses of buildings and installations, road cave-ins, pipeline bursts, rail ruptures, etc. By detecting these zones, it is possible to forecast technogenic accidents and disasters (this is what geological engineering was created for).

Studying and mapping areas of tectonic faulting have become possible with the development of the spectral seismic survey profiling method [10], which can detect such areas through increased values of the Q-factor in the harmonic components of the seismic signal. It is known [11] that the notion of the Q-factor can be applied only to decaying harmonic signals. A signal can be either harmonic or generated by interference of nonharmonic components. If it does not contain a decaying harmonic process, i.e., it was generated by interference, its Q-factor equals 1. The slower a decaying harmonic signal fades out, the bigger its Q-factor is. Conversely, the longer (slower) a harmonic signal fades out, the more destructive its consequences are. The Q-factor describes both the decaying harmonic signal and the vibrating system that generated it (if there is a signal with a Q-factor, the vibrating system that generated it has the same factor).

It is known that resonance is a coincidence of the frequency of periodic external impact with the frequency of the vibrating system that is experiencing such impact. Such interaction generates a gradual, from period to period, increase in the amplitude of vibrations. The amplitude can increase Q times. Real values of the Q-factor in areas of tectonic faulting range between 50 and 100, but the growth of vibration amplitude is restricted by elastic deformation limits and the ensuing accident. Usually, destruction takes place before the maximum amplitude has been reached, and it occurs instantaneously as if struck. This phenomenon is well known and is called a mine shock or a technogenic earthquake.

The safety of many engineering systems is assessed in terms of resistance to resonances. Bridges are known to have collapsed under marching troops due to resonance when the frequency of the rhythmically stepping people coincided with the own frequency of the bridge's elements as vibrating systems with a high Q factor. However, the possibility of resonance in the "engineering system – earth strata as a foothold" has never been considered. Many technogenic disasters have never got explanation due to this reason. For example, accidents at facilities that have vibrational impact on soil (different power plants, pumping stations, etc.). Entering into resonance is a transitional process that can be caused by a change in the work of a vibrating object, such as a change in the frequency of vibration due to a higher or lower generator speed (as was the case at the Chernobyl nuclear power plant) or the passage of a train at a certain speed through an area of tectonic faulting [12]. This category of technogenic accidents can be effectively forecasted using the spectral seismic profiling method based on seismic location. So, with time, after resonance phenomena have been better studied, it will be possible to avoid technogenic accidents similar to those described above.

But another cause of destruction – planetary pulsation – is much more troublesome for engineering systems. Some of its manifestations are known. For example, it causes significant surveying errors in certain places. Planetary pulsation occurs exclusively in areas of tectonic faulting [8] and has a rather small frequency (tenths and hundredth of a Hertz). The amplitude of such vibrations can reach 10 cm, but it is not constant and can dwindle to zero from time to time. The vector of an alternating shift in the pulsing surface changes with time both in value and direction. The frequency of pulsation changes as well.

Planetary pulsation by itself causes engineering systems to collapse. This can be easily understood if one imagines that one part of the foundation under an installation stands on stable soil, while the other part, being located in an area of tectonic faulting, starts swaying. Naturally, neither a reinforced concrete nor even metal (pipelines, railway tracks, etc.) construction can withstand such loads, even though the latter can, unlike the former, absorb bending stresses. The generally accepted point of view is that the stronger the foundation, the safer the engineering system is. However, practical experience indicates that the strengthening of the foundation does not always increase safety, especially if the foundation is designed without taking into account planetary pulsation (it's like a young flexible tree that can survive a storm that can break an old and unbending one). A break in reinforced concrete plates in the so-called "floating base" caused the collapse of the water park in Moscow's Yasenevo district, killing and injuring dozens of people.

The impact of planetary pulsation on certain vertical constructions (posts, supports, etc.) causes them to incline toward the center of the stress zone (this explains why posts and trees tilt in the same direction). As a rule, the objects that tilt are located on the fringe of the zone, while those in its center sway. Clearly, if vertical posts holding a pyramidal roof sway, the risk of its collapse increases. Pulsation causes piles driven into the ground to sway in much the same way. If they are built into the foundation frame or floating base, either a pile itself will break at the place of attachment or the construction to which it is affixed will fall apart. When planetary pulsation increases, the number of such incidents increases too. This happened in the winter of 2005–2006 when the roofs of several reinforced concrete buildings in Russia, the Czech Republic, Switzerland, and Germany collapsed at the same time.

The sudden collapse of pumping stations standing on piles built into a floating base occurs quite frequently in the Russian oil and gas industry. After studying the construction of the piled foundation at one of the enterprises in Surgut, experts came to a paradoxical conclusion: the use of the most up-to-date technologies and materials does not make a building stronger and may even increase the risk of its subsequent collapse if vibratory properties of the lithosphere are not taken into account.

An observation of piling operations showed that it takes dozens of times fewer strikes (!) to drive some of the piles into the ground, suggesting lesser friction between a pile and the soil, which usually happens in tectonic stress zones due to the highly loose rock mass that makes up the upper layer of the lithosphere. Figure 2.11 shows the results of measuring the rate of friction between piles and the ground, using the spectral seismic surveying method. What makes these

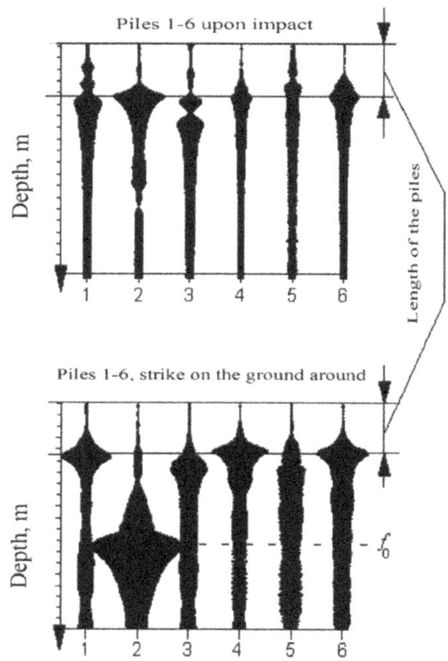

Fig. 2.11 Study of friction between piles and soil using spectral seismic surveying method

measurements distinct is that six piles driven into the ground and standing next to each other were examined twice: the first time, when the pile was struck and the second time when the soil around it was struck (in both cases, a seismic receiver was affixed to a pile).

A common feature shown by all piles was that their length on the ultrasonic image was not very clear with the exception of pile No. 2, the size of which upon impact appeared to be much clearer than that of the other piles, but it did not register at all when the ground around it was struck. This can be explained by lesser friction between pile No. 2 and the soil when, on the one hand, its own lengthwise vibrations are not dampened, and, on the other hand, there is a reduced acoustic contact with the ground. There was yet another interesting effect: when the ground was struck, the above-mentioned harmonic vibration process with a colossal Q factor was registered at a certain frequency f_0.

Although it is not quite clear what exactly generated this vibration process in the "pile–soil" system, such a vibrating system with a high Q factor is the reason for the resonance with the vibration object, for which the foundation was essentially built, and the cause of a mine shock. This leads to explosion-like and instantaneous destruction of the foundation as in the case of gas pumping stations.

Sometimes, it takes a very small number of strikes to drive a pile into the ground. It is also believed that it can't hurt if a certain number of piles simply hang upon the plate rather than hold it. However, such hanging piles will sway to break loose of the plate, and if there are many of them, the plate can simply snap. So, if a facility with vibrating devices is built in an area of tectonic stress, there are two factors causing the plate to snap: vibration and planetary pulsation. If a facility happens to be in a hazardous zone, a piled foundation with a floating base runs a greater risk of sudden collapse than when a floating base and piles are not used at all.

Therefore, there are two types of alternating impact on the lithosphere: vibration from working mechanisms and planetary pulsation. The former, being a higher frequency one, can cause resonance with the own frequencies of adjacent rocks, while the latter, as a low frequency one, can cause resonance with the underlying structures (for example, the pulsation frequency of 1 Hz corresponds to a 2.5-kilometer stratum, and a frequency of 0.1 Hz corresponds to a 25-kilometer stratum). If the frequency of planetary pulsation is close to the own frequency of the lithosphere structure, and if its Q factor as a vibrating system is big enough, the risk of resonance phenomena, i.e., earthquakes, increases immensely.

The pattern of a natural earthquake differs from that of a technogenic one. It happens even if the frequency of planetary pulsation changes and nears the frequency of resonance. The amplitude of soil vibrations increases, as has been registered by seismologists immediately before an earthquake. If the frequency of planetary pulsation does not coincide with the own frequency of the relevant vibrating system but is close to it, so-called beats occur.

The earth's crust in areas of tectonic faulting appears to be quite mobile and using it without taking this factor into account increases the frequency of technogenic accidents, and the oftener the ground comes under static or dynamic impact, the more intensive they will be. A transition to the perception of earth strata as a combination of vibrating systems both in the general methodological and purely practical terms is as important as is the transition from geocentric to heliocentric views, and helps to make the operation of engineering and technical systems safer. Such a change of paradigm will make it possible to reduce considerably the number of technogenic accidents, rule out technogenic seismic processes and effectively forecast natural earthquakes.

2.5 Forecasting of Man-Made Disasters and the Role of Space Systems in its Practical Implementation

Natural and technogenic processes often have close and sometimes paragenetic connection. But while natural calamities, especially of space and endogenous nature, can in principle be forecasted but cannot be practically prevented, technogenic accidents and disasters can be both forecasted and prevented. Much or everything depends on the will and the quality of work, the comprehensiveness and promptness of monitoring systems (both ground-based and aerospace systems), and means of protection for personnel, constructions, and facilities. There is a tendency to move from fundamental studies focusing on the physical nature of the environment and space and time changes in it to forecasting and preventing natural and technogenic disasters.

Scientists say that deterministic approaches have no fundamental physical basis for offering clear short-term forecasting criteria. Statistical probability methods of forecasting do not always justify themselves, apparently because of fundamental aspects of the transient and heterogeneous nature of space and time, and the unclear quantitative role of each of these factors in the probability of certain devastating processes. The search for evidence as well as necessary and sufficient conditions for natural and technogenic disasters and the optimal number of measureable factors is one of the vital aspects of further work.

The current state and dynamics of space and geosphere, as has been repeatedly pointed out, is largely assessed as critical not only because of the natural processes inside the planet and in the surrounding physical space, but also because of the scale of human interference in the physical and ecological state of the earth. So not only do natural processes and their dynamics cause technogenic accidents and disasters, but technogenic processes themselves bring about such changes in the natural state of the planet that eventually produce catastrophic impact on industry and society.

Dangerous exogenous and endogenous processes coupled with the lack or absence of protective measures or facilities, imperfect technologies and natural aging of infrastructure, incompetence or temporary inability of the servicing personnel can cause considerable financial losses and socio-economic damage. This warrants further efforts to explore ways for forecasting interaction between dynamically changing properties of infrastructure and continuously changing geological environment, taking into account external factors. In so doing, it is necessary to use both contact-type and remote sensing methods. For example, one of the key current processes is the growing capacity of permanently operating units and facilities, as a result of which technogenic vibration and noise already exceed the planet's maximum natural micro seismic background level by one order or two orders of magnitude, and change the engineering, geological, and seismic parameters of hitherto "quiet" places. Stresses exceeding the endurance strength of materials build up as a result of long-term operation of such facilities. Failures and breakdowns caused by wear and tear, and end of service life can be forecasted using statistical probability methods. Changes in the environment are studied using geophysical methods.

Summing up the aforesaid, we should emphasize the diversity of factors that cause technogenic accidents which require unique analysis and forecasting methods, as well as contact-type and remote prevention techniques. Man-made emergencies caused by wear and tear, end of service life, technogenic effects on the geodynamic environment, increased concentrations of hazardous and explosive substances, gases and dust can be forecasted using physical and mathematical modeling. For example, continuous mathematical monitoring of potentially hazardous facilities using a dynamic model of safe operation has been proposed [13].

Aerospace monitoring facilities have certain capabilities to detect, forecast, and prevent technogenic processes, even though the majority of them (see above) have no reliable harbingers that could be registered by aerospace systems, with the exception of anthropogenic disasters caused (provoked) by:

- Geological and meteorological calamities;

- Extreme resonance processes in mechanical systems caused by geodynamic vibrations in the earth's crust;
- Increased concentrations of potentially hazardous gases and/or dust;
- Cascading failures of geotechnical systems (trunk pipelines, railroads, etc.) as a result of interaction with the environment;
- Use of equipment for unusual purposes associated with gross violations of technical system operating rules.

Aerospace monitoring provides a tool for early detection and prevention of destructive processes in geotechnical systems, oil spills and gas leaks, siphoning of petrol products, etc., for mapping areas intended for the construction of large technical facilities and monitoring their construction in order to detect erroneous or intentional deviations (in pipeline routes, in the construction of power transmission lines, railroads, etc.); for geodetic and environmental monitoring, and early warning of emergencies (fires, gas outbursts, hazardous spills, etc.).

The Federal Agency for Mining and Industrial Supervision (ROSGORTECH NADZOR), the Russian Federal Inspectorate for Nuclear and Radiation Safety (ROSATOMNADZOR), and the relevant bodies within federal executive agencies monitor the state of industrial facilities and forecast possible accidents at them. Similar bodies exist at the regional level as well as within enterprises and organizations, and are known as industrial safety units. Monitoring is conducted using ground-based and aerospace systems of the relevant ministries, agencies, regional authorities, and organizations (enterprises) depending on their jurisdiction. The main element of the system is ground-based sensor monitoring networks and laboratory control capabilities of the federal civil defense organizations and its key elements subordinated to the Federal Agency for Hydrometeorology and Environment Monitoring, the Emergencies Ministry, the Agriculture Ministry, the Ministry of Health and Social Development, and the Ministry of Natural Resources.

Russian space-based monitoring means are currently used mainly for detecting and specifying slowly evolving large-scale natural phenomena and processes. Aviation capabilities are used for the same purpose, as well as for monitoring the state of trunk pipelines, radiation, road, snow, and ice in areas of extensive destruction, etc. They have broad possibilities both in terms of coverage and promptness of information delivery, which makes them useful for a number of monitoring services that operate within their respective jurisdictions.

General operational terms of Russia's national forecasting and monitoring system are set forth in Regulation on the System of Monitoring, Laboratory Inspection and Forecasting of Natural and Technogenic Emergencies No. 483 of the Ministry for Emergency Situations of November 12, 2001, and in the regulations of its units and elements, in the regulatory documents of the relevant federal ministries and agencies, regional and territorial authorities. The results of emergency monitoring and forecasting serve as the starting point in developing long-term, medium-term, and short-term target programs and plans, and in making relevant decisions to prevent and respond to disasters, including technogenic ones.

The Russian Federal Space Program calls for developing national remote earth sensing system up to 2015, including the creation and/or development of certain space assets that can jointly monitor natural calamities and technogenic disasters. These include remote sensing and monitoring, navigation and hydrometeorology, communications and relay systems. All of these space systems should, despite of their departmental disunity, constantly interact with each other through ground-based infrastructure designed for controlling space missions and for receiving, processing, and distributing space data. A combination of these space systems should be a multilevel hierarchic and constantly evolving intellectualized structure. Its ground-based elements should include centers for controlling space missions and for receiving, processing, and distributing space data, centers for thematic processing of monitoring data and developing methods of decoding and identifying observable processes and objects, an aviation surveillance system, test ranges for validating space data and grading onboard equipment used for monitoring natural and technogenic objects and processes, and situation centers for analyzing and gathering monitoring information as well as ground-based telemetric data necessary for making operational and strategic decisions.

Although Russia has launched only two remote earth sensing satellites – medium-resolution "Monitor-E" and ultra-high-resolution "Resurs-DK1" – in the past several years, new space optical and electronic monitoring satellites "Resurs-P" and "Arkon-Viktoria," radar observation satellites "Arkon-2" and "Kondor-E," a space system based on "Ekola" mini satellites, the above-mentioned "Smotr" gas industry monitoring system, the Arctic subpolar observation system, and other systems are expected to become operational by 2015 with the help of off-budget funding.

Many projects related to building of environment and/or hazardous facility monitoring space systems have been published in the world lately, but they have no clearly defined prognostication value. The analysis of such projects (see details in Chap. 3) showed that effective monitoring of technogenic facilities would require a space system incorporating a wide range of observation means, such as ultraviolet, visible, and infrared optical and electronic equipment (multispectral and hyperspectral), multichannel microwave equipment, multifrequency radar equipment (in the super-high and ultra-high frequency band), laser and lidar multifrequency equipment, as well as powerful onboard information systems with a memory of hundreds of gigabytes. Given the current level of technology, its combined weight will be no less than two tons. A satellite furnished with such equipment could weigh about ten tons, and its power consumption would exceed 10 kV (the ENVISAT satellite weighing about eight tons and carrying only some of the above-mentioned equipment weighing 1,200 kg could serve as an example of such a solution). In order to ensure hourly information updating, it would be necessary to have six to twelve satellites in such an orbiting system, the cost of creating, testing, and deployment of which would amount to billions of U.S. dollars, let alone the creation of ground-based infrastructure, which would take at least 10–12 years. Such a program can hardly be implemented by any, even the most industrialized, nation alone.

A different approach is used now: space systems are created for big groups of similar tasks facing different customers and information users. Therefore, technogenic disaster monitoring can also be conducted using various space systems.

For example, there is a large remote earth sensing orbiting system consisting of dozens of optical and radio monitoring satellites, including military and dual-purpose ones. Most of them can perform not only civilian monitoring and mapping but also surveillance and target-marking functions during armed conflicts (the U.S. and NATO use both their own civilian high-resolution remote earth sensing satellites and similar systems owned by European countries, Israel, India, Japan, Canada, and others, thus conducting almost continuous monitoring of conflict zones, planning combat operations, and using high-precision weapons). It is possible that similar policy can be applied during periods of increased natural and/or techno risks, with the use of military (dual-purpose) space systems for monitoring of natural disasters and emergencies.

Weather satellites which perform global climate and weather monitoring functions make it possible to forecast and warn of natural disasters caused by atmospheric disturbances: storms, floods, cyclones, torrential rains, droughts, etc., which often cause transport accidents on land and at sea. Weather satellite constellations are located on two levels. The upper echelon of such systems is located in geostationary orbit and forms an integral part of the global monitoring system under the World Meteorological Organization. The number of satellites in geostationary orbit should be sufficient for conducting continuous Earth monitoring in an area stretching from 70° North latitude to 70° South latitude and for downloading fresh weather data updated every 30 min or more frequently. This used to be achieved effectively by a system of five geostationary satellites: two American, one European, one Russian, and one Japanese. In 2007, this orbital constellation consisted of the following satellites: a U.S. satellite (GOES), a Japanese satellite (MTSAT-IR), two Indian satellites (Metsat-1, Insat-3A), and one Chinese satellite (FY-3 C). The lower echelon of weather satellites is deployed at low geosynchronous subpolar orbits for the monitoring of the earth's cloud cover and other weather phenomena; measuring vertical atmospheric temperature and humidity profiles and sea surface temperature, surface wind, ice and snow cover; and gathering information from environmental geophysical monitoring platforms, etc. The low-orbit echelon currently consists of national weather satellites: NOAA-K, DMSP 5D-3 (U.S.), Metop (Western Europe), FY-1D, FY-3 (China), and Meteor-M (Russia).

The availability and broad capabilities of the domestic and foreign navigation systems (GLONASS, GPS, and GALILEO) make it possible to promptly position mobile objects and solve space surveying tasks as part of geodynamic studies and for the purpose of monitoring dangerous natural and technogenic objects.

Space communications and relaying system becomes increasingly important. There are dozens of communications and data transmission satellites owned by the U.S., Russia, China, India, Japan, Italy, and another 16 countries and currently operating in geostationary orbit.

Therefore, the space infrastructure needed for monitoring of natural and technogenic disasters is already in place. It is only necessary to focus on creating a system of satellites carrying special equipment that is not available in traditional monitoring spacecraft. Such equipment can be made and put into orbit aboard small microsatellites and multipurpose orbiting platforms that alongside other monitoring, communications, and navigation systems could effectively solve monitoring and forecasting tasks for the benefit of all interested parties using integrated international information, navigational and telecommunication capabilities. The fourth chapter in this book explores a possible design for such a system – the International Global Aerospace Monitoring System (IGMASS).

References

1. Статистика страха http://www.raexpert.ru/editions/panorama2001-1/part1/
2. А. Беззубцев-Кондаков Почему это случилось? Техногенные катастрофы в России история самых крупных техногенных катастроф в СССР и России Издательство: Питер, 2010г.: с. 288, (Russian)
3. Вегенер А. Происхождение материков и океанов / Современные проблемы естествознания. – М. – Л.: Геоиздат, 1925. – 145 с. (Russian).
4. Кузьмин Ю.О. Современная геодинамика и оценка геодинамического риска при недропользовании. – М.: Агентство экологических новостей, 1999. – 220 с. (in Russian).
5. Sashourin, A.D., Panzhin, A.A., Kostrukova, N.K., Kostrukov, O.M.: Field investigation of dynamic displacement in zone of tectonic breaking. Rock mechanics – a challenge for society: Proceedings of the ISRM regional symposium EUROK 2001, pp. 157–162. Espoo, Finland 3–7 June 2001/Balkema 2001. (Russian)
6. Гликман А.Г. Физика и практика спектральной сейсморазведки. Internet. http://www.newgeophys.spb.ru/ru/book/index.shtml. (Russian).
7. Сашурин А.Д. Явления изостазии при разработке месторождений полезных ископаемых // Приложение результатов исследования полей напряжений к решению задач горного дела и инженерной геологии. – Апатиты: Кольский фил. АН СССР, 1985. – С. 27–31. (in Russian).
8. Николаев Н.И. О применении исторических, археологических и палеосейсмических методов при изучении исторических проявлений землетрясений // Влияние инженерной деятельности на сейсмический режим. – М.: Наука, 1977. – С. 179–182. (Russian).
9. Sashourin, A.D.: On the parameteres of the first invariant of the stress state in the upper part of the earth crust: The 1-st. International Seminar "Stresses in the lithosphere (global, regional, local)". Abstract of papers, Moscow, p. 162 (1994)
10. Гликман А.Г. О принципах спектральной сейсморазведки // Геофизика XXI столетия: 2003–2004 годы. Сборник трудов V и VI геофизических чтений имени В.В. Федынского. – Тверь: ООО « Издательство ГЕРС», 2005. – с. 370–375. (Russian).
11. Харкевич А.А. Спектры и анализ. – М. – Л.: Геоиздат, 1952. (Russian).
12. Гликман А.Г. О некоторых разновидностях аварий на железных дорогах // Строительный журнал. – Ноябрь 2007. (Russian).
13. Hegai, V.V., Kim, V.P., Nikiforova, L.I.: A possible generation mechanism of acoustic-gravity waves in the ionosphere before strong earthquakes. J. Earthquake Predict. Res. **6**, 584–589 (1997)
14. Николаев Н.И. Новейшая тектоника и геодинамика литосферы. – М.: Недра, 1988. – 491 с. (Russian).
15. Теркот Д.Л., Шуберт Дж. Геодинамика: Геологические приложения физики сплошных сред / Пер. с англ. – М.: Мир, 1985. – Ч. 1. – 376 с. (in Russian).
16. Лурье А.И. Пространственные задачи теории упругости. – М.: Наука, 1955 (in Russian)

17. Королев В.И. Упругопластические деформации оболочек. – М.: Машиностроение, 1971 (in Russian).
18. Гликман А.Г. О структуре поля упругих колебаний при сейсмоизмерениях // Седьмые геофизические чтения им. В.В. Федынского 3–5 марта 2005 года. Тезисы докладов. – М.: Центр ГЕОН. –2005. – С. 9–10. (Russian).
19. Гликман А.Г. Эффект акустического резонансного поглощения (АРП) как основа новой парадигмы теории поля упругих колебаний // Геофизика XXI столетия: 2003–2004 годы. Сборник трудов V и VI геофизических чтений имени В.В. Федынского. – Тверь: ООО « Издательство ГЕРС», 2005. – С. 293–299. (Russian).
20. Sashourin, A.D., Panzhin, A.A., Kostrukova, N.K., Kostrukov, O.M.: Field investigation of dynamic displacement in zone of tectonic breaking. Rock mechanics – a challenge for society: Proceedings of the ISRM regional symposium EUROK 2001, pp. 157–162. Espoo, Finland 3–7 June 2001/Balkema 2001.
21. Гликман А.Г. О резонансной природе землетрясений // В мире науки. – 2007. – № 3. – С. 59–63. (Russian).
22. Техногенная и экологическая безопасность и управление риском. – М.: МЧС России, РАН, 2008 (Russian)

Further Reading

1. Меньиков, В.А., Перминов, А.Н., О.М. рличич Глобальные проблемы человечества и космоса – М.: Н КС имени А.А. Максимова, 2010. (Russian) – 570с.
2. V. A. Menshikov "Directions of the international technological cooperation in remote automated control of stationery and mobile objects field.". – \"European journal of natural history". – 2006. – № 3
3. Sandau R. "Small satellite earth monitoring missions potential and trends." by book "Space for Security and Prosperity of the Peoples.", editors Contant J.-M., Menshikov V., IAA,- 2010, p. 128–154

Global Risks and Threats to Humanity from Outer Space: Prospects of Warning and Parrying

3

On October 8, 2009, an unobserved asteroid approaching the Earth exploded in the upper atmosphere (at height of 15–20 km) directly over South Sulawesi province in Indonesia. According to NASA, this explosion of a 10-m-sized stone object, which entered the dense atmosphere at a speed of more than 20 km/s, realized an energy of 50,000 tons TNT equivalent (three times more powerful than the Hiroshima nuclear blast). The event was detected by the West Ontario University Observatory, at a distance of 16,000 km away from its epicenter.

The Earth has been struck by asteroids and comets (Near-Earth Objects, NEOs) many times throughout its history. The largest of these impacts have led to large-scale extinctions of life, such as the one 65 million years ago which caused the disappearance of the dinosaurs. Humans possess technological tools that could deflect NEOs and avoid such catastrophic impacts if we were to develop the necessary plans and facilities. This chapter addresses the nature of the threat from outer space, expected future dangerous rendezvous with various sizes of NEO, and their consequences. It reviews current programs to detect, track, and characterize asteroids and comets, and the future developments required in order to take responsible and timely action. It identifies a number of techniques that could alter an incoming NEO's orbit so as to avoid impact (Fig. 3.1).

It addresses the organizational aspects that will have to be dealt with if a serious international capability is to be developed and employed to mitigate the threat. Lastly, the chapter examines some of the principal international policy implications that must be dealt with if the world is to act in a timely, unified, and effective way in relation to the very real threat from NEOs. In addition to the hazard from asteroid and meteoroids the authors discuss the problem of space debris as well as the effects of increased solar activity in terms of global threats to modern civilization.

V.A. Menshikov et al., *Global Aerospace Monitoring and Disaster Management*, DOI 10.1007/978-3-7091-0810-9_3, © Springer-Verlag/Wien 2012

Fig. 3.1 Danger from outer space: is it a reality?

3.1 The Risk and Threat from Asteroids and Comets: Are They Really Dangerous?

Hazardous nature of comets represents an idea that dates back at least to the seventeenth century, when Edmond Halley is said to have addressed the Royal Society and speculated that the Caspian Sea might be an impact scar. The first NEO (Eros) wasn't discovered until 1898 and the first NEO that actually crosses the Earth's orbit (Apollo) wasn't found until 1932. By 1940s, three Earth-crossing NEOs were known, their basic rocky nature and relationship to meteorites was appreciated, and it was possible to crudely estimate their impact rate.

Later, some scientists (who understood both orbital dynamics and impact physics) proposed that NEO impacts might account for large scale extinctions in the Earth's paleontological record. Around the same time, Shoemaker and his colleagues firmly established the impact origin of Meteor Crater in Arizona. The first decade of planetary exploration revealed that Mars and Mercury were heavily cratered.

Not until 1980/1981 did it begin to be realized in the scientific community that NEOs and impact craters carried major implications for the history of life on Earth, both in the past and possibly in our own times.

Even after the discovery of the Chixchulub impact structure in Mexico and its temporal simultaneity with the Cretaceous–Tertiary (K-T) boundary and large-scale extinctions, it has taken some Earth scientists a while to recognize and accept the statistical inevitability that the Earth is struck by asteroids and comets (Fig. 3.2). Each impact, typically spaced 50–100 million years, liberates tens of millions to billions of tons (MT, BT TNT-equivalent) of energy into the fragile ecosphere, which must have had dramatic consequences every time. Some skeptics still consider the Chixchulub impact to be only one of several contributing factors to

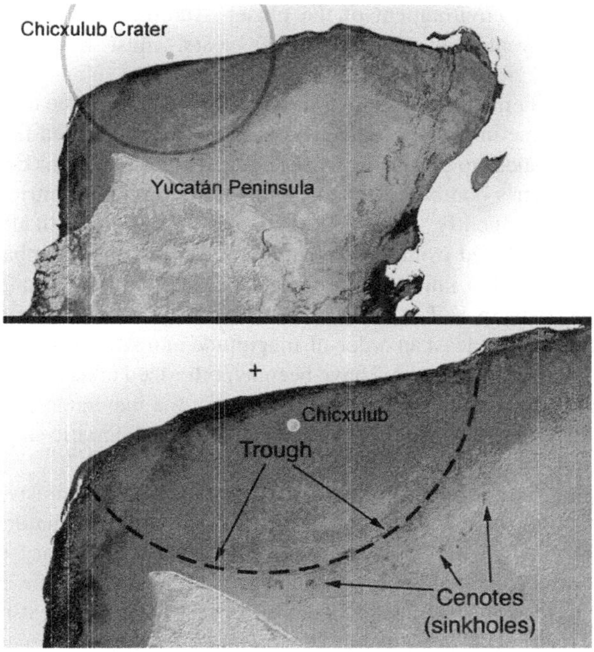

Fig. 3.2 Chixchulub impact scheme

the K-T extinctions. They also point out that direct evidence firmly linking other, older large-scale extinctions to impacts is so far either more equivocal than for the K-T, or altogether lacking – but this is a natural result of the ongoing tectonic resurfacing of our planet. If the great large-scale extinctions can somehow be explained by forces that are much less sudden and powerful than impacts (e.g., episodes of volcanism or sea regressions), one must ask how the huge impacts that must have occurred failed to leave dramatic evidence in the fossil record.

So, we'd like to draw attention of the readers to the role impacts have played in the geological and biological history of our planet, which sets the stage for the modern impact hazard. About 170 well-known impact craters have been recognized on the Earth, and this number may double according to private, commercial records. They range from recent, small (tens to hundreds of meters in size) impact craters to multi-100 km-structures expressed in the geologic record although no longer retaining crater-like morphology, which has been eroded away. Published ages for some craters are of varying reliability (raising doubts about alleged periodicities in impact rates). The Earth's history is increasingly incomplete for older epochs, but the virtual total loss of datable rocks back toward 4 billion years is consistent with the inferences from the lunar period that Earth was pummeled by a couple of lunar-basin-forming projectiles every million years for 50–100 million years, which would have boiled away any oceans and completely transformed the atmospheric, oceanic, and crustal environment of the planet. Additionally, thousands of K-T boundary level events, one every 10,000 years, must have had profound repercussions.

The late heavy bombardment period must have "frustrated" the origin of life on the Earth while some impacting projectiles might have contributed life-enhancing, volatile-rich substances to our planet. Any simple, extant life forms could conceivably survive a crater's bombardment by being ejected into geocentric or heliocentric orbits, and subsequently "re-seed" life upon re-impacting Earth after terrestrial environments had relaxed from the violent aftereffects of such bombardments. As noted above, the Earth's impact environment became similar to todays by ~ 3.5 billion years. Dozens of K-T level impacts would have happened since that time, several of which were at least an order-of-magnitude more devastating. Momentous events, like "The Snowball Earth," have been hypothesized to have occurred in pre-Paleozoic times (i.e., before 570 million years ago); the inevitable cosmic impacts must be considered as plausible triggers for such dramatic climatic changes, or their cessation, during those eons.

For example, during the Paleozoic Era, there must have been several K-T (or greater) impact events, roughly equaling the number of major large-scale extinctions recorded in the fossil record. Only the K-T boundary extinction is now accepted as being largely, or exclusively, due to impact (the formation of Chixchulub). Evidence accumulates that the greatest large-scale extinction of all, the Permian-Triassic (P-T) event, was exceptionally sudden; one recent study that argues for a gradual P-T extinction is invalidated by its faulty methodology. It is possible that the K-T impact was exceptionally efficient in causing extinction (e.g., because of the composition of the rocks where it hit, or if it were an oblique impact or augmented by accompanying impacts). However, straightforward evaluations of the expected physical and biological repercussions of massive impacts suggest that any such impact should result in such extreme environmental havoc that large-scale extinction would be plausible, although conditions may cause consequences to vary among impacts of similar magnitude. It is plausible that the difficulty of finding incontrovertible proof of the impact origin of earlier large-scale extinctions is because of the much poorer preservation and quality of the more ancient geological records.

According to some experts [1] ancient asteroid impacts must be exceptionally more lethal than any other proposed terrestrial causes for large-scale extinctions because of two unique features: their environmental effects happen essentially instantaneously (on timescales of hours to months, during which species have little time to evolve or migrate to refuges) and there are multiple environmental consequences (e.g., broiler-like skies as ejecta re-enter the atmosphere, global firestorm, ozone layer destroyed, earthquakes and tsunami, months of ensuing "impact winter," centuries of global warming, poisoning of the oceans). Not only the rapidity of changes, but also the cumulative and synergistic consequences of the compound effects, makes asteroid impact overwhelmingly more difficult for species to survive than alternative, Earth-generated crises. Volcanism, sea regressions, and even sudden effects of hypothesized collapses of continental shelves or polar ice caps are far less abrupt than the immediate (within a couple of hours) worldwide consequences of impact; life forms have much better opportunities in longer-duration scenarios to hide, migrate, or evolve.

Other hypotheses for large-scale extinctions lack the diverse, compounding negative global effects of impacts. Only the artificial horror of global nuclear war or the consequences of the very tiny possibility of a stellar explosion near the Solar System could compete with impacts for immediate, species-threatening changes to the Earth's ecosystem. What other process could possibly be so effective? And even if one or more extinctions do have other causes, the largest asteroid/comet impacts during ancient eras cannot avoid having left traces in the fossil record. By analogy, in the modern world, the very tiny possibility of an impact by a large comet or asteroid, exceeding several km in diameter, is the largest conceivable natural disaster that humanity confronts.

But how could we estimate the risks and consequences of asteroids and comet impacts for society in the twenty-first century? Less well understood are the physical and environmental consequences of impacts of various energies. Some researches have elucidated the previously poorly understood phenomena of impact-generated tsunami. There has also been recent argumentation that the ozone layer might be largely destroyed by NEOs as small as 0.5 km, smaller than previously estimated.

The most comprehensive recent analysis of the risks of NEA impacts is that of the NASA NEO Science Definition Team (SDT) [2]. The SDT evaluated two other sources of mortality due to NEO impacts smaller than those that would cause global effects: impacts onto land, with local and regional consequences analogous to the explosion of a high power bomb and impacts into an ocean, resulting in inundation of shores by the resulting tsunamis [3]. The SDT evaluated fatalities for land impacts using (a) a model for the radius of destruction by impactors >150 m diameter that survive atmospheric penetration with most of their cosmic velocity and a map of population distribution across the Earth, along coastlines, in particular. A thorough analysis of the tsunami hazard, based on reanalysis of wave and run-up physics, provided an estimated number of "people affected per year" by impact-generated tsunami. The SDT notes that, historically, only ~ 10% of people in an inundation zone die, thanks to advance warning and evacuation. Since a similar level of advance warning from an ocean impact could be expected, we suggest that the actual fatality fraction might be similar.

Experts separated out each of the general classes of hazard: land impacts, which are dominated by small-sized bodies in the range from 50 to 150 m diameter; ocean impacts causing tsunami, dominated by bodies in the 150–700-m size range; and the large-sized asteroids, 1.5 km or larger that can cause global climatic catastrophes affecting people anywhere in the world, and hence are more or less independent of the impact site.

Figure 3.3 presents the risk versus size of impactor in histogram form, using the 2008 population model, for the intrinsic risk before any NEAs were discovered, and the residual risk at the current (mid-2008) level of completion. In both cases, we have used the SDT model of tsunami risk. If we had used that level divided by

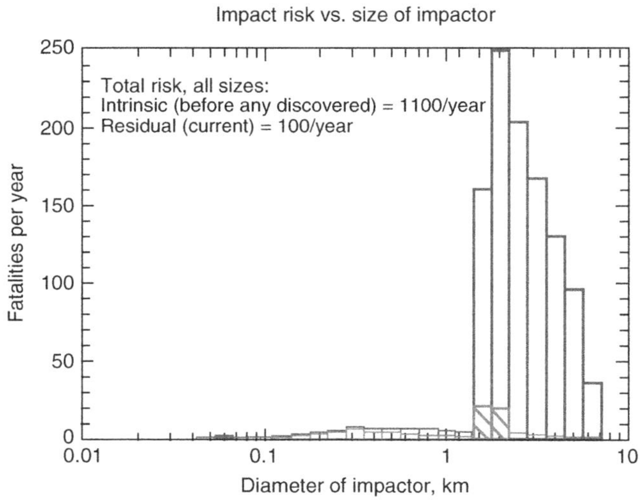

Fig. 3.3 Estimates of NEA impact risks

ten, the risk level in the mid-range from 150 m to 1.5 km diameter would be a factor of several times lower still.

The first set of estimates is essentially copied from the 2003 SDT report, using that population model and their calculated values of F. The next two columns use the same "kill curve" but impact rates derived from the more recent population model of Harris (2008). The next pair of columns show the "residual" impact risk, given that recently discovered (as of June, 2008) NEAs have been found to have essentially zero probability of impact in the next half century or so.

Individual human beings and society itself reacts in subjective ways to comparisons of the impact hazard with other societal hazards, especially because of the inherently low-probability high-consequence character of the impact hazard. We consider mortality rather than property damage as being more central to fears of impacts. But neither mortality nor economic loss estimates provide a good forecast of how societies may respond in the future to different kinds of hazards. The ~ 3,000 deaths from the terrorist attacks on September 11, 2001, had dramatic national and international consequences (involving economics, politics, war, etc.) while a similar number of U.S. highway fatalities during the same month were hardly noticed, except by family members and associates of the deceased. But asteroid impacts have many elements of a "dreadful" hazard (being perceived as being involuntary, fatal, uncontrollable, catastrophic, and increasing in news reports, anyway), like terrorism or nuclear threats, in contrast with more mundane hazards that may be more serious as measured by objective criteria.

Contemporary society often spends much – even orders of magnitude – more per life saved to reduce "dreadful" hazards than mundane ones. For this reason, efforts to reduce the impact hazard and to plan for mitigation (e.g., evacuation of ground zero, storing food supplies in order to survive a global agricultural disaster, or developing capabilities to deflect a threatening NEA), may be perceived by many citizens as money well spent. Conversely, public opinion polls show that many others regard the impact hazard as being trivial.

At the World Summit on Sustainable Development in Johannesburg in 2002 were presented three interesting scenarios, which illustrated the breadth of issues that must be confronted in managing potential consequences of NEA impacts. For each impact disaster scenario, were presented the nature of the devastation, the probability that the event will happen, the likely warning time, the possibilities for postwarning mitigation, the nature of issues to be faced in after-event disaster management, and – of most practical interest – what can be done now to prepare in advance (see Table 3.1).

1. "Civilization Destroyer" Scenario.
   The chances of such an event are probably <1-in-100,000 during the century. The warning time would almost certainly be longer term in the case of an NEA, but with current technology telescopes might be only months in the case of

Table 3.1 Asteroid impact threat scenarios

2–3-km diameter asteroid impact ("Civilization Destroyer")	Tunguska atmospheric explosion	Near-term impact scenario
A million MT impact, even though ~ 100 times less energetic than the K-T impact, would probably destroy our civilization. The dominant immediate global effect would be sudden cooling, lasting many months, due to massive injection of dust into the stratosphere following impact. Moreover, the ozone layer would be destroyed. Agriculture would be largely lost, worldwide, for an entire growing season. Combined with other effects (e.g., a firestorm the size of India), it is plausible that billions might die from collapse of social and economic institutions and infrastructure. No nation could avoid direct, as well as indirect, consequences of unprecedented magnitude. Of course, because civilization has never witnessed such an apocalypse, predictions of consequences are fraught with uncertainty.	It would explode ~ 15 km above the ground, releasing the energy of ~ 00 Hiroshima-scale bombs. Some researchers consider that such an event would be spectacular to witness but would not have lethal consequences. Our review of the literature suggests, however, that weak structures might be damaged or destroyed by the overpressure of the blast wave out to 20 km. The death toll might be hundreds; although casualties would be far higher in a densely populated place.	The problem, which can develop within hours in the 24-h global news media, is that something possibly real about an NEA is twisted by human fallibility and/or hyperbole. Hypothetical examples include: (a) a prediction, a few days in advance, of an actual near-miss ("just" 60,000 km from the Earth) by a >100 m asteroid, which might be viewed with alarm by a distrustful public who would still fear an actual impact; (b) the reported (or mis-reported) prediction by a reputable (but mistaken or misquoted) astronomer that a huge impact will occur on a specific day in the future in a particular country, resulting in panic for several days until the report is withdrawn; or (c) a prediction, officially endorsed by an entity of a few percent chance of impact by a multi-100-m NEA on a specific date decades in the future.

a comet. With years or decades of advance warning, a technological mission might be mounted to deflect an NEA so that it would miss the Earth (and also possibly a comet should new technologies enable similar warning times for them). Moving such a massive NEA would be very challenging. In any case, given sufficient warning, many immediate fatalities could be avoided by evacuating ground zero and longer-term casualties could be minimized by storing food supplies to survive the agricultural catastrophe. Susceptible infrastructure (transportation, communications, and medical services) could be strengthened in the years before impact.

However, no preparation for mitigation is warranted for such a rare possibility until a specific impact prediction is made and certified. The only advance precaution that might make sense would be at the margins of disaster planning developed for other, "all-hazards" purposes: considering such an NEA apocalypse might foster "out-of-the-box" thinking about how to define the outer envelope of disaster contingencies, and thus prove serendipitously useful as humankind faces an uncertain future.

2. "Tunguska Atmospheric Explosion" Scenario.

   Even with Spaceguard Survey, it is unlikely that such a small object would be discovered in advance – the impact would occur without warning. Since it could occur literally anywhere, there are no location-specific kinds of advance measures that could or should be taken, other than educating people (perhaps especially military forces that might otherwise mistake the event as an intentional attack) about the possibilities for such atmospheric explosions. In the lucky circumstance that the object is discovered years in advance, a relatively modest space mission could deflect such a small body, preventing the impact (Fig. 3.4).

3. "Near-Term Impact" Scenario.

   This scenario is the one most likely to become an urgent issue for public officials, indeed, such events have already happened. The last case took place around Christmas 2004, involving an asteroid then designated as 2004 MN4, except that the fortuitous location of prediscovery observations rendered the impact moot within a few days rather than months; because this prediction happened over the holidays and was then overshadowed by the Indian Ocean tsunami, media hyperbole was muted.

Ways to eliminate instances of hype and misunderstanding involve public education about science, critical thinking, and risk; familiarizing science teachers, journalists, and other communicators with the impact hazard might be especially effective. One approach that has evolved since a 1999 conference in Torino (Turin), Italy, is promulgation of the Torino Scale [1] (Fig. 3.5), which attempts to place impact predictions into a sober, rational context (on a 10-point Richter-like scale, predicted impact possibilities usually rate a 0 or 1, and are unlikely to exceed 4 during our lifetimes).

We have to note the primacy of psychological perceptions in characterizing the impact hazard. Since the consequences of an impact (other than the spectacle of meteors, and occasional meteorite falls) have never been experienced by human

Fig. 3.4 Tunguska meteorite consequences

Fig. 3.5 NEA impact prediction

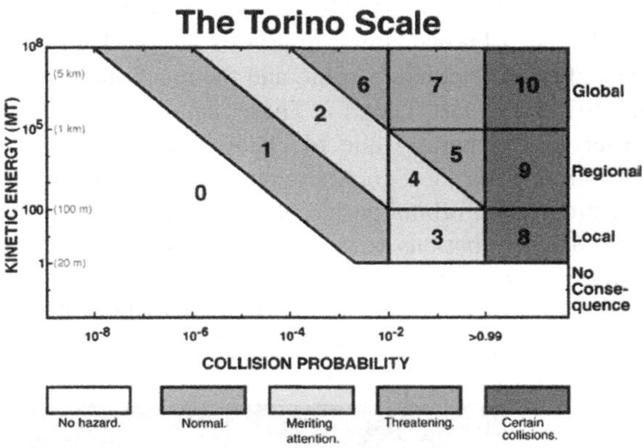

beings now alive, we can relate to this hazard only theoretically. Because society fails to apply objective standards to prioritizing hazard mitigation funding, it is plausible that the residual risks of this hazard might be altogether ignored (the Spaceguard Survey has been cheap, but it becomes increasingly costly to search for the remaining, smaller sized NEAs); or society may instead over-react and give "planetary defense" more priority than battling such clear-and-present dangers as influenza. Yet the impact hazard can be mitigated in much more concrete ways than is true for most hazards. An impact can be predicted in advance in ways that remain imperfect but are much more reliable than predictions of earthquakes or even storms, and the components of technology exist – at affordable costs given the consequences of an actual impact – to move any threatening object away and avoid the disaster altogether.

3.1.1 Historical Background of the Hazards of Near-Earth Objects: Some Estimates of Collision Risks

When the epoch of planetary accretion was over, numerous small-sized planets remained in orbit around the Sun. Most of them were still on planet-crossing orbits, causing their intense bombardment. According to some estimates [4] the population of the planetesimals on planet-crossing orbits should have decayed with a half-life of at most ~ 100 million years, being driven into the Sun or ejected into interstellar space by the combination of planetary perturbations. If this is true, the bombardment rate of the terrestrial planets should have declined accordingly. However, the geological record on the Moon bears witness to an extremely intense bombardment between 3.9 and 3.8 Giga years ago (i.e., 6 to 7 hundred million years after planet formation), when about a dozen huge impact basins formed. This period erased most of the previous geological record. It is still controversial whether it was a sudden spike in impact rate – possibly because of an abrupt change in the planetary orbital configuration that destabilized a distant reservoir of the small-sized planets – or instead the final phase of the intense bombardment, which for some reason declined much more slowly than the computer models predict (Fig. 3.6).

The well-known comet reservoirs are the Kuiper Belt and associated scattered disk, beyond Neptune's orbit, and the much more distant spherical halo of comets, called the "Oort Cloud." These small-sized bodies slowly leak from these reservoirs, generally due to chaotic dynamics near planetary resonances (e.g., distances from the Sun where such a body has an orbital period that is a simple fraction of the orbital period of a planet), facilitated by collisions and other minor orbital perturbations (e.g., the Yarkovsky Effect, which is a force on a small, spinning body due to the asymmetric re-radiation of absorbed sunlight on the body's warmer "afternoon" side). Some dislodged bodies soon arrive in the terrestrial planet zone, becoming NEOs. The latter are in comparatively transient orbits,

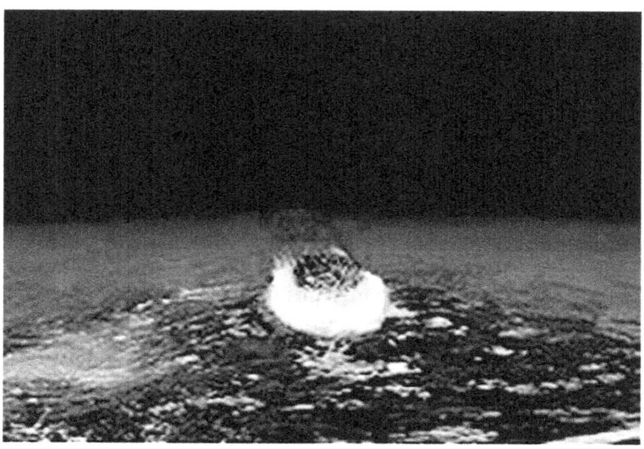

Fig. 3.6 Asteroid earth impact (art feature)

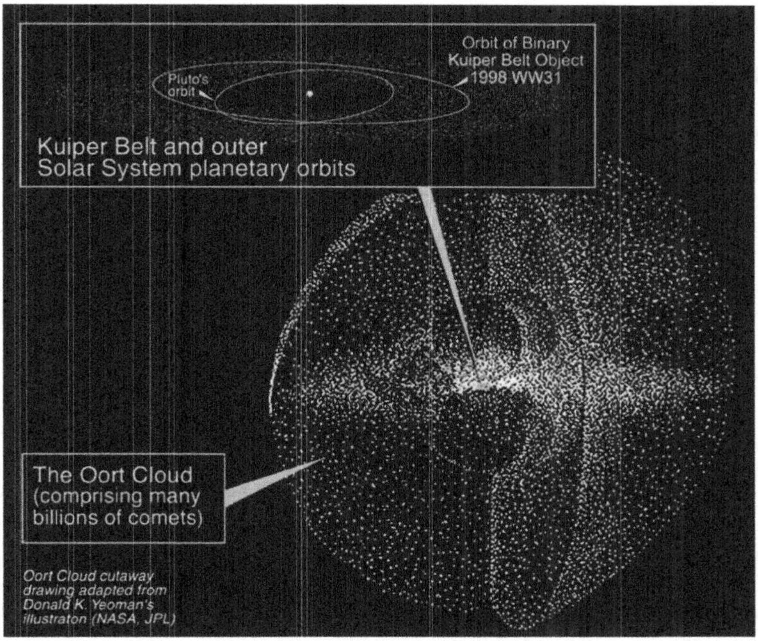

Fig. 3.7 Oort cloud – the birthplace of danger to the Earth?

typically colliding with the Sun, or more unusually with a terrestrial body, or being ejected from the solar system on hyperbolic orbits, on timescales of a few million years; however, being continually replenished from the reservoirs, the NEO population remains in a sort of steady state. Indeed, over the past 3.5 Giga years, the lunar and the terrestrial crater records show that the average Earth/Moon impact rate has varied little more than a factor of two during that time, although brief, moderate spikes in cratering rate must have happened (Fig. 3.7).

At this low modern impact rate, impacts nevertheless happen often enough to affect profoundly the evolution of life (e.g., the Chixchulub impact 65 million years ago, responsible for the K-T large-scale extinction). Because of the comparatively short time span of human lives and even of civilization, the importance of impacts as a modern hazard is debatable. The impact hazard is comparable, in terms of deaths and damage averaged over very long time periods, with other man-made and natural hazards that society takes seriously.

Therefore, NEOs are defined as bodies whose perihelia (closest orbital distance to the Sun) are <1.3 Astronomical Units (1 AU = the mean distance of the Earth from the Sun). About 20% of NEOs are currently in orbits that can approach the Earth's orbit to within <0.05 AU; these are termed Potentially Hazardous Objects (PHOs). PHOs are physically no different from other NEOs; they just happen to come close enough to Earth at the present time so that planetary perturbations could conceivably modify their orbits so as to permit an actual near-term collision, hence they warrant careful tracking.

The observed distribution of the NEOs, concerning both orbits and sizes, is not representative of the true distribution. Each survey is affected by observational biases. Obviously, it is easier to find large NEOs than small NEOs, as the latter are detectable only when passing close to the Earth. Thus, the observed size distribution is strongly skewed towards large objects. But also, NEOs on moderate eccentricity, low inclination orbits with period longer than 1 year are more easily discovered than NEOs on orbits with short periods, high inclinations, or large eccentricities. In fact the former are much more likely to pass close to the opposition point in the sky, where most of the NEO surveys are concentrated, while the latter spend most of the time at small solar elongation or far from the ecliptic.

Nevertheless, taking advantage of the growing number of observations, it has been possible to build models of the true orbital and size distributions of the NEO population. According to the two most recent models [5, 6] the estimated number of NEAs >1 km in diameter (the size for which NASA established Spaceguard's 90% completeness goal by 2008) is ~ 1,100 ±200, of which about 70% have been found to date.

Traditionally, the NEO population is subdivided into three groups [6]: the Apollos (the Earth-crossing objects with an orbital period larger than 1 year), the Amors (non-Earth-crossing NEOs, with perihelion distance between 1 and 1.3 AU), and the Atens (the Earth-crossing objects with an orbital period shorter than 1 year). According to the NEO distribution model (see [7]), about 32% of the NEOs are Amors, 62% are Apollos, and 6% are Atens. Forty-nine percent of the NEOs should have orbital periods shorter than 2.8 years, which is the minimal orbital period of main belt asteroids. This model also shows that, in addition to the NEO population, there is a population of non-Earth-crossing objects with an aphelion distance smaller than 1 AU, called IEOs. There are about 50 NEOs for every IEO.

The models of the NEO orbital and size distribution have been used to evaluate the effectiveness of the current surveys in discovering the remaining (i.e., not yet detected) NEO population. It has been predicted that the discovery rate of new objects would have started to drop off in 2003 to lower and lower rates (this has indeed happened). The reason is not simply that it is statistically less probable to find one object, if fewer remain. It is also that the NEOs which are still to be discovered are the most difficult ones, as they are small (e.g., faint) and reside on orbits whose geometry relative to the Earth maximize the observational biases against discovery. Thus, according to [8], if there were no improvements to the current facilities the Spaceguard goal would not be reached before 2030, at best. It was also shown that the current surveys are completely inadequate to discover a large fraction of the population of NEOs that is smaller than 1 km in diameter. Despite their small size, these objects could still constitute a significant hazard for human civilization. This has motivated NASA to mandate the SDT to study how the search for NEOs could be extended to the smaller objects.

According to some recent estimates [1] a collision liberating energy of 1,000 MT should happen on average only once every 65,000 years. They are caused by NEOs of about 250–300 m in size. Such energies should produce consequences at a regional level, possibly with global implications on the world-wide economy (for comparison, the December 2004 disaster due to the tsunami in South Asia was caused by an earthquake liberating energy of 10,000 MT). The UK Task Force on Potentially Hazardous NEOs, defined 1,000 MT as a lower limit for "dangerous" impacts. There is only one chance out of 650 that such an impact will occur in the next century.

From these statistics, one might conclude that the NEO hazard is a nonproblem. It is strange, therefore, that one recently discovered NEO (2004 MN4/Apophis) was calculated as having a better than 5% chance of impacting the Earth on April 13, 2029 until prediscovery images were found and radar detection of the object was accomplished, ruling out an impact in 2029. The pass will be very close, however, below the distance of geostationary Earth satellites, and there remains a 1-in-10,000 chance that the asteroid will pass through one of several "keyholes" and enter a near-resonant orbit resulting in an Earth impact in one of several years during the 2030s. As of the writing of this chapter, there was yet another "dangerous" asteroid with an estimated diameter of 580 m (impact energy of 15,000 MT), that has a probability of 1-in-10,000 of colliding with the Earth before the end of the twenty-first century. So it is not entirely clear that the statistics are giving us a complete understanding of the hazard from these "dangerous" NEOs (Fig. 3.8).

Setting the threshold of "dangerous" at 1,000 MT is arbitrary. The Tunguska explosion in 1908 is widely estimated at 10–15 MT, and possibly less. It struck in a barren location and killed few people, maybe none at all. Yet the world population is increasing and it is increasingly likely that the impact of a small NEO would be genuinely harmful. As evidenced by the South Asia tsunami, there is explosive population growth along coastlines in countries around the world, and an NEO >150 m diameter might create a tsunami with disproportionate consequences.

It is difficult to pre-judge the threshold of what constitutes a "dangerous" NEO impact or what level of priority should be given to trying to mitigate the potential

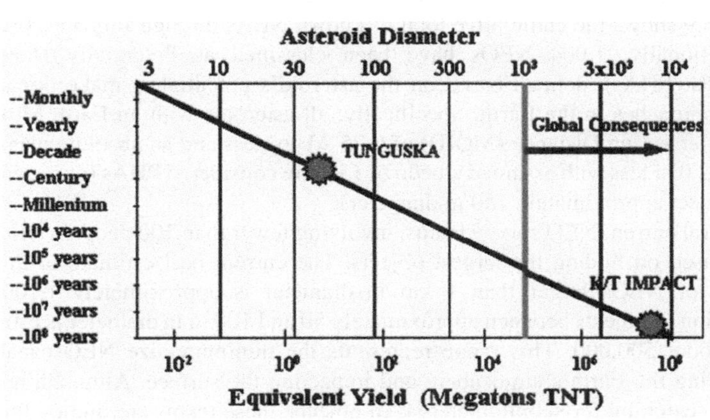

Fig. 3.8 Scale of danger from asteroids

consequences of a predicted close pass of an NEO with a specific (and ever-changing) probability of striking. The odds are that, once all 250 m NEOs are discovered and catalogued, none will turn out to be on a collision trajectory with our planet during the next century. But, given the possibility that we can deflect an NEO in the unlikely event one is found to be on an impact trajectory, we are obliged to improve the current NEO survey capabilities in order to find all NEOs larger than 200 m or so. This will make us definitely sure (not only at a probabilistic level) that the Earth is "safe." Whether it makes sense to extend, at considerable additional expense, the search down to 150 m or even smaller remains a matter of discussion.

3.1.2 NEO Detection, Impact Predictions, and Early-Warning Efforts

It is well known that to successfully avert a future asteroid or comet impact with the Earth the dangerous celestial body must first be discovered and the object's orbit and physical properties must be characterized sufficiently. Additionally, these functions must be successfully performed with enough lead time to provide early warning time to avert the collision or mitigate the effects on the Earth and its inhabitants. Since we cannot predict with certainty the direction from which a threatening asteroid or comet will come, we must observe the entire sky. Finally, if a mission to deflect or disrupt an impacting NEO is accomplished, a reassessment of that object must be completed to determine the effectiveness of that mission. Accomplishing these critical tasks inherently requires international cooperation. We'll try to describe the current capabilities to provide these functions as well as improvements that can be reasonably expected in the near future.

The total population of NEOs larger than 1 km in diameter is currently estimated to be between 1,000 and 1,200 [1]. As of December 18, 2008, a total of 5,901 NEOs have been discovered, with 761 of these approximately 1 km or larger in diameter. Figure 3.9 shows the cumulative total of known NEAs through July 17, 2008.

Additionally, 1,004 NEOs have been classified as Potentially Hazardous Asteroids (PHAs), defined based on the asteroid's potential to make threatening close approaches to the Earth. Specifically, all asteroids with an Earth Minimum Orbit Intersection Distance (MOID) of 0.05 AU or less and an absolute magnitude (H) of 22.0 or less with assumed albedo of 13% are considered PHAs (corresponding to an object approximately 150 m diameter).

Several current NEO survey teams, involving fewer than 100 people worldwide, are focused on finding the largest objects. The current best estimate of the total number of NEOs larger than 1 km in diameter is approximately 1,100. The population of objects between approximately 50 and 100 m in diameter is estimated to be about 500,000. This range represents the minimum-size NEO capable of penetrating the Earth's atmosphere and impacting the surface. Although not efficient at searching for subkilometer-sized objects, these teams are finding far more NEOs smaller than 1 km in diameter (Fig. 3.10).

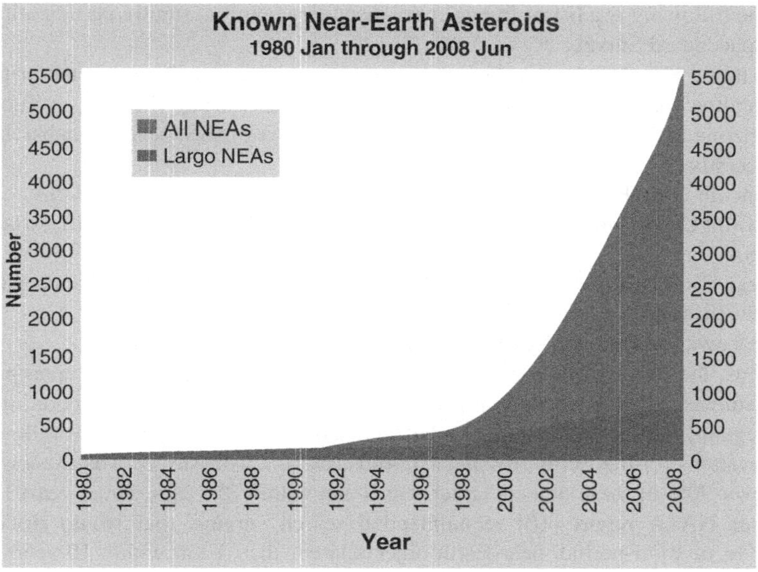

Fig. 3.9 Cumulative total of known near-earth asteroids since 1980 (Courtesy NASA/JPL-Caltech)

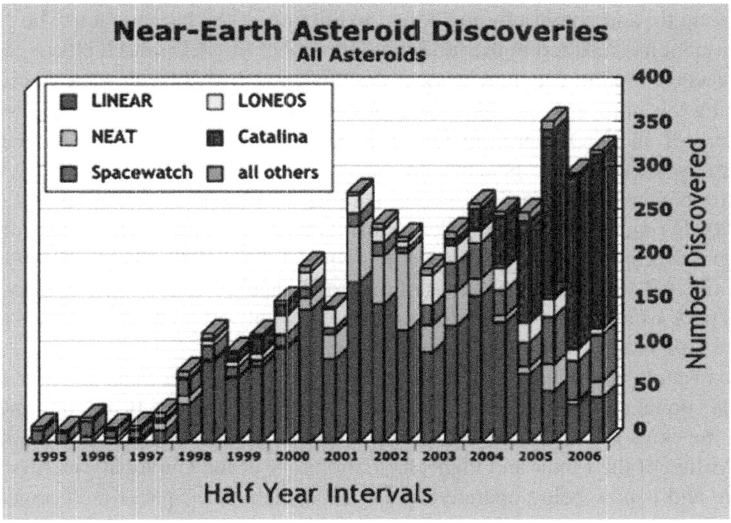

Fig. 3.10 Annual NEA discoveries by team (Courtesy NASA/JPL-Caltech)

The following is a list of five primary discovery teams currently participating in the Spaceguard Survey:

- Catalina Sky Survey – A consortium of three cooperating surveys: the original Catalina SkySurvey (CSS) and the Mt. Lemmon Survey (MLSS) in Tucson, Arizona, USA, along with the Siding Springs Survey (SSS) near Coonabarabran, Australia
- Lincoln Near-Earth Asteroid Research (LINEAR) in New Mexico, USA
- Lowell Observatory Near-Earth Object Search (LONEOS) in Flagstaff, Arizona, USA
- Near-Earth Asteroid Tracking (NEAT) at the Maui Space Surveillance Site in Hawaii, USA
- Spacewatch at the University of Arizona in Tucson, Arizona, USA

Thus, the role of "main player" in the field of NEO detection and appropriate assessments belongs to the USA. In a 1992 report to NASA [9] a coordinated Spaceguard Survey was recommended to discover, verify, and provide follow-up observations for the Earth-crossing asteroids (ECAs). This survey was expected to discover 90% of these objects larger than 1 km within 25 years. Three years later, another NASA report [10] recommended search surveys that would discover 60–70% of short-period, near-Earth objects larger than 1 km within 10 years and obtain 90% completeness within 5 years more.

In 1998, NASA formally embraced the goal of finding and cataloging, by 2008, 90% of all near-Earth objects with a diameter of 1 km or larger that could represent a collision risk to Earth. The 1-metric km diameter was chosen after considerable study indicated that an impact of an object smaller than 1 km could cause significant local or regional damage but is unlikely to cause a worldwide catastrophe [9]. The impact of an object much larger than 1 km diameter could well result in worldwide damage up to, and potentially including, extinction of the human race. The NASA commitment has resulted in the funding of a number of NEO search efforts that are making considerable progress toward the 90% goal by 2008. NASA is close to achieving this goal, and should achieve it within a few years. However, as the 2009 discovery of an NEO approximately 2–3 km in diameter demonstrates, there are still large-scale objects to be detected.

Astronomers have been conducting surveys to locate the NEOs, many (as of early 2007) funded by NASA's Near-Earth Object program office as part of their Spaceguard program. One of the best-known is LINEAR that began in 1996. By 2004 LINEAR was discovering tens of thousands of objects each year and accounting for 65% of all new asteroid detections [11]. LINEAR uses two 1-m telescopes and one half-m telescope based in New Mexico [12].

Spacewatch, which uses a 90 cm telescope sited at the Kitt Peak Observatory in Arizona, updated with automatic pointing, imaging, and analysis equipment to search the skies for intruders, was set up in 1980 by Tom Gehrels and Dr. Robert S. McMillan of the Lunar and Planetary Laboratory of the University of Arizona in Tucson, and is now being operated by Dr. McMillan. The Spacewatch project has acquired a 1.8 m telescope, also at Kitt Peak, to hunt for NEOs, and has provided the old 90 cm telescope with an improved electronic imaging system with much greater resolution, improving its search capability [13].

Other near-earth object tracking programs include Near-Earth Asteroid Tracking (NEAT), Lowell Observatory Near-Earth-Object Search (LONEOS), CSS, Campo Imperatore Near-Earth Objects Survey (CINEOS), Japanese Spaceguard Association, and Asiago-DLR Asteroid Survey [14]. Pan-STARRS is expected to complete telescope construction by 2012.

"Spaceguard" is the name for these loosely affiliated programs, some of which receive NASA funding to meet a U.S. Congressional requirement to detect 90% of near-earth asteroids (NEAs) over 1 km diameter by 2008 [15]. A 2003 NASA study of a follow-on program suggests spending US$250–450 million to detect 90% of all near-earth asteroids 140 m and larger by 2028 [16].

What results of the detection, impact prediction, and early warning of dangerous NEO do we have presently? Research published in March, 2009, describes how scientists were able to identify an asteroid in space before it entered Earth's atmosphere, enabling computers to determine its area of origin in the solar system as well as predict the arrival time and location on the Earth of its shattered surviving parts. The 4-meter-diameter asteroid, called 2008 TC3, was initially sighted by the automated Catalina Sky Survey (CSS) telescope, on October 6, 2008. Computations correctly predicted the impact would occur 19 h after discovery in the Nubian Desert of northern Sudan [17]. A number of potential threats have been identified, such as 99942 Apophis (previously known by its provisional designation 2004 MN4), which had been given an impact probability of $\sim$ 3% for the year 2029. This probability has been revised to zero on the basis of new observations [18] (Fig. 3.11).

However, at the present time, there are no dedicated space-based assets for NEO surveying and discovery. They are limited to telescopes that can provide serendipitous discovery and limited support observations, particularly when higher resolution observations are required. For example, on November 8, 2007, the House Committee on Science and Technology's Subcommittee on Space and Aeronautics held a hearing to examine the status of NASA's Near-Earth Object survey program. The prospect of using the Wide-field Infrared Survey Explorer (WISE) was proposed by NASA officials [19]. WISE will survey the sky in the infrared band at a very high sensitivity. Asteroids that absorb solar radiation can be observed through the infrared band. NASA officials told Committee staff that NASA plans to use WISE to detect NEOs, in addition to performing its science goals. It is projected that WISE could detect 400 NEOs (roughly 2% of the estimated NEO population of interest) within the 1-year mission.

A recent example of this is the set of observations provided by NASA's Spitzer and Hubble Space Telescopes during the break-up of Comet 73P/Schwassman-Wachmann 3 into over 60 fragments during April and May of 2006. The Canadian Space Agency (CSA) and Defence Research and Development Canada (DRDC) are currently developing the Near-Earth Observation Surveillance Satellite (NEOSSat), which will use a 15-cm optical telescope to search for Atens and inner-Earth objects near the ecliptic within 45° of the Sun that may not be visible to ground-based

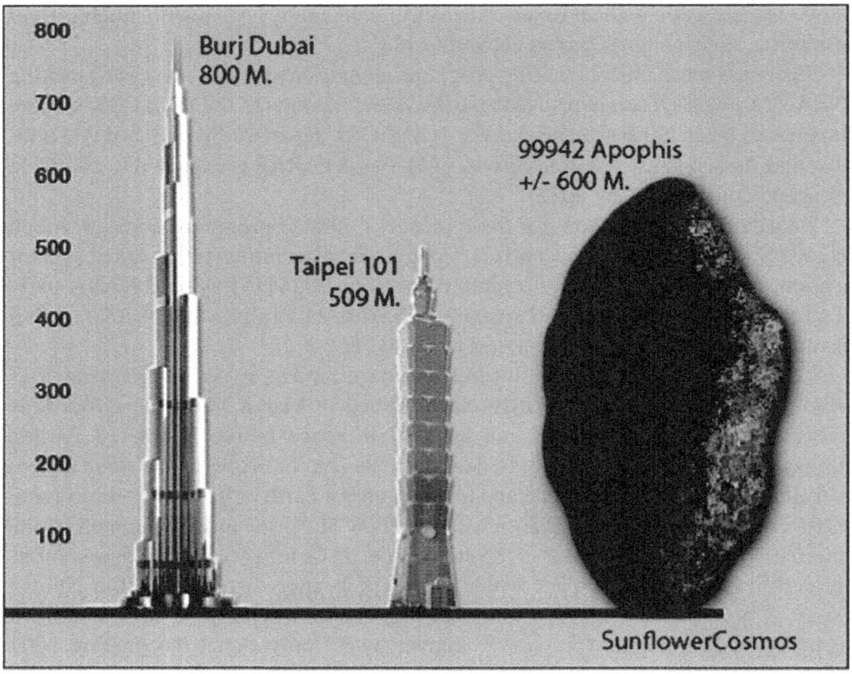

Fig. 3.11 The real size of NEA 99942 Apophis compared against two well-known Towers (Courtesy sunflowercosmos.org)

observatories. This microsatellite will be the first space-based asset deployed specifically to search for NEOs and was launched in 2009. The objective of the mission is to detect at least 50% of all IEOs having diameters greater than 1 km.

In 2003 NASA issued the Report of the Near-Earth Object Science Definition Team titled "Study to Determine the Feasibility of Extending the Search for Near-Earth Objects to Smaller Limited Diameters" [1]. Motivated by the substantial progress being made toward the 2008 goal of discovering and cataloging 90% of all NEOs with diameters 1 km or larger, this study specifically addressed what, if anything, should be done concerning the much more numerous smaller NEOs that pose a potential impact threat. The team recommended that a search system be constructed to discover and catalog 90% of the potentially hazardous objects, which is a subset of the NEOs, larger than 140 m in diameter. PHOs approach the Earth's orbit within 0.05 AU (~ 7.5 million km). The study evaluated a broad range of technology and search systems including ground-based and space-based optical and infrared systems across a credible range of optics and detector sizes. Ground-based telescopes with apertures of 1, 2, and 4 m were considered along with space-based telescopes with apertures of 0.5, 1, and 2 m. Various geographic locations were considered for the ground-based systems. Space-based telescopes in low-Earth orbit (LEO), the Earth-Sun L2 Lagrange point, and a Venus-trailing orbit

(0.7 AU semimajor axis) were also assessed. According to the cost/benefit assessment performed for the study, the benefits associated with significantly reducing the risk of an impact with a smaller object justified substantial investment in PHO search systems. If the survey was to be completed within 10 years space-based systems were preferable, but at a higher cost than ground systems. If as many as 20 years were permitted to achieve the 90% completion goal, ground-based systems could perform the survey at a significantly reduced cost. Additionally, the ground-based systems could be repaired and upgraded more easily than the space-based systems.

In 2004, European Space Agency (ESA) established an international panel called NEOMAP (Near-Earth Object Mission Advisory Panel) and the final NEOMAP report stated "The current consensus amongst the impact-hazard community is that future NEO search telescopes should be made sensitive enough to achieve near completion for objects significantly smaller than 1 km." According to Harris (2004), systems such as the U.S. Pan-STARRS (Panoramic Survey Telescope and Rapid Response System), the DCT (Discovery Channel Telescope) and LSST (Large Aperture Synoptic Survey Telescope) should be capable of discovering 90% of the population of NEAs with diameters of around 200 m or more after 10 years of operation. However, some of these facilities will serve general astrophysical research and will not be dedicated to NEO searches.

The future ground-based observatories identified in the NEOMAP report will greatly improve the search for NEOs and are currently at various stages of planning or implementation. Pan-STARRS is an innovative wide-field imaging facility being developed at the University of Hawaii's Institute for Astronomy. Pan-STARRS will consist of four 1.8 m telescopes and will have a 1.4 gigapixel CCD camera (38,400 × 38,400). The PS1 prototype telescope is essentially one quarter of Pan-STARRS, and "first light" occurred in June 2006. It will have the same optics design and camera design as anticipated for the full version of Pan-STARRS, which is currently in the site selection phase. The DCT is Lowell Observatory's project to design and construct a powerful, 4.2 meter telescope. It is anticipated that the DCT, with its powerful wide-field capability, will be able to discover 10–20 NEAs per hour, which is approximately ten times the discovery rate of all current survey programs combined. The Large Synoptic Survey Telescope is planned to be the world's most powerful survey telescope, consisting of an 8.4-m aperture telescope with a 10 square-degree-field telescope. The LSST will be capable of scanning the entire visible night sky in a matter of days rather than years as is the case with current telescopes. A Chilean mountain peak has been chosen as the location of the telescope with construction to begin in 2009 and the facility to achieve "first light" in 2014.

Future space-based observatories that are not dedicated to NEO detection will indirectly have a significant impact on the survey and cataloging of NEOs. For example, the European Space Agency's Global Astrometric Interferometer for Astrophysics (GAIA) mission is planned for launch by 2012. The GAIA spacecraft will observe the entire sky down to a visual magnitude of approximate V = 20 and down to solar elongations of 35°. The GAIA spacecraft, which is anticipated to operate for 5 years, will be placed in orbit about the Earth-Sun L2 Lagrange Point. GAIA consists of two telescopes that will continually scan the sky and record every visible object that crosses its line of sight down to its limiting magnitude. Among its other astronomical objectives, GAIA will contribute to the search for NEOs due to its unprecedented sensitivity to faint, moving objects. The spacecraft is expected to detect tens of thousands of minor planets, many of which will be NEOs, main-belt asteroids, and Kuiper Belt objects.

Russian researchers at one point had elaborated a project for an asteroid and comet threat monitoring system. According to preliminary estimates, an effective warning of 3–5 days for unidentified celestial bodies needs the creation of a dedicated space system (the so-called "space patrol") comprised of three large spacecraft with optical IR telescopes onboard (Fig. 3.12). Instead of surveillance of the entire space the space-based telescopes will survey a certain narrow "barrier zone" where all potentially dangerous objects approaching Earth from all

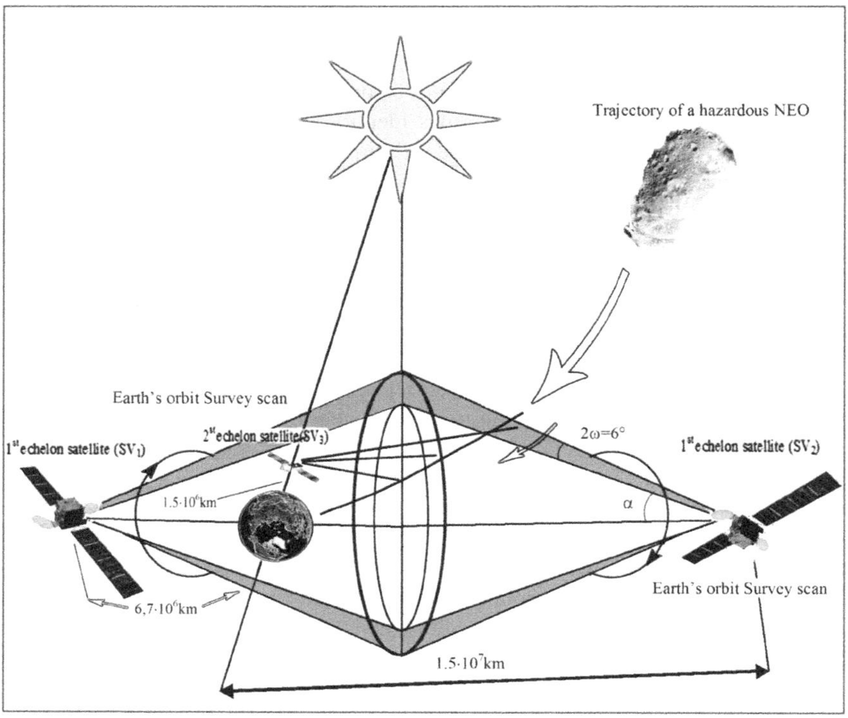

1st echelon satellite (SV1), 1st echelon satellite (SV2) – space vehicles of the asteroids detection and preliminary their ephemerids determination;
2st echelon satellite(SV3) - space vehicles of the precision ephemerids determination

Fig. 3.12 Scheme of the "Space Patrol" developed by Russian scientists

directions, including from the Sun, i.e., invisible to Earth-based telescopes, will be registered.

"The barrier zone" will embrace Earth from distances sufficient for timely (several days) warning about the fall of a dangerous object on Earth. According to the designers, the zone depth will provide roughly a 5-h warning for the dangerous object. This will make it possible to determine fairly accurately the parameters of its movement in orbit. In compliance with this requirement, the telescope's instant viewing angle is chosen. The viewing method must provide a capability of using a Charge-Coupled Device in the charge temporary delay and accumulation mode. By such means the legitimate signal from a distant asteroid is amplified a thousand times thanks to which the subsystem allows the potentially dangerous objects to be detected.

It is expected that the space patrol will consist of two echelons. The first echelon of two spacecraft telescopes is to be placed in the orbit of Earth's rotation around the Sun at points before Earth and behind it at equal distances of, respectively, ~0.1 AU and ~0.7 AU. This ensures observation at large angles between the optical axis of the telescope and possible directions of movement of dangerous asteroids towards the Earth. This will resolve problems of detection and determination of parameters of movement of objects no smaller than 50 m. An orbital telescope on each such spacecraft with a constant angular speed scans the barrier zone by using rotation of the platform satellite around the spacecraft–Earth axis (the optical axis of the telescope and this direction make the angle α) (Fig. 3.13).

The narrow field of view of the telescope of ~6° will be sufficient to view over several sessions a dangerous asteroid as it moves through the barrier zone and is fixed in consecutive cycles of scanning of the barrier zone. Telescopes with such viewing fields have a rotating focal surface which is almost flat, measuring ~0.3 m over which the image of a dangerous asteroid moves in a radial direction whereas the image of nondangerous celestial bodies move with the formation of tracks registered in each scan of the rotating field of view. To determine the relevant parameters of the movement of a dangerous object it is sufficient to have several such images obtained in different scans during its presence in the barrier zone of several hours.

To increase the accuracy of determination of parameters of asteroids, the first echelon is supplemented with the second echelon consisting of a single platform satellite featuring a long-focus telescope installed in the Lagrangian point of libration between the Earth and the Sun. Thus, with the help of target designation supplied by the detection echelon, a narrow (~40 angle min.) field of view is provided for detecting dangerous asteroids and their subsequent tracking. The optical axes of spacecraft detection and tracking telescopes form considerable angles with the direction of movement of the dangerous asteroid and between themselves. The small error in determination of the angular position of the asteroid relative to the optical axis of the tracking telescope is attained by means of a large focus distance (~17 m). The detection telescopes with a diameter of the entry pupil of 1.5 m will be able to register with a more than 0.95 probability asteroids larger than 50 m. The approach time to Earth after the asteroid has passed the registration

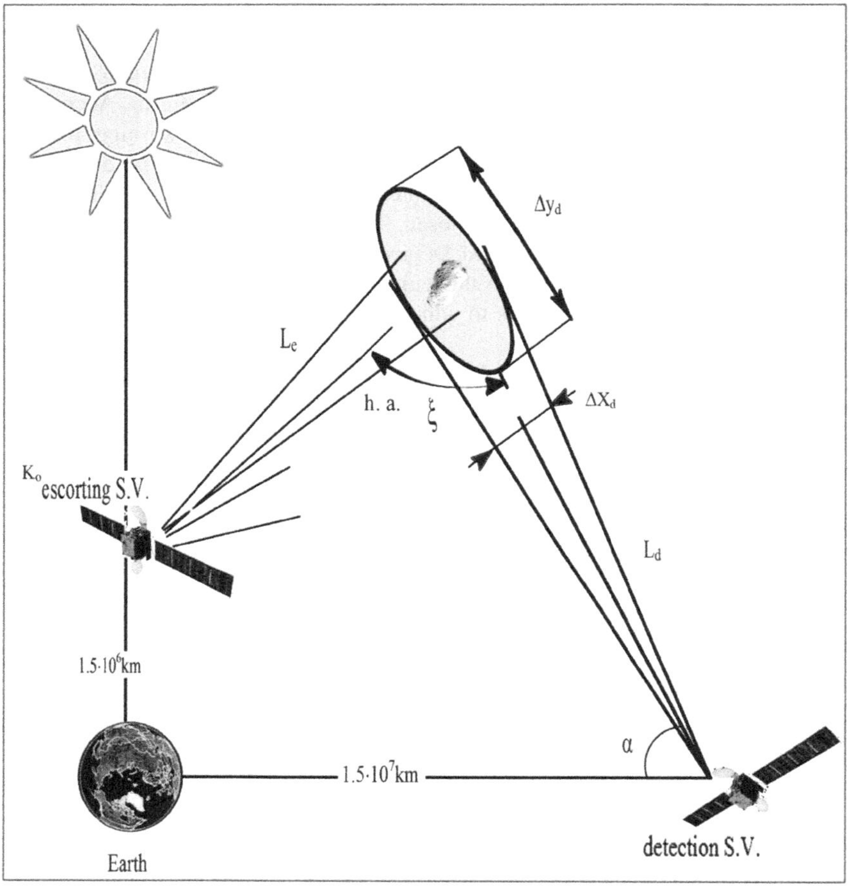

ξ - angle between optical axis detection and escorting telescopes at the covering of a hazardous asteroid with both field of view, $\xi > 25°$

α- angle between optical axis detection telescope and the direction "detection S.V. – Earth", $\alpha = 35°$

$\Delta y_d, \Delta x_d$

$\Delta y_e, \Delta x_e$ - errors in the measurements of asteroid positions

$\Delta y_d \sim 2 \cdot 10^4$ km $\quad$ $\Delta y_e, \Delta z_e \sim 10$km

$\frac{\Delta x_d}{L_d} \sim 1.5$ ang.sec. $\quad$ $\frac{\Delta z_e}{L_e} \sim \frac{\Delta y_e}{L_e} \sim 0.2$ ang.sec.

Fig. 3.13 Scheme of asteroid detection

barrier zone is 3 days and more, depending on the direction of its movement (Fig. 3.14).

The platforms of Russian spacecraft "Koronas-Foton" and "Spektr" or of U.S. Hubble Space Telescope (see Fig. 3.15) may become analogs of a similar system.

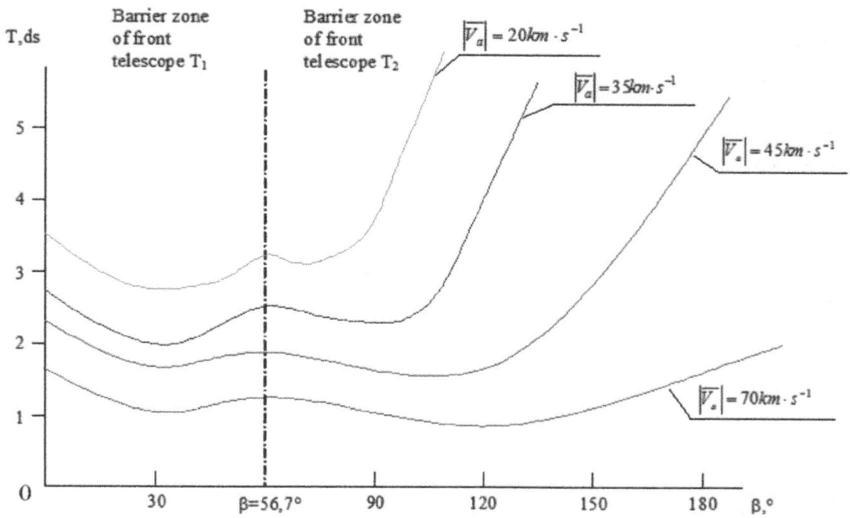

Fig. 3.14 Barrier zone parameters

Fig. 3.15 Hubble space telescope

In this respect, as already mentioned above, American space IR telescope WISE, launched into near-Earth orbit in December 2009 may be of great interest. During January–February 2010 alone, this satellite detected some 1,500 new asteroids, 15 of which are approaching the Earth, and two new comets. Unfortunately, the limited service life of the satellite (only 10 months because of the consumption of liquid hydrogen used as the coolant) makes it impossible to use it for long-term

Table 3.2 Capabilities of the "Space patrol" system

Characteristics	Characteristic value
Minimum size of a detected asteroid	Not smaller than 50 m
Maximum detection range	1.5 million km
Time of warning at asteroid speed of 20 km/s	≥3 days
Predicted impact area	3 days in advance
Probability of detecting an asteroid from an arbitrary direction	>0.999

Table 3.3 Main characteristics of a typical "Space patrol" system

Spacecraft subsystem name	Consumed power, W	Weight, kg
Telescope with a shade and glare shield	1,500	1,800
Space platform	1,400	2,000
• Onboard control system	400	200
• Orientation and stabilization system	300	350
• Power supply system	200	450
• Thermal mode maintenance system	400	80
• Mechanical systems	–	300
• Onboard cable network	–	120
• Integrated propulsion system	100	500
Total	3,000	3,800

monitoring. In addition, the precise determination of parameters of the movement of a dangerous asteroid necessitates employment of several spacecraft.

Delivery of heavy spacecraft to distant points in outer space is possible with the help of Russian heavy-lift carrier rockets like Proton or Zenith. The capabilities of the system for detection and timely warning of the approach of previously unknown asteroids are shown in Table 3.2, while approximate characteristics of spacecraft of the proposed detection system are shown in Table 3.3.

Discovery of a new NEO is a necessary first step, but by itself is not sufficient. Securing a NEO's orbit is a critical aspect of the discovery and cataloging process, and refinement of the orbit sufficiently to determine if the object is probably on an impacting trajectory requires many highly accurate observations and sufficient observational arc lengths. This can be performed fairly well with the current NEO survey teams combined with the multitude of individuals (professional and amateurs) that provide follow-up observations to the Minor Planet Center. This effort is international in scope, but as the surveys become more numerous and more capable, follow-up observations become increasingly difficult, particularly for amateur observers. Additionally, determining the physical characteristics of an asteroid or comet is crucial in the design of a mission to deflect or disrupt an Earth threatener as well as planning any precursor missions. Of primary importance are the characterization of the object's mass, gravity field, spin state, surface topography and roughness, surface gravity field, and density distribution. Astrometric measurements (ground or space-based) can improve the accuracy of the orbit knowledge, estimate the body size, and potentially identify the existence of

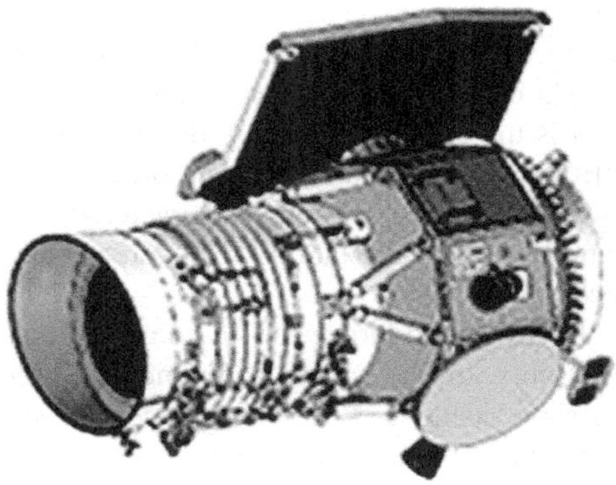

Fig. 3.16 WISE space telescope

co-orbitals. Spectral imaging using filters can be used to determine asteroid type including composition, grain density, surface albedo, and size. Finally, intensity fluctuations can be used to estimate spin period, spin state, body shape, and rotation angular momentum (Fig. 3.16).

However, telescopic observations have their inherent limitations. Radar measurements provide a unique and highly capable source of information about a NEO's physical properties and orbit. General methodology for a radar observation is to transmit a well-characterized signal and compare it with the return echo to analyze the object's properties. Radar can be used to determine body shape, rotation state, co-orbitals, rotation angular momentum, improve heliocentric orbit prediction, and place constraints on surface density and roughness. Radar systems are limited in their range, but given sufficiently strong return signals they can permit two-dimensional spatial resolutions on the order of meters.

The accuracy of orbit determination can be improved greatly with measurements of range and range rate obtained from radar instruments. Radar measurements can determine the orbit well enough to prevent "loss" of a newly discovered asteroid and reduce the positional uncertainty by several orders of magnitude compared to optical astrometric observations only. Predictions based solely on optical data typically contain significant errors, even with long data arcs. Radar measurements can make the difference between estimating that an object will pass several Earth–Moon distances away from the Earth and realizing that the NEO will actually impact the Earth. However, because radar is an active technique it has very limited range capability compared to optical observations, and thus can only add to optical observations when the object passes relatively close to the Earth. This limits its ability to provide ephemeris-refining information, which could improve warning times when optical observations are not sufficiently accurate. Nonetheless radar is an extremely powerful detection tool and an indispensable observatory for detection and impact prediction of NEOs.

Radars have another vital capability, which is to help characterize the NEO. Given sufficient directional coverage of the object, the measurements can be combined to provide detailed three-dimensional models of the NEO and accurately define its rotation state. Detailed models of the object can provide tremendous insight into many of the areas critical to averting an impact with Earth. One critical area is the effect of the mass distribution on the stability of an orbit close to the NEO, which is required for any sort of rendezvous mission or landing. In addition radar observation can help to characterize the makeup of the object so that estimates can be made of its stability if it were subjected to various deflection or disruption techniques.

Existing or future space-based assets to assist in NEO characterization in close proximity to the target body include robotic flyby, rendezvous, and landing missions using a variety of spacecraft such as orbiters, landers, and surface penetrators. Many instruments (visible cameras, synthetic aperture radar, lidar, mass spectrometers, seismographs, alpha particle X-ray spectrometers, etc.) currently exist and could be utilized to take extensive measurements of a particular asteroid or comet target. Many of these instruments have extensive flight heritage and have flown on various missions to asteroids and comets. Additionally, a radio transponder could be placed on a threatening NEO allowing for its orbit to be precisely determined. Robotic missions for the sole purpose of averting an Earth impact would only be possible if sufficient time existed before the impact and if the NEO's orbit was known well enough to be able to classify it as a probable threat. In general all scientific missions to investigate asteroids and comets provide valuable information and insight that can be useful in conjunction with principal means of observation in preventing an Earth impact by a NEO.

Till now, only a small number of missions demonstrating the ability to flyby a NEO or rendezvous with the object and perform proximity operations have already been conducted by Japan and the United States. The NASA Near-Earth Asteroid Rendezvous – Shoemaker (NEAR Shoemaker) mission also demonstrated similar capabilities. NEAR Shoemaker rendezvoused with asteroid Eros on February 14, 2000, and returned many images, some of which are shown in Figs. 3.17 and 3.18.

Of particular note is the tremendous Japan Aerospace Exploration Agency (JAXA) "Hayabusa" mission to asteroid Itokawa. The Hayabusa spacecraft observed Itokawa with a suite of instruments including a multispectral telescopic imager, a laser altimeter, a near-IR spectrometer, and an X-ray fluorescence spectrometer. In addition to the data collected, the Hayabusa mission demonstrated the capability to characterize NEAs during rendezvous and docking missions (including return a sample of the asteroid Itokawa to the Earth for laboratory analysis).

In addition to these missions, ESA conducted a spectacularly successful, unprecedented flyby of the comet Halley with spacecraft Giotto in 1986, and is directing the current Rosetta mission, which is due to orbit and land a probe on comet 67P/Churyumov-Gerasimenko in 2014 for long-term observations. Comet Halley was also visited by the Soviet Union by spacecraft "Vega 1" and "Vega-2," by Japan with spacecraft Sakigake and Susei, and by the United States with spacecraft ISEE/ICE.

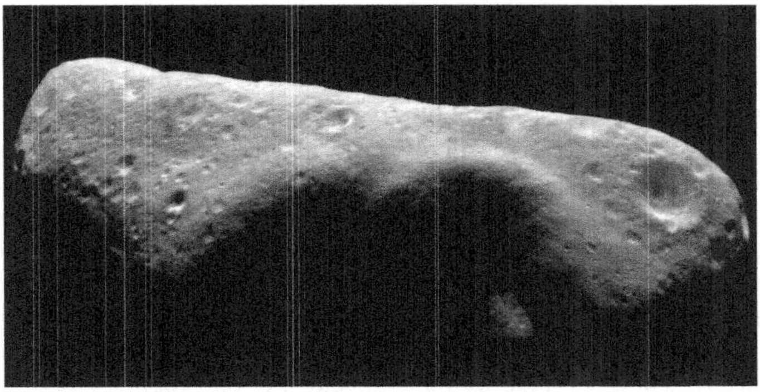

Fig. 3.17 Image of the EROS asteroid from the spacecraft

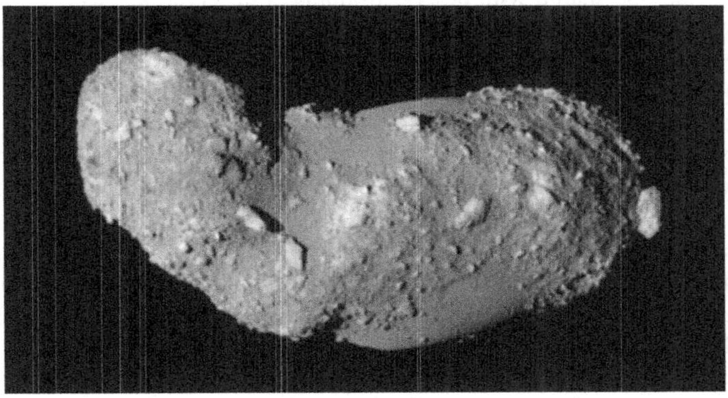

Fig. 3.18 Asteroid Itokawa imaged by the Hayabusa spacecraft (Credit and Copyright: ISAS/JAXA)

A separate issue, which has to be mentioned, is the detection of long-period comets. These celestial bodies tend to be ignored in NEO studies at this time because the probability of impact is believed to be very much smaller than that of asteroids. However, virtually all NEOs larger than a few kilometers are comets rather than asteroids, and such large NEOs are the most destructive, and potentially the "civilization killers." Additionally, the Earth regularly passes through the debris field of short-period comets (SPCs) giving us the annual meteoroid showers such as Leonids and Taurids. These are very predictable, but thankfully benign impact events. If the Earth were to encounter sizable objects within the debris field of a long-period comet, we would likely have very little warning time and would potentially be confronted with many impactors over a brief period of time. Although this type of event is currently speculative, this is a conceivable scenario which humanity could face. While the risk of a cometary impact is believed to be small, the destruction potential from a single large, high-velocity comet is much greater than from a NEA. Therefore, it is important to address their detection and potential methods for deflecting, disrupting, or mitigating the effects before one impacts the Earth.

Ground-based telescopes cannot observe comets during local daytime or when there are clouds, and their sensitivity is seriously affected by moonlight and atmospheric effects. One or more dedicated robotic observatory telescopes of greater capability than the upcoming NEOSSat instrument could readily be placed into Earth orbit, in solar orbit near the Earth, or at the above-mentioned Earth-Sun (E-S) Lagrangian point. Such telescopes would be able to observe nearly continuously, and their angular coverage would be almost spherical except for angles near the Sun. At the L2 point the telescope would also be continuously shielded from interference by sunlight. There would be no cloud or day–night effects either, and therefore a telescope in orbit could have 10–18 times the observation time on NEOs compared to any one ground observatory, and could be dedicated to NEO observation rather than be shared with other astronomical observations, as is typical with large ground telescope facilities.

The sensitivity obtained from a space-based observatory like this could be sufficient even with an aperture diameter of approximately 1 m, which is easily achievable with today's technology. Future facilities will support space telescopes whose primary mirrors consist of thin membranes, and thus will be orders of magnitude lighter and cheaper than space telescopes such as the Hubble or James Webb.

The last part of our analysis is devoted to NEO impact facts and warning time prediction. This is a very complicated issue, first of all due to comparison of impact threats. According to [1] threatening near-Earth objects are typically divided into three classifications based on their orbital characteristics and telescopic appearances: the near-Earth asteroids, the short-period comets, and the long-period comets. Many publications also use the terms Earth-crossing asteroids and Earth-crossing comets (ECCs) to describe objects whose orbits can intersect the Earth. Although the primordial population of NEAs has long been cleaned out from the solar system by collisions and gravitational ejections, it is important to recognize that the population of NEOs is constantly being replenished through a variety of mechanisms.

As was mentioned above, there are two main sources of NEAs. The first is the main asteroid belt, which is believed to replenish NEAs through collisions and chaotic orbital dynamics. The second source of asteroids is believed to be extinct comet nuclei. Short-period comets are thought to originate from the Kuiper belt, a vast population of small bodies orbiting the Sun beyond Neptune and extending 30–1,000 AU from the Sun. It is estimated to contain from 35,000 to 70,000 objects larger than 100 km residing in the region between 30 and 50 AU. SPCs may also originate as LPCs with planetary interactions (primarily with Jupiter) perturbing them into short-period orbits. LPCs are believed to originate from the vast reservoir of comets known as the Oort cloud (however there is no direct evidence of the Oort cloud because no comet has ever been observed at this great distance). It is believed that the inner Oort cloud begins approximately 1,000 AU from the Sun and may extend out to 100,000 AU (almost halfway to the Sun's nearest stellar neighbor). Although comets and asteroids have become the typical classifications applied to NEOs, there are asteroids that exhibit some amount of comet-like behavior, and some extinct comet nuclei may be classified currently as asteroids.

The size and structural integrity of the NEO are important when considering whether or not it should be considered an impact threat. Objects between 50 and 100 m in diameter are capable of penetrating the Earth's atmosphere. Precise orbit knowledge is the paramount factor for identifying a near-Earth object as an impactor. More rapidly determining an impacting object's orbit allows more warning time and opportunity to divert the object or mitigate against impact effects. Assuming that the NEO is of sufficient size and composition to pose a hazard, the following *three categories generally define the impact threat* from a warning time standpoint using the above classifications:

1. Well-defined NEO orbits (warning time – decades, NEOs detected are the Earth-Crossing Asteroids);
2. Uncertain NEO orbits (warning time – years, NEOs are newly discovered and include the Earth-Crossing Asteroids and Short-Period and Long-Period Comets);
3. Sudden NEO orbits ("immediate threat": warning time – from months to none; NEOs are Small Earth-Crossing Asteroids, Long-Period Comets (if no large aperture space-based telescopes exist, or previously undiscovered Earth-Crossing Asteroids).

From the point of view mentioned above we can approach the subject of estimating impact probability and warning time. So, in order to provide sufficient warning time to avert an impact, or at the very least mitigate the effects of the impact, precise orbit determination calculations must be made. For the vast majority of NEOs whose initial orbit calculations indicate a high impact probability, the probability will reduce to zero as additional observations are included. In the unlikely event that a high probability of an Earth impact is confirmed, the next step would be to define the likely location of impact, along with an estimate of the object's size and physical properties, to determine the likely consequences. The accuracy of an impact prediction is a function of both the observational data and the computational methods. Classical methods of orbit determination have been largely replaced with statistical techniques due to the ever-increasing processing capabilities of modern computers (Fig. 3.19).

Relatively not long ago, NEO researchers have explored complex orbital dynamics of NEO close encounters with the Earth. Although an asteroid might miss the Earth on a close approach, the object may pass through one of several small regions in space (which could be as small as on the order of only 0.5 km) called resonance "keyholes" which causes the Earth's gravity to change its solar orbit period so that it is a multiple or submultiple of the Earth's, and thus strike the Earth on one of perhaps many subsequent encounters. The size of these keyholes will only be a small fraction of the target plane error ellipse, and correspond to different resonances when the object returns to the same region in space where the first encounter occurred, causing a second encounter to take place. There can be several

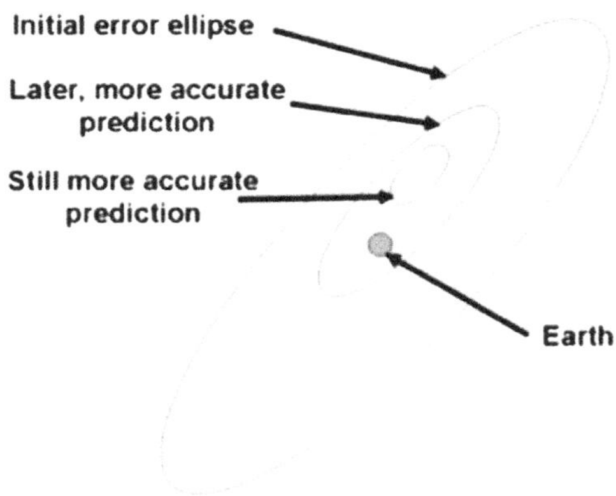

Fig. 3.19 Scheme of a typical asteroid impact prediction

keyholes corresponding to different resonances which results in possible impacts on different dates in the future. In order to prevent an impact during a subsequent encounter the trajectory of the NEO must be modified only enough to miss the keyhole and thus avert a collision in the next resonance encounter (although it may not be a permanent solution since there could be many resonant encounters, though it certainly would gain years of additional time for defensive efforts).

Various measurements will need to be repeated once the NEO's deflection or disruption mission has been completed to assess the effectiveness of the technique of planetary defense. It is going to be vital, particularly depending on the warning time permissible with a particular threat, that the instruments and resources are rapidly available to reassess the threat and recalculate the impact probability of the object (or potentially multiple objects if the original NEO becomes fragmented during the mission). The presence of local assets such as a measurement spacecraft orbiting or hovering near the NEO would permit accurate velocity change measurements. In addition, a radio transponder could be placed on the object. All of these techniques might be required to reassess the object orbit in a timely manner and recalculate its new orbit. Additionally, depending on the deflection/disruption method that was employed it might be highly desirable to send additional instruments to the NEO. This will be vital even if the object was successfully diverted to facilitate the refinement of engineering models and to characterize how successful the mitigation was, in order to be applied to the inevitable future NEOs encounter.

A common approach for the NEO's impact probability calculation pattern is the following. The ellipses in the diagram Fig. 3.19 show the likely asteroid position at its closest approach to the Earth. At first, with only a few asteroid observations, the error ellipse is very large and includes the Earth. Further observations shrink the error ellipse, but it still includes the Earth. This raises the impact probability, since the Earth now covers a larger fraction of the error region. Finally, yet more observations (often radar observations, or discovery of a previous sighting of the same asteroid on archival images) shrink the ellipse until the Earth is outside the error region, and the impact probability returns to near zero [20].

3.2 Contemporary NEO Collision Avoidance Strategies

Asteroid mitigation strategies are "planetary defense" methods by which the near-Earth objects could be diverted, preventing potentially catastrophic impact events. A sufficiently large impact would cause massive tsunamis and/or, by placing large quantities of dust into the stratosphere blocking sunlight, an impact winter. A collision between the earth and a $\sim$ 10 km object 65 million years ago is believed to have produced the Cretaceous–Tertiary extinction event. While in theory the chances of such an event are no greater now than at any other time in history, recent astronomical events have drawn attention to such a threat, and advances in technology have opened up new options.

3.2.1 Common Approaches to Parrying the Threat of NEOs

Almost any Earth defense effort requires years of warning, allowing time to build a slow-pusher or explosive device to deflect the object. As it was mentioned above, an impact by a 10 km asteroid on the Earth is widely viewed as an extinction-level event, likely to cause catastrophic damage to the biosphere. Depending on speed, objects as small as 100 m in diameter are historically extremely destructive. There is also the threat from comets coming into the inner Solar System. The impact speed of a long-period comet would likely be several times greater than that of a near-Earth asteroid, making its impact much more destructive; in addition, the warning time is unlikely to be more than a few months [21]. Therefore, determination of the material composition of the object is also necessary before deciding which strategy is appropriate.

U.S. Representative George E. Brown, Jr. (D-CA) was quoted as voicing his support for planetary defense projects in Air & Space Power Chronicles, saying "If some day in the future we discover well in advance that an asteroid that is big enough to cause a large-scale extinction is going to hit the Earth, and then we alter the course of that asteroid so that it does not hit us, it will be one of the most important accomplishments in all of human history." Because of Congressman Brown's long-standing commitment to planetary defense, a U.S. House of Representatives' bill, H.R. 1022, was named in his honor: The George E. Brown, Jr. Near-Earth Object Survey Act. This bill "to provide for a Near-Earth Object Survey program to detect, track, catalogue, and characterize certain near-earth asteroids and comets" was introduced in March 2005 by Rep. Dana Rohrabacher (R-CA) [22]. It was eventually rolled into S.1281, the NASA Authorization Act of 2005, passed by Congress on December 22, 2005, subsequently signed by the President, and stating in part:

U.S. Congress has declared that the general welfare and security of the United States require that the unique competence of NASA be directed to detecting, tracking, cataloguing, and characterizing the near-Earth asteroids and comets in order to provide warning and mitigation of the potential hazard of such near-Earth objects to the Earth. The NASA Administrator shall plan, develop, and implement a Near-Earth Object Survey program to detect, track, catalogue, and characterize the physical characteristics of the near- Earth objects equal to or greater than 140 m in diameter in order to assess the threat of such near-Earth objects to the Earth. It shall be the goal of the Survey program to achieve 90% completion of its near-Earth object catalogue (based on statistically predicted populations of near-Earth objects) within 15 years after the date of enactment of this Act. The NASA Administrator shall transmit to Congress not later than 1 year after the date of enactment of this Act an initial report that provides the following: (a) An analysis of possible alternatives that NASA may employ to carry out the Survey program, including ground-based and space-based alternatives with technical descriptions. (b) A recommended option and proposed budget to carry out the Survey program pursuant to the recommended option. (c) Analysis of possible alternatives that NASA could employ to divert an object on a likely collision course with Earth.

The result of this directive was a report presented to Congress in early March 2007. This was an Analysis of Alternatives (AoA) study led by NASA's Program Analysis and Evaluation (PA&E) office with support from outside consultants, the Aerospace Corporation, NASA Langley Research Center (LaRC).

Along with these examples of risk assessment approaches and appropriate potential NEO mitigation options we have to briefly describe the physics of interaction between them and typical technical facilities which are widely discussed by the scientific or political circles concerned.

Firstly, potentially hazardous celestial bodies orbit the Sun roughly in the same plane as the planets. Their orbits range from nearly circular to highly elliptical and their orbital velocities range upward to many tens of kilometers per second. Therefore, there are fundamentally two options for mitigating NEOs: changing their orbit so as to cause them to miss the Earth, or fragmenting them so most of the damaging fragments miss the Earth or are small enough so as to do little damage. For objects for which there is sufficient warning time the former is by far the preferable option since the fragmentation process may not be adequately predictable.

Changing the orbit of a NEO may be accomplished basically by two techniques. The first is a "fast" or kinetic change in which the interceptor collides with the NEO and the resulting momentum transfer changes the NEO's velocity; or explodes a warhead within it, or on or above its surface so that the impinging effects of the explosion cause the velocity change. The second technique is a "slow" change in which the interceptor rendezvous with the NEO and applies force to it over a substantially long time period, the integrated effects of which cause the velocity change. Each of these options has its advantages and disadvantages, which will be discussed in the following sections. Both are capable of imparting the extremely large energy and/or momentum that is required to cause the desired orbit change to occur.

As discussed in the previous chapters no two NEO's are identical and will have different shapes, sizes, rotational or tumbling rates, and may consist of rocks, metals, ice, or some combination. The objects may be solid with various degrees of density, or may be a loosely aggregated pile of rocks. Thus, the design of the preferred mitigation technique may well be very different depending on the characteristics of the specific NEO in order to accomplish the deflection and avoid fragmentation.

The fundamentals of orbit mechanics dictate that the orbit of an object may be modified with the least expenditure of energy if its velocity along its orbital path is changed, as opposed to trying to change its velocity at an angle to its velocity vector. If that much time is available, we can take advantage of the fact that a very small change in the NEO's orbital velocity will propagate with time to become a very large positional difference at the time it intersects the Earth's orbit. If applied later the velocity change imparted must be proportionally larger. Thus, whether the velocity change is applied in one short event or gently over months, and whether applied a long or shorter time before impact, the trajectory of a NEO can be modified so that it arrives sufficiently earlier or later to completely miss the Earth.

The velocity change that is required to accomplish this NEO miss is typically a very small fraction of its total velocity and typically of the order of centimeters per second; but the energies required are extremely large due to the enormous mass of even modestly sized NEOs. While the deflection required to miss the Earth completely is at most one Earth's radius plus a few hundred kilometers to avoid an atmospheric shock, it is likely that substantially larger miss distances will be required in practice due to technical uncertainties and a desire for large margins to ensure a confident outcome.

We have already mentioned that many NEOs could well be in orbits that may miss the Earth yet pass through one of a number of very small regions of space known as "keyholes" at specific but relatively near distances from the Earth, such that their orbit is altered just enough by the Earth's gravitational field that they enter a resonant orbit – one characterized by repeating encounters and thus potential collision threats to the Earth. Hence, in this scenario, the Earth could be threatened by the same NEO periodically over many years or decades. For this special case, a NEO which is predicted to miss the Earth but pass through such a keyhole for resonant return need be deflected only sufficiently that it misses the small keyhole so that, in many cases, it would not be a threat again. Since the dimensions of some keyholes are a small fraction of the size of the Earth the requirement to deflect it just enough to miss the keyhole involves changing its velocity by proportionally smaller amounts than those needed to prevent an impact were the NEO headed for a collision with the Earth, and can involve velocity changes as small as micrometers per second. This is properly illustrated in the graph of Fig. 3.20.

The accomplishing of the actual velocity change of the NEO is more complex than it would appear at first, particularly if the means is a kinetic impactor, because the composition, structure, and cohesive strength of asteroids varies considerably.

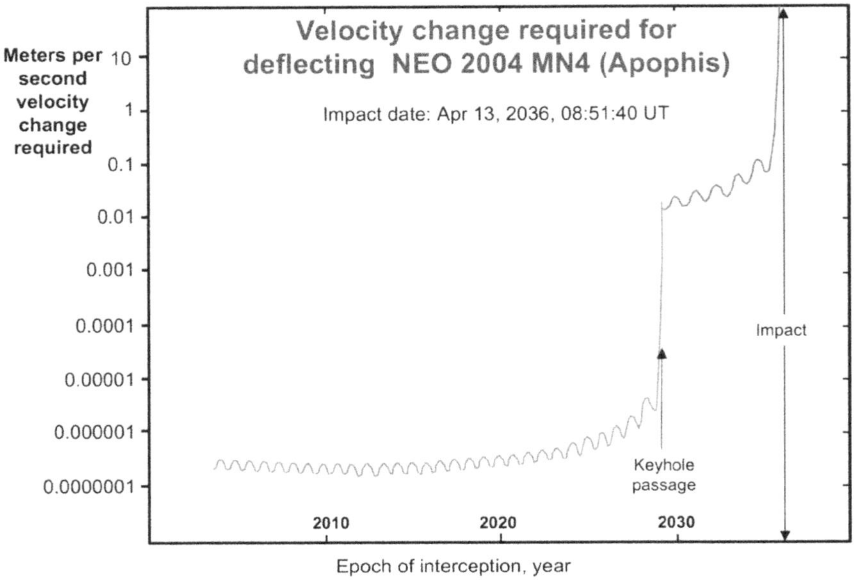

Fig. 3.20 Velocity changes required as a function of time for Apophis (Credit: B612 foundation)

It is quite possible that the force or energy applied to the NEO will fragment it as well as result in ejecta, particularly for NEOs that are characterized as loose rubble piles. If the energy of the impact is relatively low the diameter of the expanding fragmentation cloud could be of the same order of magnitude as the diameter of the Earth at the time of impact, and a single large impact would have been replaced by very many smaller impacts, possibly leading to a Hobson's choice. If it is assumed that some fragments would be relatively large their impact might cause worse damage on the Earth than one very large impact.

Besides that mentioned above, for all NEOs intervention planning must take into account orbit prediction uncertainties, mitigation system delivery uncertainties, as well as additional adequate safety factors. In addition, since the lives of millions and vast property damage may be at stake no single mitigation system will have sufficient reliability given the state of the technology. Thus, multiple missions in a coordinated and carefully planned campaign must be mounted to mitigate even a small NEO because the consequences of failure or even partial success will be horrendous and unconscionable.

Various "NEO – Earth" collision avoidance techniques have different trade-offs with respect to metrics such as overall performance, cost, operations, and technology readiness. There are various methods for changing the course of an asteroid/comet. These can be differentiated by various types of attributes such as the type of mitigation (deflection or fragmentation), energy source (kinetic, electromagnetic, gravitational, solar/thermal, or nuclear), and approach strategy (interception, rendezvous, or remote station). Strategies fall into two basic sets: destruction and delay.

Destruction concentrates on rendering the impactor harmless by fragmenting it and scattering the fragments so that they miss the Earth or burn up in the atmosphere. This does not always solve the problem, as sufficient amounts of material hitting the Earth at high speed can be devastating even if they are not collected together in a single body. The amount of energy released by a single large collision or many small collisions is essentially the same, given the physics of kinetic and potential energy. If a large amount of energy is transmitted, it could heat the surface of the planet to an uninhabitable temperature.

Collision avoidance strategies can also be seen as either direct, or indirect. The direct methods, such as nuclear devices or kinetic impactors, violently intercept the bolides' path. Direct methods are preferred because they are generally less costly in time and money. Their effects may be immediate, thus saving precious time. These methods might work for short-notice, or even long-notice threats, from solid objects that can be directly pushed, but will probably not be effective against loosely aggregated rubble piles. The indirect methods, such as gravity tractors, attaching rockets or mass drivers, laser canons, etc., will travel to the object then take more time to change course up to 180° to fly along side, and then will also take much more time to change the asteroids path just enough so it will miss the Earth.

Many NEOs are "flying rubble piles" only loosely held together by gravity, and a deflection attempt might just break up the object without sufficiently adjusting its course. If an asteroid breaks into fragments, any fragment larger than 35 m across would not burn up in the atmosphere and itself could impact the Earth. Tracking the thousands of fragments that could result from such an explosion would be a very daunting task. Many small impacts could cause greater devastation than one large impact. Against some rubble piles, a nuclear device may be delivered to it and dock with it, then it could penetrate to its center, and explode sending fragments in all directions, thus reducing the amount of material reaching the Earth. The explosion can also increase the surface area of the threat enough so that more pieces will burn up harmlessly high in the atmosphere.

Delay exploits the fact that both the Earth and the impactor are in orbit. An impact occurs when both reach the same point in space at the same time, or more correctly when some point on the Earth's surface intersects the impactor's orbit when the impactor arrives. Since the Earth is approximately 12,750 km in diameter and moves at approx. 30 km per second in its orbit, it travels a distance of one planetary diameter in about 425 s, or slightly over 7 min. Delaying, or advancing the impactor's arrival by times of this magnitude can, depending on the exact geometry of the impact, cause it to miss the Earth. By the same token, the arrival time of the impactor must be known to this accuracy in order to forecast the impact at all, and to determine how to affect its velocity.

Recent investigations into NEO intervention by many researchers [23] has resulted in identification of a number of different feasible approaches (Table 3.4). These can be divided roughly into a first class for which there is ample early warning of an impact and for which the objects' orbits have been well determined and their composition has been well characterized; and into a second class of objects that have been detected relatively much later before

Table 3.4 NEO mitigation options (projects)

Class of warning time availability	NEO mitigation options (projects)
Long warning time is available	1. Conventional rocket motor
	2. Gravitational tractor
	3. Kinetic impact facilities
	4. Laser ablation
	5. Mass driver
	6–7. Tug boat & mother ship with tug boat concepts
	8. Solar sail
Long warning time is not available	1. Non-nuclear deflection technologies
	2. Nuclear deflection technologies
	3. Other exotic concepts

predicted impact or whose orbits and characteristics have not been well determined until much later. These two classes will be addressed separately. It should be noted at the outset, however, that given the state of the art or near term projected technologies a successful mitigation may be very much more difficult if a NEO is large enough, regardless of whether it is discovered with lots of warning time. Fortunately the number of NEOs decrease exponentially as their size increases and so does the probability of an impact threat materializing in any given time interval; the warning time also generally increases with their size because they are easier to detect when far away, and technology advancement should allow continued capability increase with time over the long term. Interestingly, mitigation missions could be designed so that the resultant orbits of the target NEOs would be convenient for mining or other uses after the threat is removed.

Let's consider the first class of the NEO's mitigation options when we have enough warning time. In this case the detection and characterization systems and process are assumed to be capable of providing many years to several decades of advance warning that a PHO is in a collision orbit and what its characteristics are, with reasonably high confidence. This includes techniques for intercepting the NEOs in their keyhole passage, should there be one, which, except for the requirements placed on the tracking and orbit determination aspects, should be as easy for comets as for asteroids. Unfortunately, relatively little work has been done in the detection of the long-period comets when very far away, even though more is known about their composition from the space probes launched to date than about asteroids. As a result of the above except for the specifics of the launch systems, delivery spacecraft, and the composition of the NEOs the mitigation concepts that follow apply in general to deflection of both asteroids and comets unless specifically addressed.

1. *Conventional rocket motor*. Attaching any spacecraft propulsion device would have a similar effect of giving a steady push, possibly forcing the asteroid onto a trajectory that takes it away from Earth. An in-space rocket engine that is capable of imparting an impulse of 106 N s (e.g., adding 1 km/s to a 1,000 kg vehicle), will have a relatively small effect on a relatively small asteroid that has a mass of roughly a million times more. Chapman, Durda, and Gold's white paper [1]

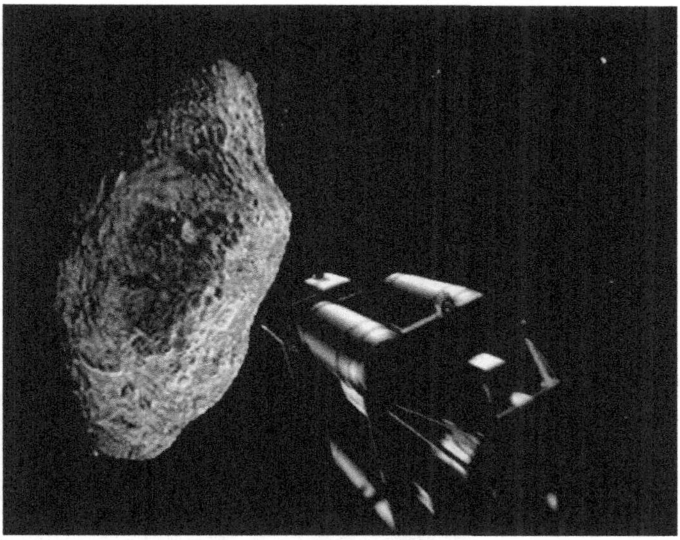

Fig. 3.21 Conventional rocket motor concept (art vision)

calculates deflections using existing chemical rockets delivered to the asteroid (Fig. 3.21).

2. *Gravitational tractor*. Another alternative to explosive deflection is to move the asteroid slowly over time. Tiny constant thrust accumulates to deviate an object sufficiently from its predicted course. Edward T. Lu and Stanley G. Love have proposed using a large heavy unmanned spacecraft hovering over an asteroid to gravitationally pull the latter into a nonthreatening orbit. The spacecraft and the asteroid mutually attract one another. If the spacecraft counters the force towards the asteroid by, for example an ion thruster, the net effect is that the asteroid is accelerated towards the spacecraft and thus slightly deflected from its orbit. While slow, this method has the advantage of working irrespective of the asteroid composition or spin rate – rubble pile asteroids would be difficult or impossible to deflect by means of nuclear detonations while a pushing device would be hard or inefficient to mount on a fast rotating asteroid. A gravity tractor would likely have to spend several years beside the asteroid to be effective.

3. *Kinetic Impact Facilities*. The hurling of a massive object at the NEO, such as a spacecraft or another near-earth object, is another violent possibility. A small asteroid or large mass in a stable high-Earth orbit would have tremendous kinetic energy stored up. With the addition of some thrust from mounted rockets (plasma or otherwise), it could be used like a stone from a slingshot to deflect the incoming threat. An alternative means of deflecting an asteroid is to attempt to directly alter its momentum by sending a spacecraft to collide with the NEO. The European Space Agency is already studying the preliminary design of a space mission able to

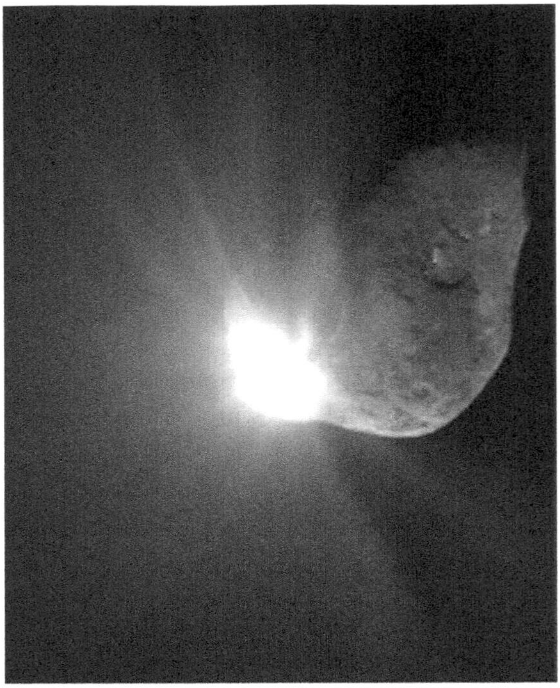

Fig. 3.22 Kinetic impact at the target

demonstrate this futuristic technology. The mission, named "Don Quixote," is the first real asteroid deflection mission ever designed. In the case of 99942 Apophis it has been demonstrated by ESA's Advanced Concepts Team that deflection could be achieved by sending a simple spacecraft weighing less than one ton to impact against the asteroid. During a trade-off study one of the leading researchers argued that a strategy called "kinetic impactor deflection" was more efficient than others (Fig. 3.22).

These kinds of mission are relatively simple and not unlike planetary and comet probes that have been launched by a number of agencies in the past, requiring only a launch vehicle (LV) with sufficient energy to place the interceptor in a collision trajectory with the NEO and a stage that can perform trajectory corrections along the way. While conceptually daunting the requisite accuracy of the guidance capability has been amply demonstrated in the planetary probes, as well as in a number of antisatellite and antiballistic missile interception tests carried out to date by several entities. Thus, it is reasonable to say that the technology to perform kinetic impact mitigation exists today, even if dedicated systems to perform such missions do not. In an emergency it is also reasonable to expect that if a scientific planetary probe were essentially ready at the time of need it could probably be modified to perform a kinetic impact mission on a NEO instead.

A special case of kinetic "interception in a keyhole" exists which requires far less energy than interception far from the Earth because the keyholes are by definition fairly close to our planet.

4. *Laser Ablation*. Another option for applying a slow push to a NEO is to use a sufficiently intense laser projection system to illuminate it, causing surface ablation and plasma ejection, whose reaction forces would result in a velocity change. Deploying the system on the Moon, in low Earth orbit (LEO), geostationary orbit (GEO), or the Earth-Sun libration points are basing options. Since all these are remote from the location of the NEO, and for physically realizable optical projector mirrors, the energy density arriving there will necessarily be low and so what surface ablation occurs will cause very little heating at depth and hence could totally avoid fragmentation even of rubble pile NEOs. In fact there is some experimental evidence that this approach might serve to actually coax some NEOs into a more structurally cohesive body.

A space-based laser could be delivered to rough vicinity of the NEO by a conventional launch vehicle and spacecraft, and could impart a greater force on the NEO than the gravity tractor, yet enjoy the same advantages of not being affected by its surface shape or composition. It would also allow much greater standoff distances, and thus eliminate the challenges of hovering near a rotating and irregular NEO, but would entail a much greater technological challenge in the laser and optical beam generator. It would also likely be considerably heavier for the same effect due to the inefficiencies of at least current high power lasers (Fig. 3.23).

A slow ablation variant similar to the laser ablation concept involves sending an accelerated beam of particles to the target, preferably neutral ones to avoid interaction with charged space plasmas. There are differences in using particle beams as

Fig. 3.23 Laser ablation of the asteroid using several spacecraft (art concept)

Fig. 3.24 Mass drivers on the surface of asteroid (art concept)

opposed to laser beams in the degree of reaction caused by their incidence on the NEO.

All such acceptable laser and beams weapon technologies were well developed in the framework of the "Space Stars" ABM Programs both in the USA and Russia at the end of the 1980s.

5. Mass Driver. A mass driver is an (automated) system on the asteroid to eject material into space thus giving the object a slow steady push and decreasing its mass. Mass driver is designed to work as a very slow Specific Impulse system, which in general uses a lot of propellant, but very little power. The idea is that when using local material as propellant, the amount of propellant is not as important as the amount of power, which is likely to be limited. Another possibility is to use a mass driver on the Moon aimed at the NEO to take advantage of the Moon's orbital velocity and inexhaustible supply of "rock bullets" (Fig. 3.24).

This is another concept which envisions imparting the requisite velocity change to a NEO by using one or more spacecraft docked to it, each of which ejects chunks of the NEO's own mass to cause reaction forces, rather than using propellants brought from the Earth. The spacecraft would use either solar or nuclear reactor energy, and since supply of the reaction mass is essentially limitless the velocity change imparted is dependent principally on the operating time.

While visions of 1970s era kilometer-long mass driver concepts come to mind more recent concepts involve emplacing independent modestly sized vehicles on the NEO surface, anchoring them, drilling for the reaction mass, and ejecting tennis-ball sized chunks normal to the surface at velocities of a few hundred meters per second using small nuclear reactors as energy sources.

6-7. "Tug Boat" & "Mother Ship with Tug Boat" Concepts. According to these concepts the Tug Boat approach envisions rendezvous and docking with the NEO with a propulsion-equipped spacecraft, attaching to its surface, and pushing it

Fig. 3.25 Tug boat on its way to the NEO (art concept)

gently along its velocity vector. Implied in this concept are the requirements that sufficient warning time must exist for the longer and more energetic rendezvous trajectories required to make such long-term operations possible; and that a hard dock sufficiently stable to transfer forces from the tug to the NEO be achieved. Furthermore, the propulsion system must operate for months to years with high reliability, and if electric, at megawatt power levels [1].

Launch vehicle energy and time of flight requirements for the tug boat mission are substantially greater than for a simple impact trajectory as in order to rendezvous it must match its orbital velocity to that of the NEO, and subsequently perform proximity and docking operations to land on the NEO and fasten itself to it. Nonzero tumbling and spin rates are typical for NEOs, and the Tug would have to match these rates in order to achieve a hard dock with the NEO. These rates also would require either thrust vector control or a time-controlled thrusting after dock so that thrust is only applied essentially along the desired inertial thrust vector coinciding with the NEO velocity vector. The difficulty of attaching the Tug Boat to the NEO must not be minimized because of the great uncertainties of surface shape, structure, and composition that could be present (Fig. 3.25).

The option of "Mother ship with tug boat" is similar to the preceding Tug Boat concept except that it overcomes some of its problems by virtue of using a separate, dedicated, robotic propulsion system which would detach from a mother ship and dock with the NEO. This would permit the mother ship to remain relatively near the NEO or orbit around it, and thus to independently measure the actual velocity change that is imparted to the NEO as it occurs, thus allowing a real-time control of

Fig. 3.26 Solar sail removing a NEO (art concept)

the propulsive action which is not possible with the simpler Tug Boat. Not only could the actual velocity being imparted be accurately controlled but if the tug boat failed to function as expected that fact would be detected in real time, as opposed to only after perhaps months of observation from the Earth. This would allow time to perform adjustments or equipment exchanges by the mother ship; the tug boat could be re-docked; or at the worst case a new mission launched.

8. Solar Sail (Using focused solar energy). H. Jay Melosh proposed to deflect an asteroid or comet by focusing solar energy onto its surface to create thrust from the resulting vaporization of material, or to amplify the Yarkovsky effect. Over a span of months or years enough solar radiation can be directed onto the object to deflect it. This method would first require the construction of a space station with a system of gigantic lens and magnifying glasses near the Earth. Then the station would be transported toward the Sun (Fig. 3.26).

Since the photon pressure from the Sun is very small the resulting increased force on even a large NEO would be measured at most in a few Newtons, and thus extremely long time would be required to effect a sufficient velocity change. In addition, since virtually all NEO rotate or tumble, the Yarkovsky effect results in the effective thrust probably being in a nonoptimal direction due to thermal delay between absorption and re-radiation of photons, and thus the effect of the thrust would be even smaller. The painting of the entire surface of even a small NEO has not even been seriously addressed. Attaching a steerable solar sail to the NEO would enhance the photon pressure, but the construction, deployment, and complexities in orienting the very large sail required would make the device complex and expensive, particularly on NEOs which are likely spinning and/or

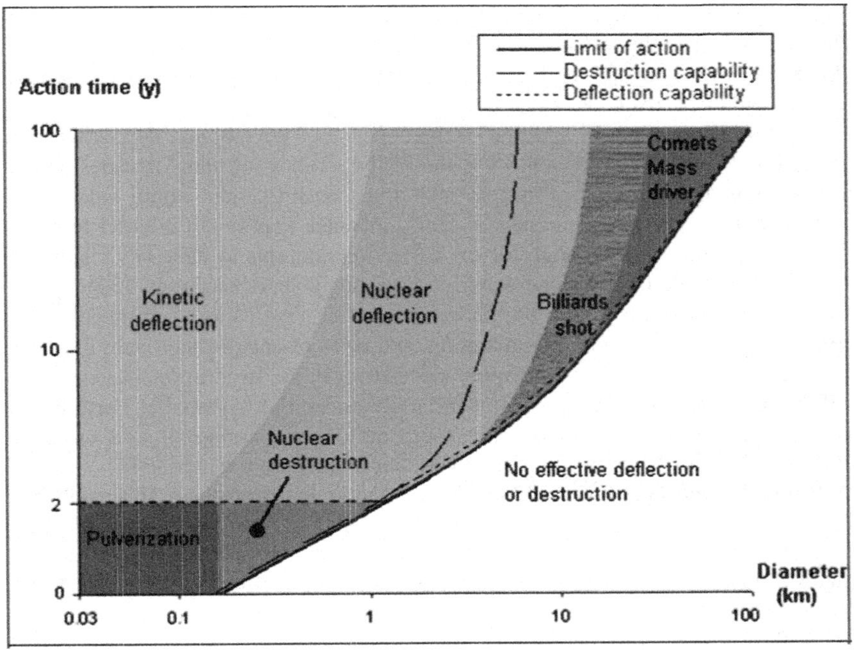

Fig. 3.27 NEO mitigations options (available time for actions)

tumbling. Though the principle is understood the magnitude of the effective thrust is tiny and a very long time would be needed for sufficient deflection. Furthermore, the uncertainties of prediction of the net thrust vector given tumbling bodies precludes this effect from being included in the higher confidence techniques.

Finally, let's pay attention to the cases of NEO's mitigation when little warning time is available (Fig. 3.27). These options are mostly considered in the public communities concerned. So, suppose a dangerously large NEO is detected late by the surveillance system, or has been known to exist before but the contacts have been insufficient to generate a confident orbit, and we now find ourselves with a high probability of collision in only a few months or at most a year or so. Can anything be done to mitigate an impact?

The only confident statement that can be made in that situation is that "slow push or pull" techniques would not be able to move the impact point significantly in the remaining time, thus narrowing practical options to those that impart velocity change in very short time intervals. These are divided into those using non-nuclear and nuclear facilities.

1. *Non-nuclear deflection techniques.* A number of sufficiently massive kinetic interceptors launched sequentially from the Earth or possibly based in the Earth orbit, on the Moon or at the Lagrangian points, targeted to impact sequentially would be able to impart sufficient velocity change to a modestly sized NEO, even

6–12 months before impact, to cause the impact point to miss the Earth. For larger NEOs or substantially reduced time available the Earth impact point might not be able to be moved beyond the Earth but may be able to be moved to an area whose impact could perhaps be less destructive.

The problem is that, in addition to the expense of developing, deploying, and maintaining such NEO interceptors at the ready perhaps for many decades before use, a command and control system including control center would have to be emplaced, fully checked out, and constantly manned, kept ready in order to make the response time short enough for the deflection attempts to have any chance of success. This undertaking alone would require many years and large budgets, and would most probably only be undertaken after international protocols and commitments covering funding, manning, and rules of engagement were in place. That process alone could easily require a decade or more. In addition these systems will be far more expensive to deploy and use than similar but smaller Earth-based kinetic systems that could take advantage of longer warning times were they available, but that is academic if such warning times do not materialize.

While this concept of multiple sequential kinetic intercepts has a finite probability of deflecting a small to modestly sized NEO sufficiently even with only 6–12 months' warning, it loses effectiveness with the inverse cube as the size of the NEO increases, and will probably prove essentially ineffective for moderate-to-large NEOs. The stationing of multiple high energy lasers on extraterrestrial outposts is also possible and might offer similar deflection potential, but would suffer from similar expenses and difficulties as the multiple impactor concepts, in addition to being extremely expensive and even more sensitive to reductions in the available time. Perhaps as important is that because of the technology of launch vehicles at least 6 months, and more probably a year or more, will be required for an interceptor to travel from the Earth to the NEO in time to act, even assuming that zero time is required for decisions and launch readiness (Fig. 3.28).

For all the above reasons non-nuclear kinetic impact techniques, though theoretically capable of deflecting small NEOs with a warning time of a year or even less, are unlikely to become a realistic and practical solution to the problem of dealing with modest-to-large sized NEOs when less than about a decade warning time is available. This leaves only the nuclear option for mitigating such threats.

2. *Nuclear Deflection techniques*. Detonating an explosive nuclear device above the surface (or on the surface or beneath it) of an NEO would be one option, with the blast vaporizing part of the surface of the object and nudging it off course with the reaction. This is a form of nuclear pulse propulsion. Even if not completely vaporized, the resulting reduction of mass from the blast combined with the radiation blast and rocket exhaust effect could produce positive results. Another proposed solution is to detonate a series of smaller nuclear devices alongside the asteroid, far enough away as not to fracture the object. Providing this was done far enough in advance, the relatively small forces from any number of nuclear blasts could be enough to alter the object's trajectory enough to avoid an impact. This is a form of nuclear pulse propulsion (Fig. 3.29).

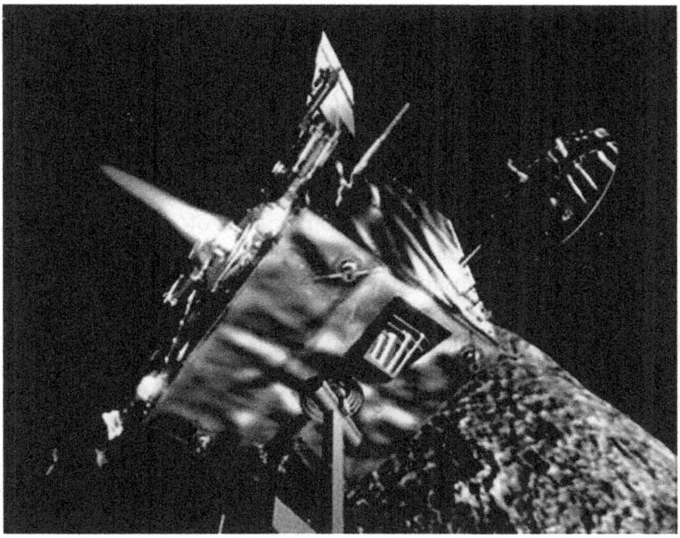

Fig. 3.28 Non-nuclear NEO deflection (art concept)

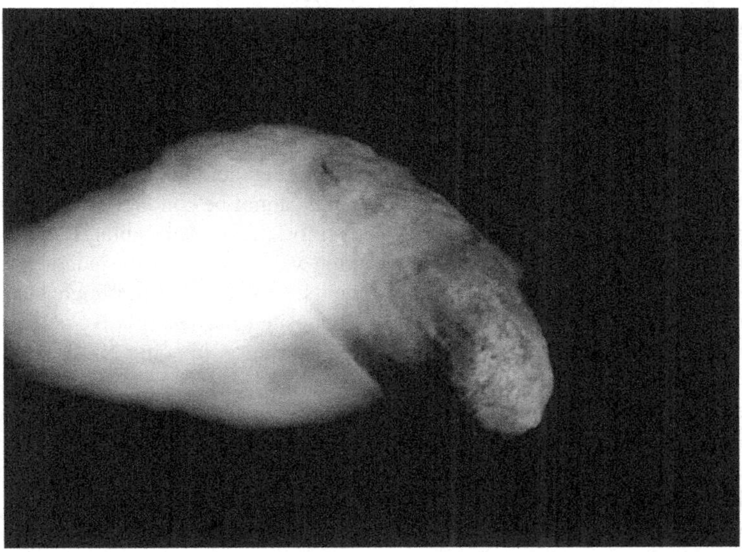

Fig. 3.29 Nuclear deflection of the NEO (art concept)

But, first of all we have to note that the placement or detonation of nuclear devices in space is currently forbidden by signed international treaties and agreements. The international treaties regarding nuclear devices in space were formulated during the cold war and were intended to minimize the chance that they would be used by some nations against others – they were clearly not formulated with an extraterrestrial threat in mind. Thus, it is not unreasonable to address the amendment of these treaties to allow employment of nuclear devices to mitigate extraterrestrial threats, even though such considerations would require extensive political and policy deliberations by the world's major powers in order to assure verifiable international safeguards and use agreements. It is also a fact that enshrined in the 1969 Vienna Convention on the Law of Treaties is a fundamental rule that states that "..if the literal interpretation of a treaty obligation would lead to a result which is manifestly absurd or unreasonable, such interpretation should not be upheld."

Clearly use of nuclear devices is less desirable than if non-nuclear means are known which could be as effective for similar investment, and therefore nuclear devices for mitigating NEO threats may be considered as a means of last resort, much as they are generally considered for use against targets on the Earth. It is recognized that there are many for whom the use of nuclear devices for any purpose is abhorrent in principle. Nonetheless given the choice of having some or all of humanity's civilization destroyed by a NEO impact versus overcoming an aversion in principle to using nuclear devices and averting the cataclysm, the former becomes the clearly preferred alternative.

Engagement of a NEO using a nuclear device is very similar to that of a non-nuclear kinetic impact deflection technique, with the device being detonated just before or at impact. Similarly the trajectory is one of simple collision at high velocity, and neither rendezvous nor docking are required. Detonation of a thermonuclear device in space near a NEO produces a tremendous radiative flux. While the total energy from the particle flux may be relatively small, the hard X-ray, infrared, neutron, and gamma ray flux is extremely large. This radiation onslaught results in extremely rapid surface heating of the NEO that cannot be conducted away rapidly enough through the material. Hence, a layer of the surface ablates into a hot expanding plasma which disperses at extreme velocities, and its reaction forces on the surface accelerate the NEO and change its velocity. With sufficient stand-off distance, the area over which the energy is deposited may be relatively large and hence the "push" gentle enough to avoid fragmentation in some NEOs; however there would still be significant fragmentation risk to susceptible NEOs. In addition the detonation at a standoff distance from the NEO reduces the coupled energy by about one order of magnitude compared to that transferred to the body were the device detonated at the surface, and by about two orders of magnitude were it buried a few tens of meters below its surface and then detonated.

Notwithstanding the above, the energy and momentum transfer imparted by a nuclear device is so large that even very large NEOs discovered late may be deflected sufficiently. While the delivery and detonation of a nuclear device above or on the NEO surface can be accomplished in a straightforward impact trajectory, the most effective use with the device deeply buried would require a rendezvous trajectory, with subsequent docking and drilling operations to emplace the device using either automated/robotic means or manned systems. The deep explosion would result in superheated ablated interior material being ejected from the NEO in a jet-like fashion which, while extremely effective in transferring momentum to the NEO, would result in such extreme interior shock pressure that the danger of partially or totally fragmenting the body is a very serious possibility. The predictability of the effects of an explosion inside the NEO is even lower than that for external ablation or even kinetic impact, and requires a much better understanding of the composition and structure of the body in order to predict the outcome with any reasonable probability.

There is a persistent and mistaken notion that the way to deal with a dangerous NEO is to simply hit it with an ICBM and vaporize it in space. Unfortunately, reality is far removed from this illusion. While it is likely that we may be able to rapidly reconfigure an ICBM computer guidance system to intercept a point or object in the near-Earth space, ICBM propulsion system performance is insufficient to enable intercept beyond a few hundred kilometers above the Earth's surface. Stages must be added to an ICBM to enable it to achieve the necessary escape velocity and to place the weapon on an intercept trajectory with a NEO. While these upper stage technologies are space qualified, such a system would have too low a reliability for the NEO intercept mission given the potentially horrendous consequences of an Earth impact, and might thus require many sequential launches of several such vehicles to have any reasonable chance of successfully deflecting a NEO. Such attempts would be part of a dedicated "campaign" utilizing several different launch vehicle types, designed with different upper stages, using different end game techniques, and different nuclear warhead types, in order to obtain a high probability of success. Furthermore, at least one failed launch attempt is likely if many are required, and with a nuclear payload this could result in serious environmental effects in and of itself. Thus, it is clear that for the nuclear concept several dedicated designs of inherently highly reliable launch vehicles and multistage interceptors would be extremely desirable to loft the nuclear warheads, and thus the use of existing ICBMs, even if outfitted with current technology upper stages, is highly undesirable if not essentially ruled out.

Finally, it must be made clear that many nuclear warheads intended for ICBMs exist that could be used with few, if any, modifications as payloads for the purpose of deflection of NEOs, whatever launch vehicle and upper stage is used to get them to the NEO. Many analyses have been carried out to compare the various techniques discussed briefly above. While much detail exists the fundamentals of orbit dynamics are simple: The longer the action time of a force applied to a NEO the lower is the energy required to move its impact point off the Earth and the larger is the NEO than can be so moved.

Conversely, the greater the energy available the shorter the action time can be and the larger is the NEO that can be moved. One comparison is shown in Fig. 3.30, which clearly shows that if the NEOs are large and if little time is available then there is no choice but the use of nuclear devices to prevent impact. It also shows that in general there will be a significant advantage to a kinetic "fast push" approach over any "slow push" technique for the same available action time, at least for the parameters considered in this comparison.

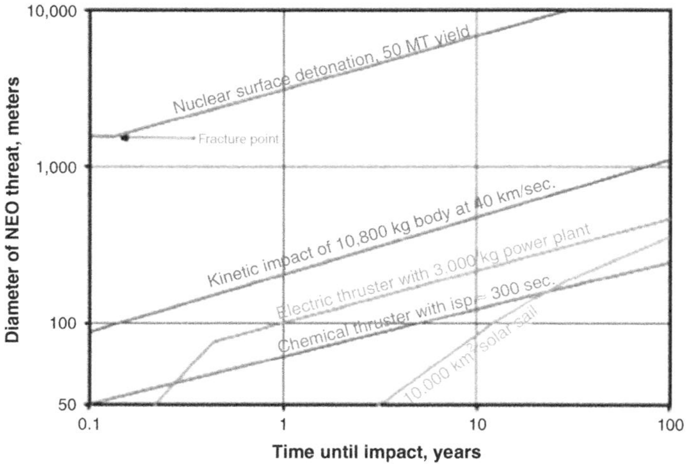

Fig. 3.30 Comparison of different NEO mitigation options

Although there are well-founded aversions to the use of nuclear devices, their use against NEOs in space is probably one of the best and most desirable applications of these devices; and is the only technique that might be able to prevent a horrendous regional or global catastrophe.

Along with the above-mentioned NEO mitigation options and impact risk assessments there are some others (sometimes "exotic" proposals), which are under consideration in scientific and public circles. There are: nonconventional engines; "painting" or dusting the object with titanium dioxide or soot to alter its trajectory via the Yarkovsky effect; and deflecting a potential threatening celestial body by releasing a cloud of steam in the path of the object, hopefully gently slowing it. In 1990 a similar idea was sketched [24], "comet aero-braking," the targeting of a comet or ice construct at an asteroid, then vaporizing the ice with nuclear explosives to form a temporary atmosphere in the path of the asteroid; attaching a tether and ballast mass to the asteroid to alter its trajectory by changing its center of mass [25]; magnetic Flux Compression etc.

Well-known scientist Carl Sagan, in his book "Pale Blue Dot" [25], expressed concerns about deflection technology: that any method capable of deflecting dangerous bodies away from the Earth could also be abused to divert nonthreatening bodies toward the planet. Considering the history of genocidal political leaders and the possibility of the bureaucratic obscuring of any such project's true goals to most of its scientific participants, he judged the Earth at greater risk from a man-made impact than a natural one. Sagan instead suggested that deflection technology should only be developed in an actual emergency situation.

Analysis of the uncertainty involved in nuclear deflection shows that the ability to protect the planet does not imply the ability to target the planet. A nuclear bomb which changed an asteroid's velocity by 10 m/s (plus or minus 20%) would be adequate to push it out of an earth-impacting orbit. However, if the uncertainty of the velocity change was more than a few percent, there would be no chance of directing the asteroid to a particular target.

According to Rusty Schweickart, the gravitational tractor method is also controversial because during the process of changing an asteroid's trajectory the point on the Earth where it could most likely hit would be slowly shifted across different countries. This means that the threat for the entire planet would be minimized at the cost of some specific states' security. In Schweickart's opinion, choosing the way the asteroid should be "dragged" would be a tough diplomatic decision [5].

After such detailed observation of the threat posed by a pending asteroid impact which is inherently international in scope, we'd like to draw the attention of the readers to some concepts of "Planetary Defense Systems," so popular in contemporary scientific and public circles. While the physical extent of an asteroid impact could range from local to regional to global, the entire world would be engaged in the unfolding drama from the announcement of a potential collision through either the successful mitigation or the disastrous consequences of impact. Fortunately, the resources of the global community would also be available to respond to the challenge. Effectively harnessing and applying these resources, however, will require unprecedented cooperation and organization. Without adequate planning and preparation before an event is underway, the challenge may overwhelm even the most enthusiastic international proponents of a coordinated response.

There is an international community of astronomers participating in surveys of the asteroid population. But beyond that, technologies to prepare, respond, and recover from asteroid impacts can also be drawn from throughout the world. Budget resources and talent are limited within individual nations, even in countries making significant contributions today. Pooling and leveraging funds and talent through wider cooperation in commonly agreed upon priority areas and more effective use of resources can substantially improve the posture of future responses. Many nations approach technological solutions differently and offer specialized areas of competence that, when shared widely, can illuminate issues and help other nations develop effective responses.

Considering and accounting for cultural differences and sensitivities in dealing with mass evacuations, establishment of relocation centers, and eventual remediation add perspective that, when applied early in the planning cycle, can save time, money, and more importantly, lives, if a call for action is necessary.

Below some historical facts about public concern about the NEO threat are presented:

- In their 1964 book, Islands in Space, Dandridge M. Cole and Donald W. Cox noted the dangers of planetoid impacts, both those occurring naturally and those that might be brought about with hostile intent. They argued for cataloging the minor planets and developing the technologies to land on, deflect, or even capture planetoids [6].
- In the 1980s NASA studied evidence of past strikes on the Earth, and the risk of this happening at our current level of civilization. This led to a program that maps which objects in our solar system both cross the Earth's orbit and are large enough to cause serious damage if they ever hit.

- In the 1990s, U.S. Congress held hearings to consider the risks and what needed to be done about them. This led to a US$ 3 million annual budget for programs like Spaceguard and the near-earth object program, as managed by NASA and USAF.
- In 2005 the world's astronauts published an open letter through the Association of Space Explorers calling for a united push to develop strategies to protect the Earth from the risk of a cosmic collision [26].
- In the year 2007 it was believed that there are approximately 20,000 objects capable of crossing the Earth's orbit and large enough (140 m or larger) to warrant concern [27]. On average, one of these will collide with the Earth every 5,000 years, unless preventative measures are undertaken [7]. It is now anticipated that by the year 2008, 90% of such objects that are 1 km or more in diameter will have been identified and will be monitored. The further task of identifying and monitoring all such objects of 140 m or greater is expected to be complete around 2020 [27].
- The CSS [28] is one of NASA´s four funded surveys to carry out a 1998 U.S. Congress mandate to find and catalog by the end of 2008, at least 90% of all near-Earth objects larger than 1 km across. CSS discovered 310 NEOs in 2005, 400 in 2006, and the record will be broken with 450 NEOs found in 2007. In doing this survey they discovered on November 20, 2007, an asteroid, designated 2007 WD5, which initially was estimated to have a chance of hitting Mars on January 30, 2008, but further observations during the following weeks allowed NASA to rule out an impact [29]. NASA estimated a near miss by 26,000 km [30].

3.2.2 Planetary Defense Concept

As the one example of joint international efforts in the field of NEO mitigation and parrying risks and threats to the Earth – the Russian Conception "Planetary Defense System "CITADEL" contributed by its ideologist – Dr.Anatoly Zaitsev, Director General of the Research Center of the same name" (Moscow) [31]. However, it must be kept in mind that the form and content of such a system would likely be very different from that below when actually implemented. This concept description is based on Russian technologies, but clearly any real concept should be implemented with the best of all the world's technologies (Fig. 3.31).

The common requirements for such a system (hereinafter named the System) would be the following:

- The System's implementation should be amenable to the defense of any country or region, and be amenable to exclusion of those that wish to be excluded from its action;
- The System must be created to efficiently and rapidly provide notification to all that a threat exists, and to prevent data suppression by anyone;
- The System has to guarantee that its capabilities will not be used for military purposes;

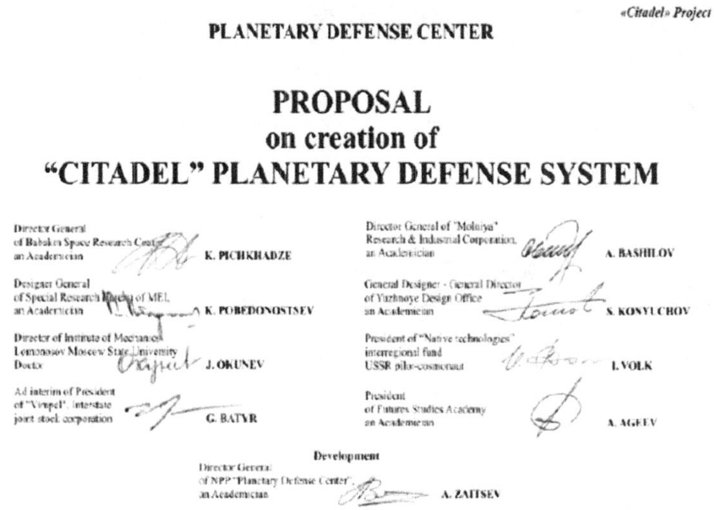

«Citadel» Project

PLANETARY DEFENSE CENTER

PROPOSAL
on creation of
"CITADEL" PLANETARY DEFENSE SYSTEM

Director General
of Babakin Space Research Center
an Academician
K. PICHKHADZE

Director General of "Molniya"
Research & Industrial Corporation,
an Academician
A. BASHILOV

Designer General
of Special Research [illegible] of MEI,
an Academician
K. POBEDONOSTSEV

General Designer - General Director
of Yuzhnoye Design Office
an Academician
S. KONYUCHOV

Director of Institute of Mechanics
Lomonosov Moscow State University
Doctor
J. OKUNEV

President of "Native technologies"
interregional fund
USSR pilot-cosmonaut
I. VOLK

Ad interim of President
of "Vimpel", interstate
joint stock corporation
G. BATYR

President
of Futures Studies Academy
an Academician
A. AGEEV

Development

Director General
of NPP "Planetary Defense Center",
an Academician
A. ZAITSEV

Fig. 3.31 "Citadel" project scientific proposals

- The System is to be able to provide maximum warning time and minimize damage from fragments of the NEO or the mitigation means themselves;
- It must allow for modernization as new technologies become available.

Since it is unrealistic to initially deploy and operate continuously a system designed to mitigate large NEOs which occur very infrequently, the system concept envisions a two-tier structure in which a subset would operate continuously against NEOs of tens to hundreds of meters size, and the capabilities against larger NEOs brought into play if needed later upon detection of a viable threat. This latter capability, though designed and understood, can be established only when a real threat materializes and thus to some extent will exist initially in virtual form.

The small-NEO initial system is envisioned to have international ground and space surveillance means and two Planetary Defense Centers, a European/Asiatic one and an American one. The observation component would use optical and radar systems sited both on the ground and in space. The Planetary Defense Centers would include interception launch facilities with both non-nuclear and nuclear devices at the ready. The observation and interception means will be connected and netted as shown in Fig. 3.32.

A typical operation sequence could be as follows: Once a threat NEO/PHA has been detected the appropriate Control Center would issue a prediction concerning the expected place and time of impact as well as expected damage. Catastrophe prevention measures would be developed and an operations plan submitted to the hierarchy of nations/consortia/global entities comprising the management of the System (Fig. 3.33).

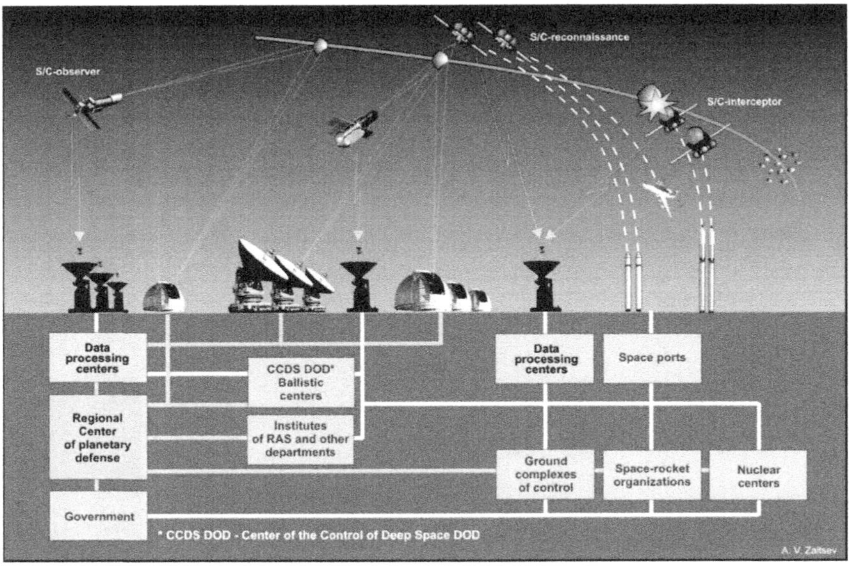

Fig. 3.32 Integrated infrastructure and dedicated planetary defense system

Once the plan has been proved at that intergovernmental level, launches of reconnaissance spacecraft on "Dnepr" or "Zenit" heavy launchers and then interceptor spacecraft on "Zenit" or "Proton" launchers can proceed. In the future multipurpose launch/intercept systems could be developed and used. The reconnaissance spacecraft would approach the NEO as fast as possible and come to a minimum distance from the NEO. It would define the NEO's trajectory, velocity of rotation, mass, mineralogical content, and dimensions, and acquire a detailed panorama of its surface and download these data to the Planetary Defense Centers. A mission design would then be undertaken to determine the best means of intercepting and mitigating the NEO threat using nuclear or non-nuclear facilities as appropriate. Nuclear devices with a yield of 1.5–5 MT would suffice to mitigate a stony NEO with a diameter up to a few hundred meters. Several interceptor modules could be assembled in Earth orbit to implement a capability to mitigate substantially larger NEOs.

A number of steps would be necessary for its implementation, including:

1. Development of top-priority measures of population relocation and property recovery for potential areas subject to damage, including protection of property and items of cultural value. This should be done even before a defense system is developed.
2. Perform experiments to understand the interaction between the interception means and the NEO to assess the practicality and effectiveness of the mitigation operation.

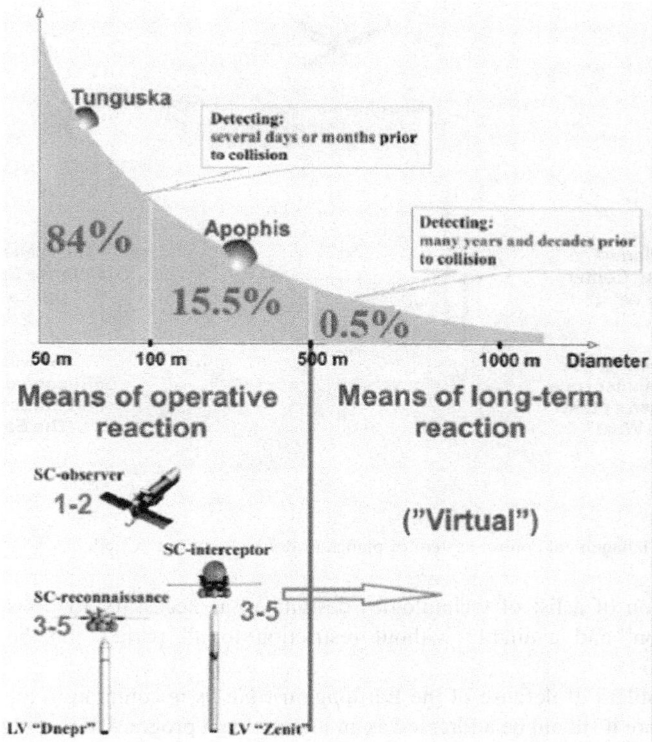

Fig. 3.33 Key points of the "Citadel" concept

3. Demonstration projects to test methods and means of NEO interception and deflection, including development of kinetic and slow deflection interceptors.
4. Deployment of the short-term system tier to respond to smaller threat NEOs, including all the surveillance, launch, communications, and command/control systems required. This could use upgraded existing facilities, develop new ones, or both. It is estimated that the first such near-term tier of a Planetary Defense System could be developed within 5–7 years from an international decision to do so.
5. Creation of a plan for development, deployment, and operation of the longer term tier to respond to much larger NEOs later in time. This could include development of new high energy launch vehicles and interceptors as needed. This plan would not be implemented until later.
6. Development of the legal regime addressing organizational, financial, political, juridical, ethical, and other questions at national and international levels necessary to create and operate a Planetary Defense System.

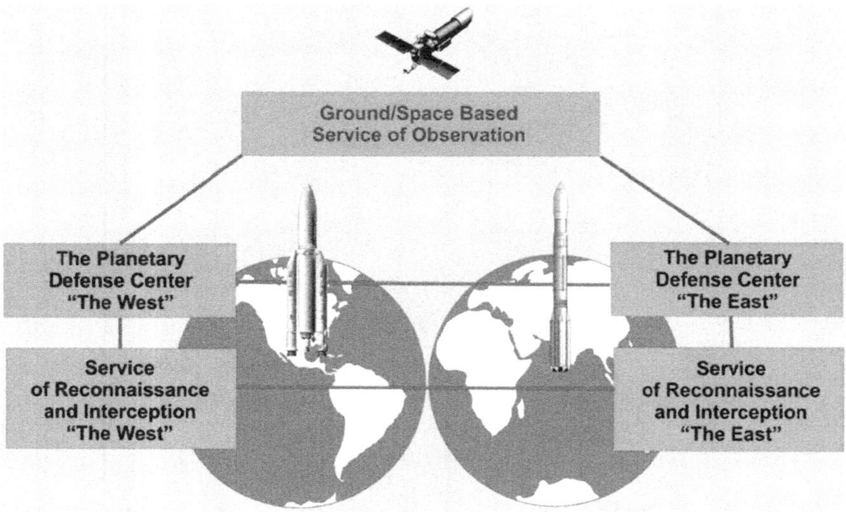

Fig. 3.34 Management control system of planetary defense concept "Citadel"

7. Definition of a list of technologies designated as necessary for its successful operation, and available without restriction to all parties for the purpose (Fig. 3.34).

The problem of defense of the Earth against NEOs is common to all mankind and therefore it should be addressed as an international program of the whole world community. The importance of the problem and its global and complex nature require resources to be pooled and managed at the intergovernmental level. One major step forward to be taken in this direction would be the creation of a "Mankind Insurance Fund" for financing the development and operation of such a Planetary Defense System. Such a fund would be established by all the more developed countries of the world with participation of government resources, banks, organizations, and individuals. In addition to financial, pooling of intellectual, technical, and other world assets would be required. Specifically the objectives should include:

- Establishing an International Coordination Council of heads of leading organizations, scientists and specialists in the NEO defense field in order to coordinate efforts to define and develop a Planetary Defense System proposal.
- Drawing up a draft of constituent documents of the "Mankind Insurance Fund" intended to finance the project and forward them, together with the proposal, to major government organizations, banks, funds, individuals, and others and solicit their participation in establishing the Fund.
- To create the Mankind Insurance Fund and when financial resources become available to proceed with development and operation of the first phase of the "Citadel" Planetary Defense System.

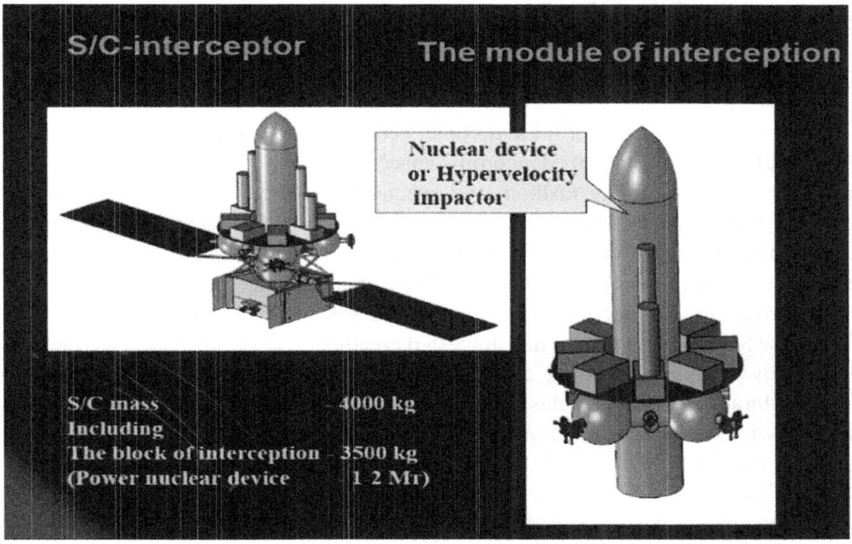

Fig. 3.35 Citadel project NEO interceptor concept

- Preliminary estimates for the costs of such a Planetary Defense System are US$ 3–5 Billion. This equates to an annual cost of only 5–10 cents per human being on the planet (Fig. 3.35).

In summary, exploration and development work already carried out show that there is a good chance that a first operational system can be developed in 3–5 years, and that it could operate successfully based on current technical assessments. The more difficult problems of organizational, political, and judicial nature need addressing, and hopefully can also be solved in this time frame. Thus, while there is little doubt that the solutions exist or can be created we need to overcome the main problem which is a moral one: to understand that there is a necessity for all mankind to realize a responsibility for its own preservation and that of the Earth biosphere, as well as that of cultural, material and other values that have been created over millennia by billions of human beings.

Thus, the problem of the protection of Earth against NEOs can be seen as a kind of test for mankind's ability to solve global problems which it faces. So this "Citadel" Planetary Defense System could become a model for the first global project of mankind in the third millennium, which could turn the Earth into an unassailable fortress with protection from space threats. This, in turn could become a catalyst for development of many industries and technologies that facilitate not only improvements in Planetary Defense but also further development facilitating unity of many nations. But to make that happen many countries will be required to pool their resources, including financial as well as specialists in both natural science and humanitarian fields.

3.2.3 Some Principles Concerning the Adaptation of Space Rocketry Facilities for Asteroid and Comet Threat Mitigation Missions (Russian Approach)

The above analysis of asteroid and comet security tasks showed that they can be effectively solved only if dangerous space objects (DSO) are detected and characterized in a timely fashion, and special devices are delivered to them on time to cause them to veer off the collision path or to disintegrate into harmless fragments. The development of specialized space devices created for this purpose will reduce environmental and functional risks for the planet and people.

A group of Russian scientists and designers from the Academician V.P. Makeyev State Rocket Center has suggested creating an asteroid and comet security system by designing, building, and deploying a set of specialized universal impact and exploratory space vehicles that will provide an acceptably prompt, environmentally harmless, and functionally safe way of sending a dangerous space object off the collision path or neutralizing it (fragmenting it into harmless pieces) (Fig. 3.36).

Following below are some of the results of the conceptual analysis and systemic synthesis of the principles of adaption of existing Russian space systems for asteroid and comet security missions. Dangerous space objects were assumed to be asteroids, comets and their fragments that cross the Earth's orbit. The authors acted on the assumption that the actual majority of such objects do not cross the Earth's orbit in three-dimensional space even though their perihelion distance d is less than 1 AU and aphelion distance Q is more than 1 AU, and therefore they create

Fig. 3.36 Director General of Acad. V.P. Makeyev state rocket center Prof. Vladimir Degtyar at the plenary session of the Cyprus symposium "Space & Security of Humanity," November 2, 2009

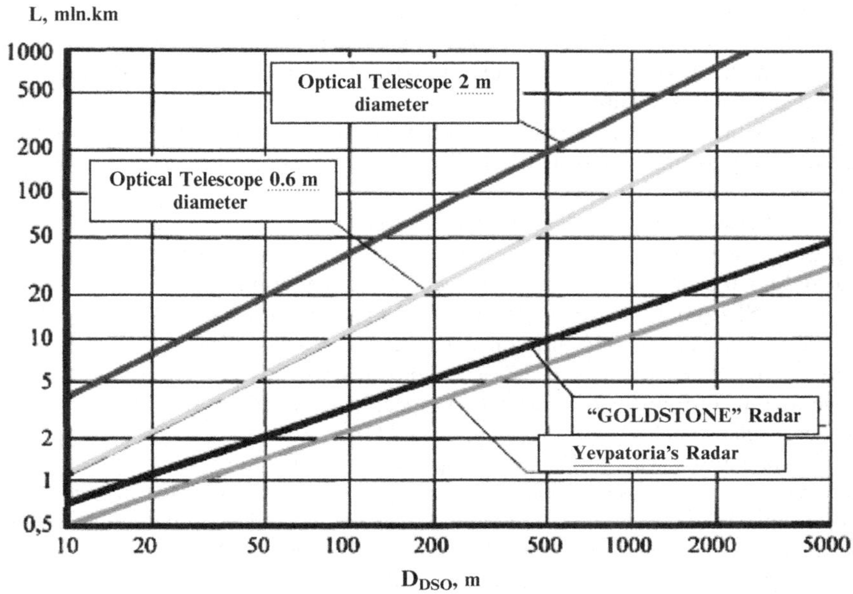

Fig. 3.37 Maximum asteroid detection range for optical telescopes and radars (D_{DSO} – object size, L – detection range)

no real threat for the planet. Astronomers use the term "minimum orbit intersection distance," which is defined as the minimum distance between a potentially hazardous object and the Earth, both occupying a "random" position with regard to each other in their respective orbits at any given moment. Potentially hazardous are space objects at a distance less than 0.05 AU (<7.5 million km) from the Earth, with changes in their orbits caused by gravitational interaction over 100 years or more leading to their collision with our planet.

Powerful optical and radio telescopes (radars) are used to detect DSOs. Detection range depends on many factors: DSO dimensions, material and surface structure, DSO sight line position in relation to the direction of the Sun, and characteristics of the detection facilities used. Figure 3.37 shows the maximum DSO detection range for the Goldstone and Yevpatoria (Ukraine) radars and optical telescopes 0.6–2 m in diameter. The asteroid phase is accepted to be "full Moon," albedo 0.1. Analysis shows that optical telescopes make it possible to detect DSOs 7.6–38 million km away from the Earth (DSO diameter is 20–100 m) and to roughly estimate their trajectories. When a DSO is 1.1–3.2-million km from the Earth, radar stations may be used to determine its trajectory within the accuracy limits required.

According to Russian specialists, a dangerous space object must be either shifted from its trajectory or disintegrated into small fragments. A 20–100 m DSO can be detected about 30–130 h before its approach to the Earth (when the DSO-to-Earth approach velocity is within 70 km/s). During this time an interceptor starts at an

11 km/s initial velocity and reaches a 300,000–700,000 km altitude. In this case, about 1.6–1.7 km/s additional velocity is required to be imparted to a DSO to deflect it from the trajectory by about 7,000 km (slightly more than the Earth's radius). However, it is not possible to impart to a DSO a more than 1–10-m/s incremental velocity without its disintegration. So, when intercepting a DSO of less than 100 m the only way to eliminate dangerous effects for ground objects is to destroy such a DSO into small fragments.

The Russian designers of the Earth protection system suggest destroying dangerous space objects using so-called "kinetic star-shaped penetrators" (KSP). According to experimental and modeling data obtained from such systems, a set of several KSPs (total mass of up to 10 tons, impact speed 30 km/s) is required to destroy a DSO of 100–150 m in diameter. As present-day launch vehicles cannot carry payloads heavier than 10–20 tons, in order to destroy larger DSOs it would be necessary to use more powerful and compact, in comparison with KSPs, attack means, such as nuclear explosive devices (NED).

By comparison with NED, KSP is capable of disintegrating DSOs with considerably lower power consumption. According to Russian specialists, in a contact NED explosion only about 10% of explosion energy penetrates a DSO (most of the depth energy is spent for DSO material heating: about 0.1% of DSO mass evaporates, 1% melts, and within 3–7% is thermally damaged) and the remaining 90% is scattered in space. The explosion seismic wave is formed with just 1% of full explosion energy. On KSP impact and deep penetration the DSO is damaged along the full length of the hole, and material evaporation and plasma initiation are minimal. Because of the above features the KSP efficiency is by more than two orders of magnitude higher than that of NED (thus, either 140 kt-power NED or 1 kT-power KSP would be required to disintegrate a DSO of 100 m diameter).

To eliminate the threat of large DSO impact on the Earth, it is necessary not only to disintegrate DSOs into small (1–3 m) fragments, but also to provide an average distance between them of no less than ten times as large as their diameter during their atmospheric entry. Table 3.5 shows the altitude required for KSP–DSO interception as a function of the DSO diameter. One percent of full KSP impact energy (KSP mass is 6 tons) is accepted to be spent for DSO fragment separation (DSO material density is 2,000 kg/m3). The same Table also gives the interceptor's initial velocity at a 200-km altitude required for the interceptor to fly along a semielliptical trajectory to the point of impact with a DSO at the specified altitude, and the interceptor's flight duration. In this case, the flight duration includes the time of interceptor injection into a 200-km reference orbit, interceptor flight along

Table 3.5

DSO diameter, m	<10	20	30	40	70	100	150	300
Minimum DSO interception altitude, thou. km	0.2	0.3	0.8	1.7	8.3	23.8	80.0	639.1
Initial speed, km/s	7.78	7.81	7.95	8.17	9.15	9.97	10.61	10.95
Duration of flight, hours	2.21	2.28	2.41	2.34	2.99	4.92	15.33	259.0

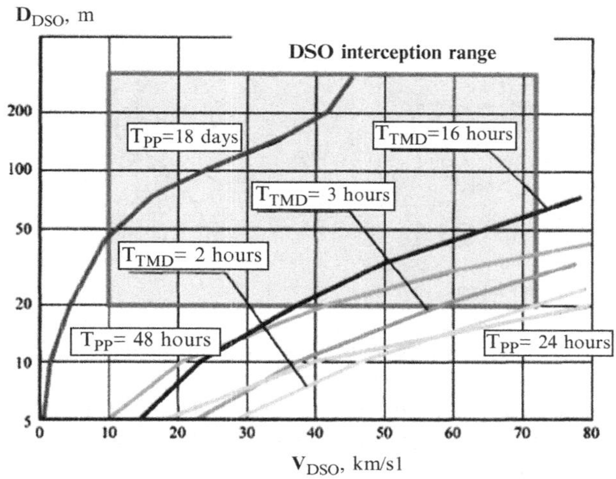

Fig. 3.38 DSO interception possibilities (T_{PP} – prelaunch preparation period; T_{TMD} – target and mission data specification period)

the reference orbit from the orbit injection point to that of transfer to the DSO interception trajectory (1.47 h per circuit), and interceptor flight along a semielliptical trajectory to the point of impact with the DSO.

The analysis of the data above indicates that DSOs of up to 150 m in diameter can by intercepted by KSP at an 80,000 km altitude. In this case, the initial velocity of the interceptor is within 10.6–10.7 km/s, and flight duration does not exceed 1 day.

Figure 3.38 illustrates estimates of a minimum diameter DSO that can be intercepted with prelaunch preparations (from the moment of DSO detection to LV launch) of 18, 2, and 1 days and a target and mission data specification period (from the moment of DSO detection with radar facilities to LV launch) of 6, 3, and 2 h. The analysis of these data shows that interception of 20–300 m DSOs at distances up to 100,000 km would require the launch vehicle to be in a high degree of readiness (prelaunch preparation of about 1–2 days).

Asteroids and comet nuclei rotate during flight, i.e., their position in space is influenced not only by gravitational interaction with the planets, but also by gyroscopic effect and conservation of the rotational momentum vector. This is why the vector of the impulse that will strike a DSO off a dangerous trajectory should pass through its center of gravity in the optimum direction. If such impulse is produced by several impact modules, their resultant force should meet this condition. The points of applying deorbiting forces should be distributed evenly across a DSO. NEDs exploding near a dangerous object ("soft" explosion) will be more effective in pushing a DSO off a dangerous path, even though their impulse will be smaller than that of a contact explosion. In order to meet these requirements, a launch vehicle and the impact modules should be provided with an effective

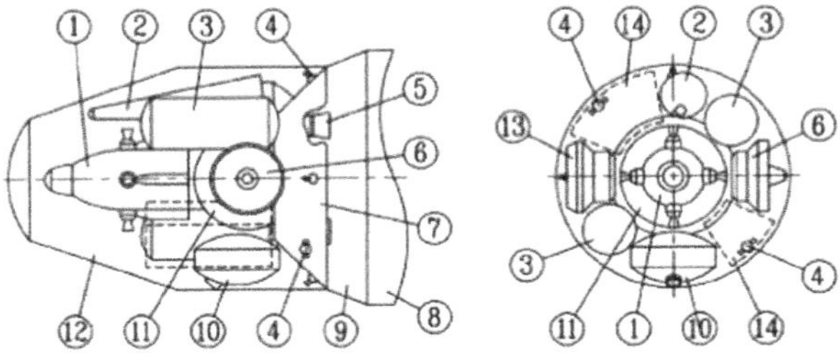

Fig. 3.39 Space explorer layout

control system that will guarantee delivery to the designated point and synchronous operation of the impact modules with high accuracy.

Research conducted by the project authors shows that the DSO interception task can be effectively solved using existing and future Russian rocket systems ("Soyuz-2" and "Rus'-M"), provided two types of specialized probes are created: a so-called space explorer (for studying the structure and characteristics of the dangerous asteroid) and a universal space interceptor (USI) for impacting the asteroid.

The space explorer (Fig. 3.39) will consist of a power plant (7) with the instrument console including an astronavigation system astro-corrector (10), control equipment and power supply system (3), liquid propellant engine made up of a combined tank (11), steering engines (4) and high-thrust engines (5); a set of optical guidance and asteroid exploration systems (6), a set of radio command and radio telemetric systems and systems for transmitting obtained data to the Earth (13). Placed near the center of gravity (1) will be a separable self-guidance impact penetrator (IP). All other equipment will be placed in the control sections (3), as part of the optical systems (6), and in free spaces (14). The space explorer will have an adapter (9) for docking with the launch vehicle (8) and will be protected by an aerodynamic fairing (12) during the flight through the atmosphere.

The space explorer will be used mainly for the following purposes: flying to the area of the dangerous asteroid along the preset trajectory with corrections effected by radio, if necessary; searching for and detecting the asteroid; correcting the approach trajectory in self-guidance mode; separating the probe and guiding it to the selected point on the surface of the DSO; receiving data from the probe after penetration into the DSO; conducting remote sensing of the surface and physical properties of the matter retrieved from the asteroid (solid fragments, vapors, plasma) and transmitting the results to the Earth.

The impact penetrator (IP) is a guided shell weighing 500 kg (400-kg copper penetrator proper, with a correction engine, special equipment, and communication means making up the rest of the weight) in the form of an elongated thick-walled pointed container complete with its own guidance system. The penetrator is made mainly of copper, the spectral lines of which can easily be distinguished from those of the asteroid. Four correction engines in cruciform configuration are placed inside the container near the IP center of gravity. The cone houses the instrument bay with

Table 3.6

Space explorer's functional tasks	Special onboard equipment
Photographing the surface of the asteroid, long-range guidance, short-range guidance for the probe to direct it to the selected area on the surface of the asteroid, documenting the approach, penetration, crater formation, and release of matter	Multispectral long-range CCD video camera
Photographing the dispersion of the matter from the asteroid and documenting the spread of the impact wave across it	Multispectral short-range CCD video camera
Registering radiation spectrums of asteroid surface matter evaporation	Image spectrometer
Measuring the asteroid's magnetic field	Magnetometer
Registering radiation on the surface of the asteroid for determining the chemical and isotopic composition of its matter	Fluoroscope and neutron detector

the self-guidance system equipment, and the rear part of the module includes sensors for registering gas (plasma) parameters after asteroid penetration and determining braking dynamics, the spin engine that stabilizes the angular position of the probe after its separation from the space explorer, as well as communication equipment.

A standard DSO exploration scenario is as follows. The impact produced by the probe striking an asteroid at a speed of 10–20 km/s releases energy that is equivalent to an explosion of 6–25 tons of TNT. Depending on the material and structure of the asteroid, this may produce a crater up to 100 m deep. The passing space explorer conducts both standard celestial object parameter registration in several spectral ranges, measuring gravitational, magnetic, and thermal fields, and monitors the probe as it approaches and strikes the asteroid, while analyzing the chemical composition of the rock retrieved from the asteroid, and the impact wave produced. For the space explorer to be able to conduct all the necessary observations and measurements, it has to be provided with special equipment as shown in Table 3.6.

Limited information on the results obtained during the Deep Impact experiment raise some doubts about the use of a heavy penetrator to study the structure of a comet nucleus because it takes different types of such probes to explore the structure of asteroids and comets. The use of lighter penetrators than those in the Deep Impact program and the Russian proposals concerning their telemetric types will make it possible to reduce the weight of the space explorer.

It has been proposed that the universal space interceptor be built by the modular principle: one command-impact module and several separable self-guidance impact penetrators. The interceptor is flown to intersect the flight trajectory of the approaching asteroid to either fragment it or change its orbit so it misses the Earth. It is assumed that objects up to 150 m in size will be destroyed by a penetrator upon single impact. Larger objects will require several simultaneous strikes by a group of impact modules, each of which will have its own optimum aiming point.

In order to determine the number of impact modules to be carried by the universal space interceptor and work out a strategy for using it, a space explorer should be sent to the dangerous asteroid in advance. It will determine the target points for the penetrators and conduct telemetric and spectroscopic analysis upon impact in order to determine the physical, chemical, and mechanical properties of the asteroid matter. The interceptor and its penetrator have engines and control systems that guarantee interception of the asteroid and acquisition of information upon impact at a set distance from the asteroid. The design of the space explorer allows it to effectively study approaching asteroids at distances of up to one million kilometers.

Figure 3.40 shows one of the possible command-impact module layouts. Length 2.3 m, diameter 1.8 m, dry mass (without fuel) 1,400 kg, and fuel capacity up to 600 kg.

Figure 3.41 shows one of the possible penetrator layouts. Length 1.5 m, diameter 3.0 m, total mass 1,625 kg (fuel capacity 225 kg).

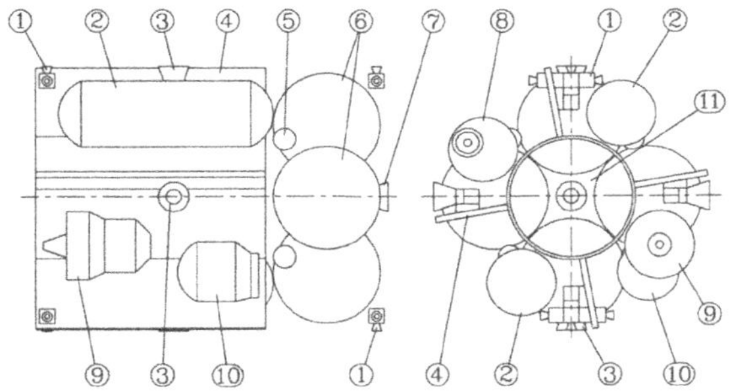

Fig. 3.40 Command – impact module layout

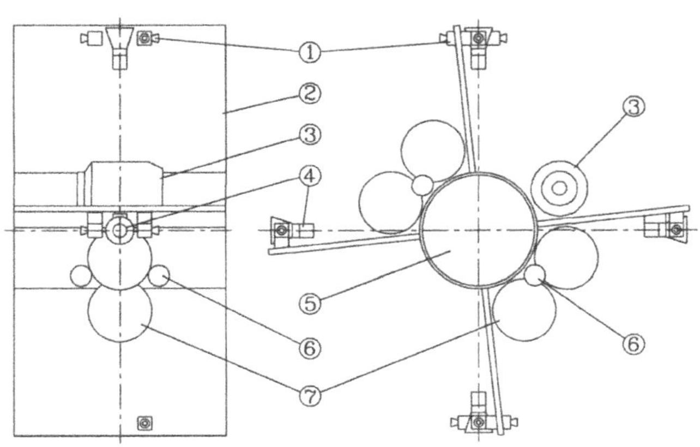

Fig. 3.41 Penetrator layout

Table 3.7

Universal space interceptor (USI)	USI-6	USI-5	USI-3	USI-1	USI-nuclear
Number of impact modules	5	4	2	none	
USI weight, kg	10,054	8,304	4,891	1,590	2,757
USI length, m	10.05	8.5	5.4	2.3	
USI diameter, m	3.0			1.8	
KSP Weight, kg	7,250	5,950	3,350	750	
NED weight, kg	non				1,127

A shaped charge placed on the impact module increases the impact on the asteroid, the upper layers of which are destroyed by an explosion at about one meter above the surface in order to ensure deeper penetration. An explosive device in the impact module may be necessary when intercepting large celestial bodies or those composed of super strong materials (iron and iron-stone asteroids). A space interceptor may carry up to five impact modules, a nuclear explosive device, and an emergency descent safety system (activated in an emergency during LV prelaunch preparations and in flight to ensure a soft (parachute) touchdown).

Table 3.7 contains the main characteristics of USIs. When assessing the USI fuel supply required, the USI-DSO delivery accuracy was accepted to be within 25 km. At an interception altitude of 80,000 km, the USI-DSO approach velocity is 70 km/s, USI-DSO guidance accuracy (defined as a deviation of the USI-DSO approach velocity vector from DSO direction) is not lower than 0.0001 radian, provided that the accuracy of self-guidance is not lower than 0.001 radian. In order to effectively neutralize the most dangerous space objects (20–300 m in diameter), a USI should carry at least two impact modules.

The system that controls USI transportation to the interception orbit employs the principles of flexible control adaption for the launch vehicle, the USI and impact module onboard systems that are integrated into one whole based on function. The command and computing core in the upper stage of the launch vehicle controls the flight in all sections of the impact vehicle flight trajectory by activating appropriate control algorithms for related systems. The use of these principles will guarantee USI delivery to the target orbits in inertial mode with accuracy 1.5–2 times higher than that for existing launch vehicles and boosters, injection reliability of 0.995, and confidence level of no less than 0.9. The use of data from star trackers in the control system will further increase (by 30% and more) injection accuracy as compared to inertial mode. The development of Russia's new "Rus-M" launch system will allow a USI to carry up to five impact modules and destroy (deflect) DSOs up to 700 m in size.

According to Academician V.P. Makeyev State Rocket Center specialists, the proposed planetary protection system will reduce the risk of DSO approach to the Earth from 0.3535 now to ~ 0.007. The cost of building such space vehicles and adapting them to existing and future rocket systems would be incomparably smaller than the potential damage that even small asteroids may cause to the Earth.

3.3 Other Global Risks and Threats in Space and from Outer Space

Among some other global risks and threats to our Planet we have to mention three types of danger from outer space: solar variations, space ecology (the *space debris* problem), and anomaly events, such as UFO phenomena. Some of them may have a futuristic character and at first glance are unconnected with the global security issues considered in this book. However, the authors intend to draw the attention of the readers to the phenomena as some of them could provoke both natural and man-made disasters in the near future.

Let's begin our review with solar activity and solar cycle variations.

3.3.1 Risks and Threats from Solar Activity and Solar Cycle Variations

Solar variation refers here to changes in the amount of total solar radiation and its spectral distribution. There are periodic components to these variations, the principal one being the 11-year solar cycle (or sunspot cycle), as well as aperiodic fluctuations (Fig. 3.42) [32]. Solar activity has been measured by satellites during recent decades and estimated using "proxy" variables in prior times. Scientists studying climate change are interested in understanding the effects of variations in the total and spectral solar irradiance on the Earth and its climate.

The variations in total solar irradiance (TSI) remained at or below the threshold of detectability until the satellite era, although the small fraction in ultraviolet wavelengths varies by a few percent. Total solar output is now measured to vary (over the last three 11-year sunspot cycles) by approximately 0.1% [33–35] or about 1.3 W/m peak-to-trough during the 11 year sunspot cycle. The amount of solar

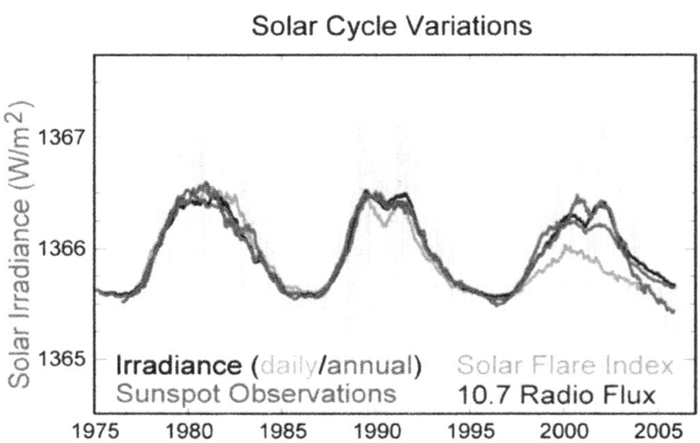

Fig. 3.42 One estimate of the last 30 years of solar variability

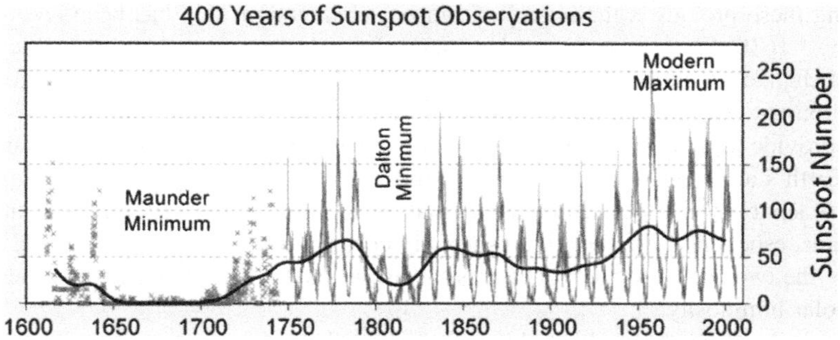

Fig. 3.43 Prehistory of sunspot numbers

radiation received at the outer surface of the Earth's atmosphere averages 1,366 watts per square meter (W/m) [36–38]. There are no direct measurements of the longer-term variation and interpretations of proxy measures of variations differ. The intensity of solar radiation reaching the Earth has been relatively constant throughout the last 2,000 years, with variations of around 0.1–0.2% [39–41]. The combination of solar variation and volcanic effects are likely to have contributed to climate change, for example during the Maunder Minimum (Fig. 3.43). Apart from solar brightness variations, more subtle solar magnetic activity influences on climate from cosmic rays or the Sun's ultraviolet radiation cannot be excluded although confirmation is not at hand since physical models for such effects are still too poorly developed [42].

Around 1900, researchers began to explore connections between solar variations and weather on the Earth. Of particular note is the work of Charles Greeley Abbot. Of note is his detection of 27 harmonic periods within the 273-month Hale cycles, including 7, 13, and 39 month patterns. He looked for connections to weather by means such as matching opposing solar trends during a month to opposing temperature and precipitation trends in cities. With the advent of dendrochronology, scientists such as Waldo S. Glock attempted to connect variation in tree growth to periodic solar variations in the extant record and infer long-term secular variability in the solar constant from similar variations in millennial-scale chronologies [43].

Statistical studies that correlate weather and climate with solar activity have been popular for centuries, dating back at least to 1801, when William Herschel noted an apparent connection between wheat prices and sunspot records [44] They now often involve high-density global datasets compiled from surface networks and weather satellite observations and/or the forcing of climate models with synthetic or observed solar variability to investigate the detailed processes by which the effects of solar variations propagate through the Earth's climate system [45].

Direct irradiance measurements have only been available during the last three cycles and are based on a composite of many different observing satellites [46, 47]. However, the correlation between irradiance measurements and other proxies of solar activity make it reasonable to estimate past solar activity. Most important among these proxies is the record of sunspot observations that has been recorded since ~ 1610. Since sunspots and associated faculae are directly responsible for small changes in the brightness of the sun, they are closely correlated to changes in solar output. Direct measurements of radio emissions from the Sun at 10.7 cm also provide a proxy of solar activity that can be measured from the ground since the Earth's atmosphere is transparent at this wavelength. Lastly, solar flares are a type of solar activity that can impact human life on the Earth by affecting electrical systems, especially satellites. Flares usually occur in the presence of sunspots, and hence the two are correlated, but flares themselves make only tiny perturbations of the solar luminosity.

Recently, it has been claimed that the total solar irradiance is varying in ways that are not duplicated by changes in sunspot observations or radio emissions. However, this conclusion is disputed. Some scientists believe that shifts in irradiance may be the result of calibration problems in the measuring satellites [48, 49]. These speculations also admit the possibility that a small long-term trend might exist in solar irradiance [50].

Sunspots are relatively dark areas on the radiating "surface" (photosphere) of the Sun where intense magnetic activity inhibits convection and cools the photosphere. Faculae are slightly brighter areas that form around sunspot groups as the flow of energy to the photosphere is re-established and both the normal flow and the sunspot-blocked energy elevate the radiating "surface" temperature. Scientists have speculated on possible relationships between sunspots and solar luminosity since the historical sunspot area record began in the seventeenth century (Fig. 3.44) [51, 52]. Correlations are now known to exist with decreases in luminosity caused by sunspots (generally <−0.3%) and increases (generally <+0.05%) caused both by faculae that are associated with active regions as well as the magnetically active "bright network" [53]. Modulation of the solar luminosity by magnetically active regions was confirmed by satellite measurements of total solar irradiance by the ACRIM1 experiment on the Solar Maximum Mission (launched in 1980) [53].

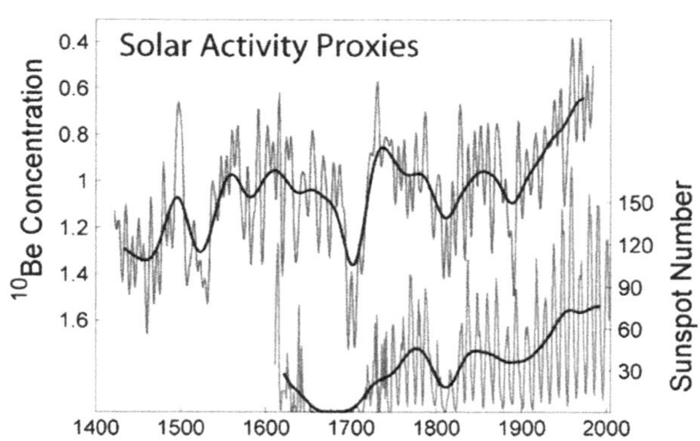

Fig. 3.44 Proxies of solar activity, including changes in sunspot number and cosmogenic isotope production

The modulations were later confirmed in the results of the ERB experiment launched on the Nimbus 7 satellite in 1978. Sunspots in magnetically active regions are cooler and "darker" than the average photosphere and cause temporary decreases in TSI of as much as 0.3%. Faculae in magnetically active regions are hotter and "brighter" than the average photosphere and cause temporary increases in TSI. The net effect during periods of enhanced solar magnetic activity is increased radiant output of the sun because faculae are larger and persist longer than sunspots.

Various studies have been made using sunspot number (for which records extend over hundreds of years) as a proxy for solar output (for which good records only extend for a few decades). Also, ground instruments have been calibrated by comparison with high-altitude and orbital instruments. Researchers have combined present readings and factors to adjust historical data. Other proxy data – such as the abundance of cosmogenic isotopes – have been used to infer solar magnetic activity and thus likely brightness.

Sunspot activity has been measured using the Wolf number for about 300 years. This index (also known as the Zürich number) uses both the number of sunspots and the number of groups of sunspots to compensate for variations in measurement. A 2003 study by Ilya Usoskin of the University of Oulu, Finland found that sunspots had been more frequent since the 1940s than in the previous 1,150 years [54].

Sunspot numbers over the past 11,400 years have been reconstructed using dendrochronologically dated radiocarbon concentrations. The level of solar activity during the past 70 years is exceptional – the last period of similar magnitude occurred over 8,000 years ago. The Sun was at a similarly high level of magnetic activity for only ~ 10% of the past 11,400 years, and almost all of the earlier high-activity periods were shorter than the present episode [55].

Different cyclic changes exist in the behavior of the Sun (Fig. 3.45). Although many possible patterns have been suggested (87, 210, 2,300, and 6,000 years) only the 11 and 22 year cycles are clear in the observations.

The most well-known *11 year cycle* is named after Heinrich Schwabe (Schwabe cycle). Obviously it is a gradual increase and more rapid decrease of the number of

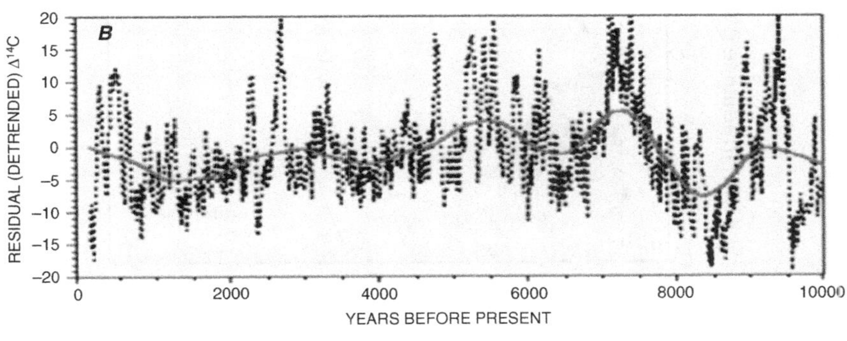

Fig. 3.45 Solar variation 2,300 year cycles

sunspots over a period ranging from 9 to 12 years. Differential rotation of the Sun's convection zone (as a function of latitude) consolidates magnetic flux tubes, increases their magnetic field strength, and makes them buoyant. As they rise through the solar atmosphere they partially block the convective flow of energy, cooling their region of the photosphere, causing "sunspots." The Sun's apparent surface, the photosphere, radiates more actively when there are more sunspots. Satellite monitoring of solar luminosity since 1980 has shown there is a direct relationship between the solar activity (sunspot) cycle and luminosity with a solar cycle peak-to-peak amplitude of about 0.1% [33]. Luminosity has also been found to decrease by as much as 0.3% on a 10 day timescale when large groups of sunspots rotate across the Earth's view and increase by as much as 0.05% for up to 6 months due to faculae associated with the large sunspot groups [53].

Another well-known, *22-year solar cycle* is the Hale cycle, named after George Ellery Hale. The magnetic field of the Sun reverses during each Schwabe cycle, so the magnetic poles return to the same state after two reversals. The sensitivity of climate to cyclical variations in solar forcing will be higher for longer cycles due to the thermal inertia of the ocean, which acts to dampen high frequencies. Scafetta and West (2005) found that the climate was 1.5 times as sensitive to 22-year cyclical forcing relative to 11-year cyclical forcing, and that the thermal inertial induced a lag of approximately 2.2 years in cyclic climate response in the temperature data [56]. Solar irradiance (Fig. 3.46) or insolation, is the amount of sunlight which reaches the Earth. The equipment used might measure optical brightness, total radiation, or radiation in various frequencies. Historical estimates use various measurements and proxies. There are two common meanings of solar irradiance of the Earth and its surface: the radiation reaching the upper atmosphere and some point within the atmosphere, including the surface. Various gases within the

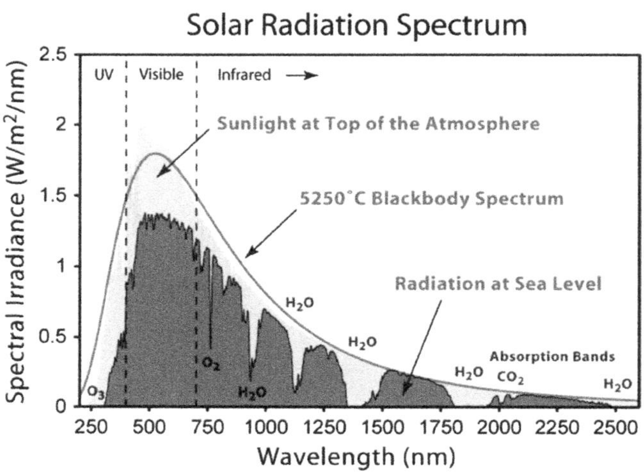

Fig. 3.46 Solar irradiance spectrum above the atmosphere and at the surface

atmosphere absorb some solar radiation at different wavelengths, and clouds and dust also affect it. Measurements above the atmosphere are needed to determine variations in solar output, to avoid the confounding effects of changes within the atmosphere. There is some evidence that sunshine at the Earth's surface has been decreasing in the last 50 years (see global dimming) possibly caused by increased atmospheric pollution, whilst over roughly the same timespan solar output has been nearly constant.

There are several hypotheses for how solar variations may affect the Earth. Some variations, such as changes in the size of the Sun, are presently only of interest in the field of astronomy:

- Total solar irradiance changes slowly on decadal and longer timescales.
- The variation during recent solar magnetic activity cycles has been about 0.1% (peak-to-peak) [33].
- Variations corresponding to solar changes with periods of 9–13, 18–25, and >100 years have been detected in sea-surface temperatures.
- Different composite reconstructions of total solar irradiance observations by satellites show different trends since 1980; see the global warming section below.
- Ultraviolet irradiance (EUV) varies by approximately 1.5% from solar maxima to minima, for 200–300 nm UV [57].
- Energy changes in the UV wavelengths involved in production and loss of ozone have atmospheric effects. The 30-hPa atmospheric pressure level has changed height in phase with solar activity during the last four solar cycles. UV irradiance increase causes higher ozone production, leading to stratospheric heating and to poleward displacements in the stratospheric and tropospheric wind systems.
- A proxy study estimates that UV has increased by 3% since the Maunder Minimum.

There are some changes in the solar wind and the Sun's magnetic flux. A more active solar wind and stronger magnetic field reduces the cosmic rays striking the Earth's atmosphere (Fig. 3.47). Variations in the solar wind affect the size and intensity of the heliosphere, the volume larger than the Solar System filled with

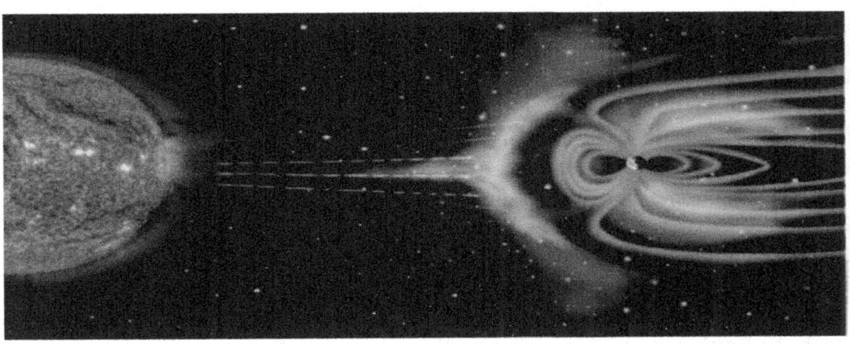

Fig. 3.47 Solar particles interact with the Earth's magnetosphere

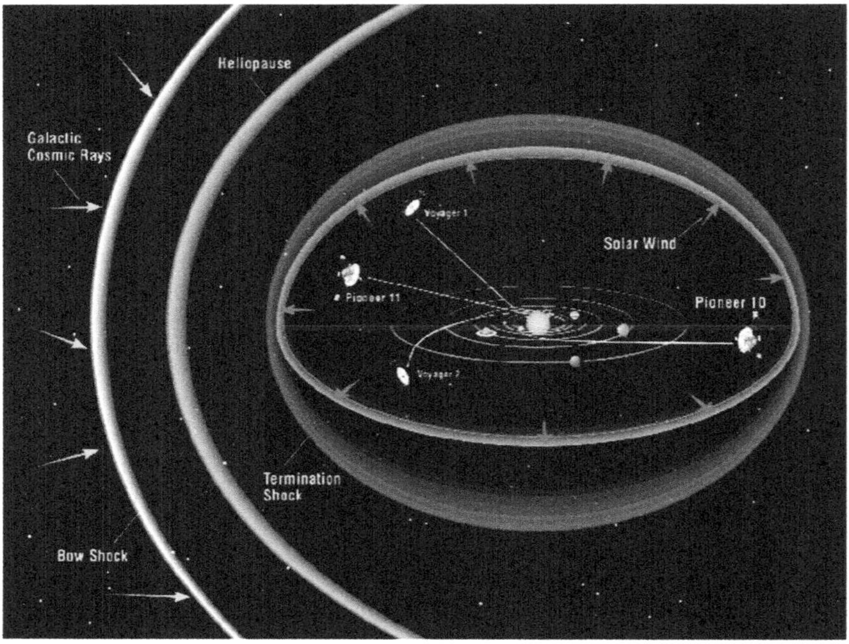

Fig. 3.48 Solar wind and the magnetic field create a heliosphere around the solar system

solar wind particles (Fig. 3.48). Cosmogenic production of ^{14}C, ^{10}Be, and ^{36}Cl show changes tied to solar activity. Cosmic ray ionization in the upper atmosphere does change, but significant effects are not obvious. As the solar coronal-source magnetic flux doubled during the past century, the cosmic-ray flux has decreased by about 15%. The Sun's total magnetic flux rose by a factor of 1.41 from 1964 to 1996 and by a factor of 2.3 since 1901.

From the point of view of disaster predictions there are some effects of solar activity on clouds: cosmic rays have been hypothesized to affect formation of clouds through possible effects on production of cloud condensation nuclei. Observational evidence for such a relationship is inconclusive. Moreover, 1983–1994 data from the International Satellite Cloud Climatology Project (ISCCP) showed that global low cloud formation was highly correlated with cosmic ray flux; subsequent to this the correlation breaks down [58].

There exist some other effects due to solar variation. These include interaction of solar particles, the solar magnetic field, and the Earth's magnetic field, and they cause variations in the particle and electromagnetic fields at the surface of the planet. Extreme solar events can affect electrical devices. Weakening of the Sun's magnetic field is believed to increase the number of interstellar cosmic rays which reach the Earth's atmosphere, altering the types of particles reaching the surface.

It has been speculated that a change in cosmic rays could cause an increase in certain types of clouds, affecting the Earth's albedo.

Among geomagnetic effects due to solar variations there are the following. First of all the Earth's polar aurorae become visible and displays signs, created by interactions between the solar wind, the solar magnetosphere, the Earth's magnetic field, and the Earth's atmosphere. Variations in any of these affect aurora displays. Sudden changes can cause the intense disturbances in the Earth's magnetic fields which are called geomagnetic storms. Moreover, energetic solar protons can reach Earth within 30 min of a major flare's peak. During such a solar proton event, the Earth is showered with energetic solar particles (primarily protons) released from the flare site. Some of these particles spiral down Earth's magnetic field lines, penetrating the upper layers of our atmosphere where they produce additional ionization and may produce a significant increase in the radiation environment.

Among the last, but not least are the galactic cosmic rays (GCR) which are deflected due to solar variations. An increase in solar activity (more sunspots) is accompanied by an increase in the "solar wind," which is an outflow of ionized particles, mostly protons and electrons, from the Sun. The Earth's geomagnetic field, the solar wind, and the solar magnetic field deflect galactic cosmic rays. A decrease in solar activity increases the GCR penetration of the troposphere and stratosphere. GCR particles are the primary source of ionization in the troposphere above 1 km (below 1 km, radon is a dominant source of ionization in many areas). Levels of GCRs have been indirectly recorded by their influence on the production of carbon-14 and beryllium-10. The Hallstatt solar cycle length of approximately 2,300 years is reflected by climatic Dansgaard-Oeschger events. The 80–90-year solar Gleissberg cycles appear to vary in length depending upon the lengths of the concurrent 11-year solar cycles, and there also appear to be similar climate patterns occurring on this time scale (Fig. 3.49).

Cloud effects also directly connected with solar variations. Changes in ionization affect the abundance of aerosols that serve as the nuclei of condensation for cloud formation [59]. As a result, ionization levels potentially affect levels of condensation, low clouds, relative humidity, and albedo due to clouds. Clouds formed from greater amounts of condensation nuclei are brighter, longer lived, and likely to produce less precipitation. Changes of 3–4% in cloudiness and concurrent changes in cloud top temperatures have been correlated to the 11- and 22-year solar (sunspot) cycles, with increased GCR levels during "antiparallel" cycles [60]. Global average cloud cover change has been found to be 1.5–2%. Several studies of GCR and cloud cover variations have found positive correlation at latitudes greater than 50° and negative correlation at lower latitudes [59]. However, not all scientists accept this correlation as statistically significant, and some that do attribute it to other solar variability (e.g., UV or total irradiance variations) rather than directly to GCR changes [61, 62]. Difficulties in interpreting such correlations include the fact that many aspects of solar variability change at similar times, and some climate systems have delayed responses.

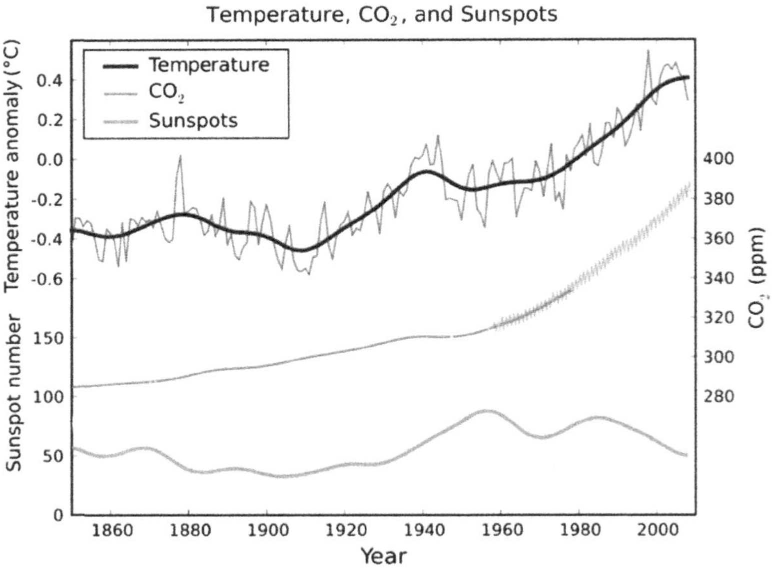

Fig. 3.49 CO_2, temperature, and sunspot activity since 1850

But one of the most interesting effects of solar activity is connected with the very popular topic of public discussion today – global warming. The scientific consensus is that solar variations do not play a major role in determining present-day observed climate change [63]. The Intergovernmental Panel on Climate Change Third Assessment Report (TAR) states that the measured magnitude of recent solar variation is much smaller than the amplification effect due to greenhouse gases [64]. Estimates of long-term solar irradiance changes have decreased since the TAR. However, empirical results of detectable tropospheric changes have strengthened the evidence for solar forcing of climate change. The most likely mechanism is considered to be some combination of direct forcing by changes in total solar irradiance, and indirect effects of ultraviolet radiation on the stratosphere. Least certain are indirect effects induced by galactic cosmic rays [32].

Crucial to the understanding of possible solar impact on terrestrial climate is accurate measurement of solar forcing. Unfortunately, such measurements of incident solar radiation have only been available since the satellite era, and even that is open to dispute: different groups find different values, because of different methods of cross-calibrating essentially the same datasets [33].

The study of sun spot cycles was generally popular through the first half of the twentieth century. Governments had collected a lot of weather data to play with and inevitably people found correlations between sun spot cycles and select weather patterns. If rainfall in England didn't fit the cycle, maybe storminess in New England would. Respected scientists and enthusiastic amateurs insisted they had found patterns reliable enough to make predictions. Sooner or later though every prediction failed. An example was a highly credible forecast of a dry spell in Africa during the sunspot minimum of the early 1930s. When the period turned out to be wet, a meteorologist later recalled "the subject of sunspots and weather relationships fell into dispute, especially among British meteorologists who witnessed the discomfiture of some of their most respected superiors." Even in the 1960s he said, ("For a young [climate] researcher to entertain any statement of sun-weather relationships was to brand oneself a crank" [65]).

Today it is well known that variations of solar activity influence both human life (including variety of social activities) and technogenic systems too. A lot of research exists that confirm that solar variations cause natural as well as man-made disasters. Moreover, solar activity influences the behavior of the humans responsible for managing and controlling complex technical systems and facilities. That is why we believe it is very important to monitor solar activity variations in connection with disaster management.

3.3.2 Risks and Threats from Space Junk (Debris)

Another matter of contemporary concern is the space debris problem. Space debris, also known as orbital debris, space junk, and space waste, is the collection of objects in orbit around the Earth that were created by humans but no longer serve any useful purpose (Fig. 3.50). These objects consist of everything from spent rocket stages and defunct satellites to explosion and collision fragments. The debris can include slag and dust from solid rocket motors, surface degradation products

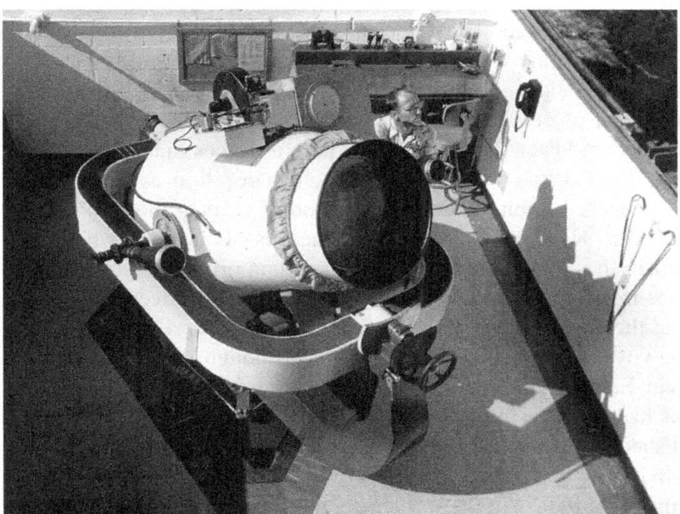

Fig. 3.50 Contemporary optical telescope for space junk observations

such as paint flakes, coolant released by RORSAT nuclear-powered satellites, clusters of small needles, and objects released due to the impact of micrometeoroids or fairly small debris onto spacecraft [66]. As the orbits of these objects often overlap the trajectories of spacecraft, debris is a potential collision risk.

The vast majority of the estimated tens of millions of pieces of space debris are small particles, like paint flakes and solid rocket fuel slag. Impacts from these particles cause erosive damage, similar to sandblasting. The majority of this damage can be mitigated through the use of a technique originally developed to protect spacecraft from micrometeorites, by adding a thin layer of metal foil outside of the main spacecraft body. Impacts take place at such high velocities that the debris is vaporized when it collides with the foil, and the resulting plasma spreads out quickly enough that it does not cause serious damage to the inner wall. However, not all parts of a spacecraft may be protected in this manner, i.e., solar panels and optical devices (such as telescopes, or star trackers), and these components are subject to constant wear by debris and micrometeorites.

The present means for spacecraft shielding, such as those used for the manned modules of the International Space Station (ISS), are only capable of protecting against debris with diameters below about 1 cm (0.39 in). The only remaining means of protection would be to maneuver the spacecraft in order to avoid a collision. This, however, requires that the orbit of the respective object be precisely known. The current equipment used to gather such information is only capable of tracking objects down to about 5 cm (2.0 in) diameter in low Earth orbit, and about 50 cm (20 in) in geosynchronous orbit. Out of the estimated 600,000 objects [66] above 1 cm (0.39 in) diameter, only 19,000 can be tracked as of today. This leads to wide uncertainties in the estimated quantities of debris, and the predicted path of their orbits.

If a collision with larger debris does occur, many of the resulting fragments from the damaged spacecraft will also be in the 1 kg (2.2 lb) mass range, and these objects become an additional collision risk. As the chance of collision is a function of the number of objects in space, there is a critical density where the creation of new debris occurs faster than the various natural forces that remove these objects from orbit. Beyond this point a runaway chain reaction can occur that quickly reduces all objects in orbit to debris in a period of years or months. This possibility is known as the "Kessler Syndrome," and there is debate as to whether or not this critical density has already been reached in certain orbital bands.

As space missions moved out from the Earth and into deep space, the question arose about the environment of the asteroid belt, which probes would have to pass through on voyages to the outer solar system. Although Whipple had demonstrated that the near-Earth environment was not a problem for space travel, the same depth of analysis had not been applied to the belt. Into this void stepped Donald Kessler, who published a series of papers, starting in late 1968, estimating the density of asteroids [67].

Since the earliest days of the space race, the North American Aerospace Defense Command (NORAD) and its Russian prototype – the Outer space Control Center (Ts K K P) – had maintained a database of all known rocket launches and the various objects that reach orbit as a result – not just the satellites themselves, but the aerodynamic shields that protected them during launch, upper stage booster rockets that placed them in orbit, and in some cases, the lower stages as well. Known as the Space Object Catalog when it was created with the launch of Sputnik in 1957, NASA later started publishing data-massaged versions of the database in the now common two-line element set format [68].The trackers that fed the database were also aware of a number of other objects in orbit, many of which were the results of on-orbit explosions. Some of these were deliberately caused as a part of 1960s antisatellite weapon (ASAT) testing, while others were the result of rocket boosters that had "blown up" in orbit as leftover fuel expanded into a gas and ruptured their tanks.

A lack of good data about the debris problem prompted a series of studies to better characterize first of all the LEO environment. In October 1979 NASA provided Kessler with additional funding for further studies of the problem. Several approaches were used by these studies. Some used optical telescopes or short-wavelength radar to more accurately measure the number and size of objects in space. The optical measurements alone demonstrated that the published population count was too low by at least 50%. Before this it was believed that the NORAD database was essentially complete and accounted for at least the majority of large objects in orbit. These measurements demonstrated that the NORAD list deliberately eliminated some objects (typically U.S. military spacecraft), could not easily account for objects under 20 cm (7.9 in) in size, and didn't bother to track many others because they were considered unimportant. In particular, the debris left over from exploding rocket boosters and several 1960s antisatellite tests were only tracked in a haphazard way in the main database [69].

Other studies used microscopes to study spacecraft that had returned to the Earth, looking for impacts that had already taken place and had gone unnoticed. Sections of Skylab and the Apollo CSMs (Command and Service modules) that had been recovered in the 1960s and 1970s were shown to be heavily pitted by debris. To everyone's surprise, every study demonstrated that the debris flux was much higher than expected, and that debris was already the primary source of collisions in space. LEO was already suffering from the Kessler Syndrome (this will be considered below), as originally defined [69]. Later refinements were added as the result of the return of Solar Max, the Long Duration Exposure Facility, numerous Space Shuttle missions, and similar spacecraft studies. Similar studies continue to this day.

One discovery that was particularly disconcerting was that 42% of all cataloged debris was the result of only 19 events, explosions of spent rocket stages, mostly from U.S. Deltas. During the same period, an experimental program was run to determine what would happen if debris collided with satellites or other debris. The study demonstrated that the process was entirely unlike the micrometeor case, and that many large chunks of debris would be created that would themselves be a collision threat [69]. This leads to a worrying possibility – instead of the density of debris being a measure of the number of items launched into orbit, it was that number plus any new debris caused when they collided. If the new debris did not decay from orbit before impacting another object, the number of debris items would continue to grow even if there were no new launches.

The space debris problem could be categorized into three regimes. With a low enough density the addition of debris through impacts is lower than their rate of decay, and the problem does not become significant. Beyond that is a critical density where additional debris can quickly upset the system and lead to additional collisions. At a high enough density the rate of production is greater than decay rates, leading to a "cascade," or chain reaction, that quickly reduces the on-orbit population to small objects of the order of a few cm in size, making any sort of space activity very hazardous. This worrying possibility became the new use of the term "Kessler Syndrome" [69]. In early 2009, D. Kessler summed up the situation bluntly:

> Aggressive space activities without adequate safeguards could significantly shorten the time between collisions and produce an intolerable hazard to future spacecraft. Some of the most environmentally dangerous activities in space include large constellations such as those initially proposed by the Strategic Defense Initiative in the mid-1980s, large structures such as those considered in the late-1970s for building solar power stations in the Earth orbit, and antisatellite warfare using systems tested by the USSR, the U.S., and China over the past 30 years. Such aggressive activities could set up a situation where a single satellite failure could lead to cascading failures of many satellites in a period of time much shorter than years.

Faced with this potentially worrying scenario, as early as the 1980s NASA and other groups within the US attempted to limit the growth of debris. One particularly effective solution was implemented by McDonnell Douglas on the Delta booster, by moving the boosters away from their payload and then venting any remaining fuel in the tanks. This eliminated the pressure build-up in the tanks that had caused them to explode in the past [70]. Other countries, however, were not as quick to adopt this sort of measure, and the problem continued to grow throughout the 1980s, especially due to a large number of launches in the Soviet Union [70].

A new battery of studies followed as NASA, NORAD, and others attempted to better understand exactly what the environment was like. Every one of these studies adjusted the number of pieces of debris in this critical mass zone upward. In 1981 when Schefter's article was published it was placed at 5,000 objects but a new battery of detectors in the Ground-based Electro-Optical Deep Space Surveillance system quickly found new objects within its resolution. By the late 1990s it was thought that the majority of 28,000 launched objects had already decayed and about 8,500 remained in orbit [70]. By 2005 this had been adjusted upward to 13,000 objects and a 2006 study raised this to 19,000 as a result of an ASAT test and a satellite collision [71].

The population growth has led to intense debate within the community on the nature of the problem and earlier dire warnings. Following Kessler's 1991 derivation, and updates from 2001 [71], the LEO environment within the 1,000-km (620 mi) altitude range should now be within the cascading region. However, only one major incident has occurred: the 2009 satellite collision between Iridium 33 and Cosmos 2251. The lack of any obvious cascading in the short term has led to a

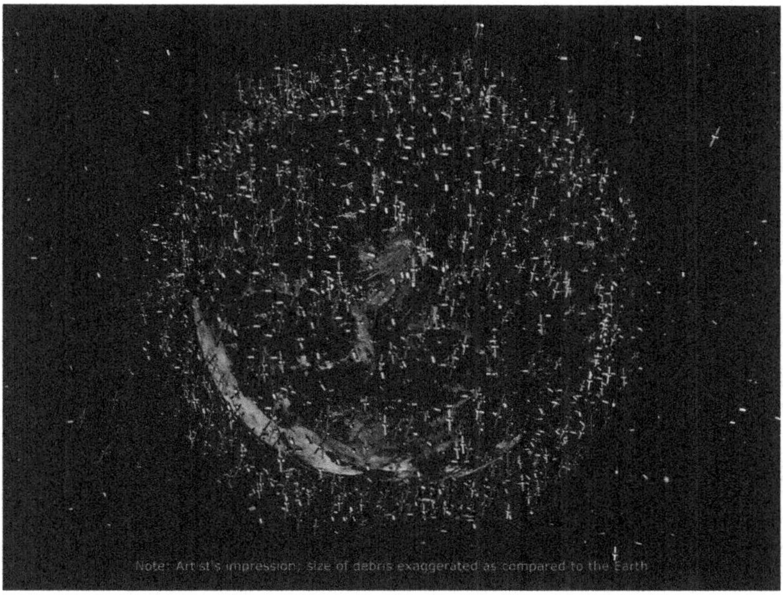

Fig. 3.51 Space ecology today and tomorrow

number of complaints that the original estimates overestimated the issue [72]. Others have pointed out that the start of a cascade would not be obvious until the situation was well advanced, which might take years (Fig. 3.51).

A 2006 NASA model suggested that even if no new launches took place, the environment would continue to contain the then-known population until about 2055, at which point it would increase on its own [73, 74]. Richard Crowther of Britain's Defence Evaluation and Research Agency stated that he believes the cascade will begin some time around 2015 [75]. The National Academy of Sciences, summarizing the view among professionals, noted that there was widespread agreement that two bands of LEO space, 900–1,000 km (620 mi) and 1,500 km (930 mi) altitudes, were already past the critical density [76].

In terms of numbers, the vast bulk of debris consists of smaller objects, 1 cm (0.39 in) or less. In mid-2009 NASA estimated the number of large debris items over 10 cm (3.9 in) at 19,000, between 1 cm and 10 cm (3.9 in) at approximately 500,000, and estimates that debris items smaller than 1 cm (0.39 in) probably exceeds tens of millions [77]. In terms of mass, the vast majority of the overall weight of the debris is concentrated in larger objects, using numbers from 2000, about 1,500 objects weighing more than 100 kg (220 lb) each account for over 98% of the 1,900 tons of debris then known in low earth orbit [78].

Since space debris comes from man-made objects, the total possible mass of debris is easy to calculate: it is the total mass of all spacecraft and rocket bodies that have reached orbit. The actual mass of debris is much lower than that, as a considerable proportion of these objects have since decayed. As debris mass tends to be dominated by larger objects, most of which have long ago been detected, the total mass has remained relatively constant in spite of the addition of many smaller objects. Using the older figure of 8,500 known debris items, the total mass is estimated at 5,500 tons [79].

The space junk problem in LEO is compounded by the fact that there are few "universal orbits" that keep spacecraft in particular rings, as opposed to GEO, a single widely used orbit. The closest would be the sun-synchronous orbits that maintain a constant angle between the sun and orbital plane. But LEO satellites are in many different orbital planes providing global coverage, and the 15 orbits per day typical of LEO satellites results in frequent approaches between object pairs. Since sun-synchronous orbits are polar, the polar regions are common crossing points [80]. Collisions there occur at very high relative velocities, typically several kilometers per second [81]. Such a collision will normally create large numbers of objects in the critical size range, as was the case in the 2009 collision. It is for this reason that the Kessler Syndrome is most commonly applied only to the LEO region. In this region a collision will create debris that will cross other orbits and this population increase leads to the cascade effect.

At the most commonly used Low Earth orbits for manned missions, 400 km (250 mi) and below, residual air drag helps keep the zones clear. Collisions that occur under this altitude are also less of an issue, since they result in fragment orbits having perigee at or below this altitude. The critical altitude also changes as a result of the space weather environment, which causes the upper atmosphere to expand and contract. An expansion of the atmosphere leads to an increased drag to the fragments, resulting in a shorter orbit life time. An expanded atmosphere for some period of time in the 1990s is one reason the orbital debris density remained lower for some time [70]. Another is the rapid reduction in launches by Russia, which conducted the vast majority of launches during the 1970s and 1980s [70].

The space junk problem is especially problematic in the valuable geostationary orbits (GEO), where satellites are often clustered over their primary ground "targets" and share the same orbital path. Orbital perturbations are significant in GEO. Active satellites maintain their station via thrusters, but if they become inoperable they become a collision concern (as in the case of Telstar 401). There has been estimated to be one close (within 50 m) approach per year [82]. On the upside, relative velocities in GEO are low, compared with those between objects in largely random low earth orbits. The impact velocities peak at about 100 m per second (330 ft/s) [83]. This means that the debris field from such a collision is not the same as a LEO collision and does not pose the same sort of risks, at least over the short term. It would, however, almost certainly knock the satellite out of operation. Large-scale structures, like solar power satellites, would be almost certain to suffer major collisions over short periods of time [84]. Additionally, the GEO orbit is too distant to make accurate measurements of the existing debris field for objects under 1 m (3 ft 3 in), so the precise nature of the existing problem is not well known [70]. Some experts have suggested that these satellites be moved to empty spots within GEO, which would require less maneuvering and make it easier to predict future motions. An additional risk is presented by satellites in other orbits, especially those satellites or boosters left stranded in geostationary transfer orbit, which are a concern due to the typically large crossing velocities.

In spite of these efforts at risk reduction, spacecraft collisions have taken place. Moreover, impacts with larger debris normally destroy any spacecraft. To date there have been several known and suspected impact events. The earliest on record was the loss of "Cosmos 1275," which disappeared on 24 July 1981 only a month after launch. Tracking showed it had suffered some sort of breakup with the creation of 300 new objects. The Soviet satellite did not contain any volatiles and is widely assumed to have suffered a collision with a small object. However, proof is lacking, and an electrical battery explosion has been offered as a possible alternative. Another mission "Cosmos 1484" suffered a similar mysterious breakup on 18 October 199 [85]. Several confirmed impact events have taken place since then. On 24 July 1996, the French microsatellite Cerise was hit by fragments of an Ariane-1 H-10 upper-stage booster that had exploded in November 1986 [70]. On 29 March 2006 the Russian "Express-AM11" communications satellite was struck by an unknown object which rendered it inoperable. Luckily, the engineers had enough time in contact with the spacecraft to send it to a parking orbit out of GEO [86].

The first major space debris collision was on 10 February 2009. The deactivated 950-kg (2,100 lb) Cosmos 2251 and an operational 560-kg (1,200 lb) Iridium 33 collided 500 miles (800 km) [87] over northern Siberia. The relative speed of impact was about 11.7 km/s (7.3 mi/s), or approximately 42,120 km/h (26,170 mph) [88]. Both satellites were destroyed and the collision scattered considerable debris, which poses an elevated risk to spacecraft.[55] The collision created a debris cloud, although accurate estimates of the number of pieces of debris are not yet available [89]. In a Kessler Syndrome cascade, satellite lifetimes would be measured on the order of years or months. New satellites could be launched through the debris field into higher orbits or placed in lower ones where natural decay processes remove the debris, but it is precisely because of the utility of the orbits between 800 and 1,500 km (930 mi) that this region is so filled with debris.

There are a number of processes and events that generate new debris objects. Events that lead to the generation of space debris are on-orbit explosions or collisions, firings of solid rocket motors, releases of RORSAT reactor coolant, the release of mission-related objects (such as covers for optical instruments, or yo-yo de-spin weights), and impacts of micrometeoroids or small space debris onto spacecraft. Processes that lead to the generation of new debris are the degradation of spacecraft surfaces and the delaminating of objects due to the influence of the radiation environment in space. In a catalog listing known launches up to July 2009, the Union of Concerned Scientists listed 902 operational satellites only [90]. This is out of a known population of 19,000 large objects and about 30,000 objects ever launched. Thus, operational satellites represent a small minority of the population of man-made objects in space. The rest are, by definition, debris.

One major source of debris in the past was the testing of antisatellite weapons carried out by both the US and Soviet Union in the 1960s and 1970s. The NORAD element files only contained data for Soviet tests, and it was not until much later that

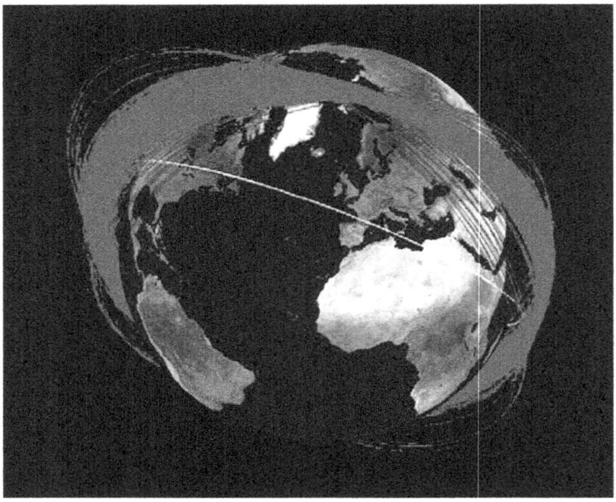

Fig. 3.52 Known orbit planes of Fengyun-1C debris 1 month after its disintegration by the Chinese ASAT

debris from US tests was identified [70]. By the time the problem with debris was understood, widespread ASAT testing had ended. However, the US re-started such programs in the 1980s. A 1985 test destroyed a 1 ton (2,200 lb) satellite orbiting at 525 km (326 mi) altitude, creating thousands of pieces of space debris larger than 1 cm (0.39 in). Because it took place at relatively low altitude, atmospheric drag caused the vast majority of the large debris to decay from orbit within a decade. Following the U.S. test in 1985, there was a de-facto moratorium on such tests [91] (Fig. 3.52).

China suffered widespread condemnation after their 2007 antisatellite missile test, both for the military implications as well as the huge amount of debris it created [92]. This is the largest single space debris incident in history, estimated to have created more than 2,300 pieces (updated 13 December 2007) of trackable debris (approximately golf ball size or larger), over 35,000 pieces 1 cm (0.4 in) or larger, and one million pieces 1 mm (0.04 in) or larger. Particularly worrying is the fact that the test took place in the most densely populated part of space, as the target satellite orbited between 850 km (530 mi) and 882 km (548 mi) [93]. Since the atmospheric drag is quite low at that altitude, the debris will persist for decades. In June 2007, NASA's Terra environmental spacecraft was the first to perform a maneuver in order to prevent impacts from this debris [94].

On 20 February 2008, the U.S. launched an SM-3 Missile from the USS Lake Erie specially to destroy a defective U.S. spy satellite feared to be carrying 1,000 lb of toxic hydrazine fuel. Since this event occurred at about 250 km (155 mi) altitude, all of the resulting debris have a perigee of 250 km (155 mi) or lower [95]. The missile was aimed to deliberately reduce the amount of debris as much as possible, and they had decayed by early 2008 [96].

The vulnerability of satellites to a collision with larger debris and the ease of launching such an attack against a low-flying satellite, has led some to speculate that such an attack would be within the capabilities of countries unable to make a precision attack like former U.S. or Soviet systems. Such an attack against a large satellite of 10 tons or more would cause enormous damage to the LEO environment [91].

Another main source of space waste leading to decline of the environmental situation in space is rocket and boosters. Upper stages, like the Inertial Upper Stage, start and end their productive lives in orbit. Boosters remain a serious debris problem and one of the major known impact events was due to an "Ariane" booster [70]. During the initial attempts to characterize the space debris problem, it became evident that a good proportion of all debris was due to the break-up of rocket boosters. Although NASA quickly made efforts to improve the survivability of their boosters, other countries did not follow suit for some time. On 11 March 2000, a Chinese Long March 4's CBERS-1/SACI-1 upper stage exploded in orbit and created a debris cloud [97, 98]. An event of similar magnitude occurred on 19 February 2007, when our Russian Briz-M booster stage exploded in orbit over Australia. The booster had been launched on 28 February 2006, carrying an Arabsat-4A communication satellite but malfunctioned before it could use all of its fuel. The explosion was captured on film by several astronomers, but due to the path of the orbit the debris cloud has been hard to quantify using radar. As of 21 February 2007, over 1,000 fragments had been identified [99, 100]. A third break-up event also occurred on 14 February 2007 as recorded by Celes Trak [101]. Eight break-ups occurred in 2006, the most break-ups since 1993 [102].

Any spacecraft in a debris field are subject to constant wear as a result of impacts with small debris. Critical areas of a spacecraft are normally protected by Whipple shields, eliminating most damage. However, low-mass impacts have a direct impact on the lifetime of a space mission, if the spacecraft is powered by solar panels. These panels are difficult to protect because their front face has to be directly exposed to the sun. As a result, they are often punctured by debris. When hit, panels tend not to produce new debris and produce a cloud of gas-sized particles that does not present as much of a risk to other spacecraft. This gas is generally a plasma when created and consequently presents an electrical risk to the panels themselves [103]. The effect of the many impacts with smaller debris was particularly notable on Mir, the Russian space station, as it remained in space for long periods of time with the panels originally launched on its various modules [104, 105] (Fig. 3.53).

So, the most dangerous consequences of space debris are connected with piloted cosmonautics. From the earliest days of the Space Shuttle missions, NASA has turned to NORAD's database to constantly monitor the orbital path in front of the Shuttle to find and avoid any known debris. At one point these simulations used up a considerable amount of the NORAD tracking system's capacity [70]. The first official Space Shuttle collision avoidance maneuver was during STS-48 in September 1991 [106]. A 7-s reaction control system burn was performed to avoid debris from the Cosmos-955 satellite. Similar maneuvers followed on missions 53, 72, and 82 [106]. One of the first events to widely publicize the debris problem was Space

Fig. 3.53 Debris impacts on Russian space station Mir's solar panels degraded their performance

Shuttle Challenger's first flight on STS-7. A small fleck of paint impacted Challenger's front window and created a pit over 1 mm (0.04 in) wide. Endeavour suffered a similar impact on STS-59 in 1994, but this one pitted the window for about half its depth: a cause for much greater concern. Postflight examinations have noted a marked increase in the number of minor debris impacts since 1998 [107].

The damage due to smaller debris has now grown to become a significant threat in its own right. Chipping of the windows became common; along with minor damage to the thermal protection system tiles (TPS). To mitigate the impact of these events, once the Shuttle reaches orbit it is deliberately flown tail first in an attempt to intercept as much of the debris load as possible on the engines and rear cargo bay. These are not used on orbit or during descent and thus are less critical to operations after launch. When flown to the ISS, the Shuttle is moved to a location where the station itself provides as much protection as possible [108].

The sudden increase in debris load led to a re-evaluation of the debris issue and today a catastrophic impact with large debris is considered to be the #1 threat to Shuttle operations on every mission [108, 109]. Mission planning now requires a thorough discussion of debris risk, requiring an executive level decision to proceed if the risk is greater than 1 in 200 of destroying the Shuttle. On a normal low-orbit mission to the ISS the risks are estimated to be 1 in 300, but the STS-125 mission to repair the Hubble Space Telescope at 350 miles was initially calculated at 1 in 185 due to the 2009 satellite collision, and threatened to cancel the mission. However, a re-analysis as better debris numbers became available reduced this to 1 in 221, and the mission was allowed to proceed [110].

In spite of their best efforts, however, there have been two serious debris incidents on more recent Shuttle missions. In 2006, Atlantis was hit by a small fragment of a circuit board during STS-115, which bored a small hole through the radiator panels in the cargo bay (the large gold colored objects visible when the doors are open) [111]. A similar incident followed on STS-118 in 2007, when Endeavour was hit in a similar location by unknown debris which blew a hole several centimeters in diameter through the panel [112] (Fig. 3.54).

The ISS uses extensive Whipple shielding to protect itself from minor debris threats [113]. However, large portions of the ISS cannot be protected, notably its large solar panels. In 1989 it was predicted that the International Space Station's panels would suffer about 0.23% degradation over 4 years, which was dealt with by overdesigning the panel by 1% [114]. New figures based on the increase in collisions since 1998 are not available.

Like the Shuttle, the only protection against larger debris is avoidance. On one occasion the crew was forced to abandon work and take refuge in the Russian "Soyuz" capsule while the threat passed [115]. This close call is a good example of the potential Kessler Syndrome; the debris is believed to be a small 10 cm (3.9 in) portion of the former "Cosmos 1275" [116], which is the satellite that is considered to be the first example of an on-orbit impact with debris. If the Syndrome comes to pass, the threat to manned missions may be too great to contemplate operations in

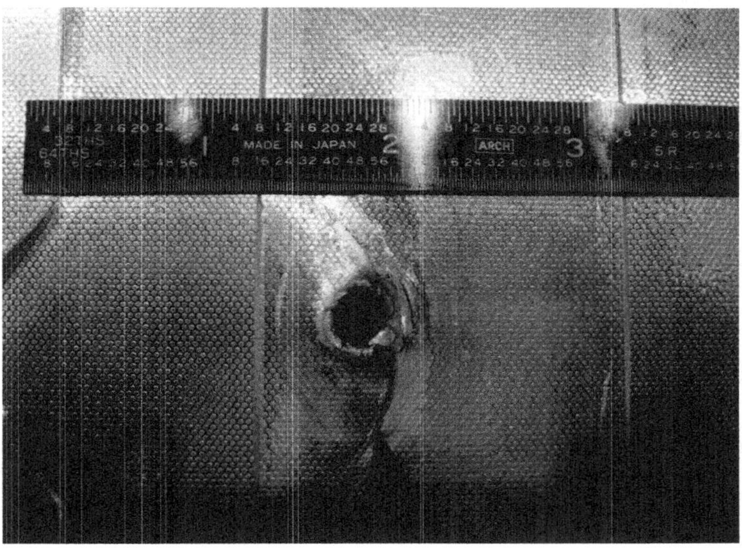

Fig. 3.54 Endeavour suffered a major hit on the radiator (less than 1/2 in) during STS-118

LEO. Although the majority of manned space activities take place at altitudes below the critical 800–1,500 km (310 mi) regions, a cascade within these areas would result in a constant rain down into the lower altitudes as well. The time scale of their decay is such that "the resulting debris environment is likely to be too hostile for future space use" [70].

Although most debris will burn up in the atmosphere, larger objects can reach the ground intact and present a risk.

For example, the original re-entry plan for "Skylab" called for the station to remain in space for 8–10 years after its final mission in February 1974. Unexpectedly high solar activity pushed the space station's orbit closer to Earth than planned. On 11 July 1979, Skylab re-entered the Earth's atmosphere and disintegrated, raining debris harmlessly along a path extending over the southern Indian Ocean and sparsely populated areas of Western Australia [117, 118]. On 12 January 2001, a Star 48 Payload Assist Module (PAM-D) rocket upper stage re-entered the atmosphere after a "catastrophic orbital decay" [119]. The PAM-D stage crashed in the sparsely populated Saudi Arabian desert (Fig. 3.55). It was positively identified as the upper-stage rocket booster for NAVSTAR 32, a GPS satellite launched in 1993. The Columbia disaster in 2003 demonstrated this risk, as large portions of the spacecraft reached the ground. In some cases entire equipment systems were left intact [120]. NASA continues to warn people to avoid contact with the debris due to the possible presence of hazardous chemicals [121].

On 27 March 2007, wreckage from a Russian recognizance satellite passed close to a Lan Chile (LAN Airlines) Airbus A340, which was travelling between

Fig. 3.55 Saudi officials inspect a crashed PAM-D module, January 2001

Santiago, Chile, and Auckland, New Zealand carrying 270 passengers [122]. The aircraft was flying over the Pacific Ocean, which is considered one of the safest places in the world for a satellite to come down because of its large areas of uninhabited water. At the same time, there has only been one recorded incident of a person being hit by human-made space debris. In 1997 an Oklahoma woman named Lottie Williams was hit in the shoulder by a 10 × 13 cm (5.1 in) piece of blackened, woven metallic material that was later confirmed to be part of the fuel tank of a Delta II rocket which had launched a U.S. Air Force satellite in 1996. She was not injured [123].

Finalizing our observations on the space debris issue we have to say a few words about its tracking and measurement. Because of the potential danger of orbital junk for both space activity and to important objects on the Earth's surface, Humanity has to pay attention to improving the environmental situation in frequently used orbits and allocate resources for the early warning and prediction of global risks and threats caused by free flying and falling artificial space bodies.

Radar and optical detectors such as lasers are the main tools used for tracking space debris. However, determining orbits to allow reliable re-acquisition is problematic. Tracking objects smaller than 10 cm (4 in) is difficult due to their small cross-section and reduced orbital stability, though debris as small as 1 cm (0.4 in) can be tracked [77, 124]. NASA's Orbital Debris Observatory tracked space debris using a 3-m (10 ft) liquid-mirror transit telescope [125]. The U.S. Strategic Command maintains a catalogue containing about 19,000 objects in the version compiled in 2009, in part to prevent misinterpretation as hostile missiles. Observation data gathered by a number of ground-based radar facilities and telescopes as well as by a space-based telescope is used to maintain this catalogue [126]. Nevertheless, the majority of debris objects remain unobserved. There are more than 600,000 objects larger than 1 cm (0.4 in) in orbit (according to the ESA Meteoroid and Space Debris Terrestrial Environment Reference, the MASTER-2005 model). Other sources of knowledge on the actual space debris environment include measurement campaigns by the ESA Space Debris Telescope, TIRA (System) [127], Goldstone radar, Haystack radar [128], and the Cobra Dane phased array radar [129]. The data gathered during these campaigns is used to validate models of the debris environment like ESA-MASTER. Such models are the only means of assessing the impact risk caused by space debris, as only larger objects can be regularly tracked.

A separate issue in relation to the processes of detecting and predicting the danger of space debris is special measurement facilities in space. For example, the Long Duration Exposure Facility (LDEF) is an important source of information on the small particle space debris environment.

Returned space debris hardware is also a valuable source of information on the (submillimeter) space debris environment. The LDEF satellite deployed by STS-41-C Challenger and retrieved by STS-32 Columbia spent 68 months in orbit. The close examination of its surfaces allowed the analysis of the directional distribution and the composition of debris flux. The EURECA satellite deployed by STS-46 Atlantis in 1992 and retrieved by STS-57 Endeavour in 1993 provided additional insight.

The solar arrays of the Hubble Space Telescope returned during missions STS-61 Endeavour and STS-109 Columbia are an important source of information on the debris environment (Fig. 3.56). The impact craters found on the surface were counted and classified by ESA to provide another means for validating debris environment models. Similar materials returned from Mir were also extensively studied (Fig. 3.57).

Fig. 3.56 Hubble space telescope on orbit

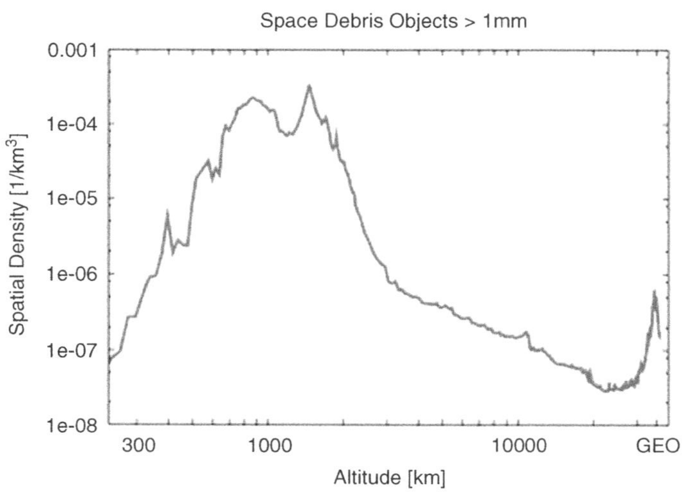

Fig. 3.57 Space debris population in the nearest future (forecast)

However, to date Humanity is far from unity in regard to the space junk issue. Discussion for more than 20 years at UN level (in the frameworks of COPUOS and its subcommittees) has seen no binding agreement between the space powers (one that carries with it strict obligations). Active discussion of the space debris problem began again after the 2009 European Air and Space Conference, at the University of Southampton, where UK researcher Hugh Lewis predicted that the threat from space debris would rise 50% in the coming decade and quadruple in the next 50 years. Currently more than 13,000 close calls are tracked each week [130]. How do we mitigate growth in space junk? Many technical, political, and management approaches exist in this context. In order to mitigate the generation of additional space debris, a number of measures have already been proposed. For example, the passivation of spent upper stages by the release of residual fuels is aimed at reducing the risk of on-orbit explosions that could generate thousands of additional debris objects.

It is already an ITU requirement that geostationary satellites are able to remove themselves to a "graveyard orbit" at the end of their lives. It has been demonstrated that the selected orbital areas do not sufficiently protect GEO lanes from debris, although a response has not yet been formulated [131]. Rocket boosters and some satellites retain enough fuel to allow them to power themselves into a decaying orbit. In cases where a direct (and controlled) de-orbit would require too much fuel, a satellite can also be brought to an orbit where atmospheric drag would cause it to de-orbit after some years. Such a maneuver was successfully performed with the French Spot-1 satellite, bringing its time to atmospheric re-entry down from a projected 200 years to about 15 years by lowering its perigee from 830 km (516 mi) to about 550 km (342 mi) [132]. Another proposed solution is to attach an electrodynamics tether to the spacecraft on launch. At the end of their lifetime it is rolled out and slows down the spacecraft [133]. Although tethers of up to 30 km have been successfully deployed in orbit the technology has not yet reached maturity [74]. It has also been proposed that booster stages include a sail-like attachment to the same end [134].

Other option – external removal of abandoned objects from the orbits. The vast majority of space debris, especially smaller debris, cannot be removed under its own power. A variety of proposals have been made to directly remove such material from orbit. One of the most widely discussed solutions is the laser broom, which uses a powerful ground-based laser to ablate the front surface of known debris and thereby produce a working mass that slows the debris in orbit. With a continued application of such thrust, the debris will eventually spiral down into a low orbit and become subject to atmospheric drag [135]. So, the U.S. Air Force worked on a ground-based design under the name "Project Orion" [136]. Although a test-bed device was scheduled to launch on a 2003 Space Shuttle, numerous international agreements, forbidding the testing of powerful lasers in orbit, caused the program to be limited to using the laser as a measurement device [137]. Finally, the Space Shuttle Columbia disaster led to the project being postponed.

Another well-studied solution is to use a remotely controlled vehicle to rendezvous with debris, capture it, and return to a central station [138]. A number of other proposals intercept the debris in a foamy ball of aero gel or even a spray of water [139]. These facilities would impact with the debris and slow it down. Some propose inflating balloons around the objects in order to increase their atmospheric drag. However, it was pointed out that a balloon could be punctured by other debris, thereby failing in its mission and actually increasing the amount of mass in orbit. But in any event, the cost of launching any of these solutions is about the same as launching any spacecraft. Johnson has stated that none of the existing solutions are currently cost-effective [74].

Therefore, this space environment issue is still so far from being solved as it was two decades ago. There is no international treaty mandating behavior to minimize space debris, but the United Nations Committee on the Peaceful Uses of Outer Space (COPUOS) did publish voluntary guidelines in 2007 [130]. As of 2008, the Committee is discussing international "rules of the road" to prevent collisions between satellites [140]. NASA has implemented its own procedures for limiting debris production [141] as have some other space agencies, such as the European Space Agency. The ISO is also in the process of preparing a new standard, dealing with space debris mitigation.

The authors therefore propose space waste questions in regard to prediction and forecasting be included in future broad international monitoring projects so that the near-Earth space environment, a unique resource for all humanity, be protected in a similar way to environmental protection of the Earth' surface to guarantee global space security by unifying all available technical, intellectual, and political recourses.

3.3.3 Risks and Threats to Humanity Connected with Abnormal Events

After lengthy discussion the authors of the book decided to include a chapter on dangerous abnormal events, connected with UFO phenomena, widely discussed in scientific and public circles over the last 60 years. This issue is generally perceived ambiguously: some people believe in UFOs, while others deny their existence assigning it to the realm of nonscientific futuristic fiction. Unidentified flying object (commonly abbreviated as UFO or U.F.O.) is the popular term for any apparent aerial phenomenon whose cause cannot be easily or immediately identified by the observer. The United States Air Force, which coined the term in 1952, initially defined UFOs as those objects that remain unidentified after scrutiny by expert investigators, [142] though today the term UFO is colloquially used to refer to any unidentifiable sighting regardless of whether it has been investigated. UFO reports increased precipitously after the first widely publicized U.S. sighting, reported by private pilot Kenneth Arnold in 1947, that gave rise to the popular terms "flying saucer" and "flying disc." The term UFO is popularly taken as a synonym for alien spacecraft and generally most discussions of UFOs revolve around this presumption [143]. UFO enthusiasts and devotees have created organizations, religious cults have adopted extraterrestrial themes, and in general the UFO concept has evolved into a prominent mythos in modern culture [144]. Some investigators now prefer to

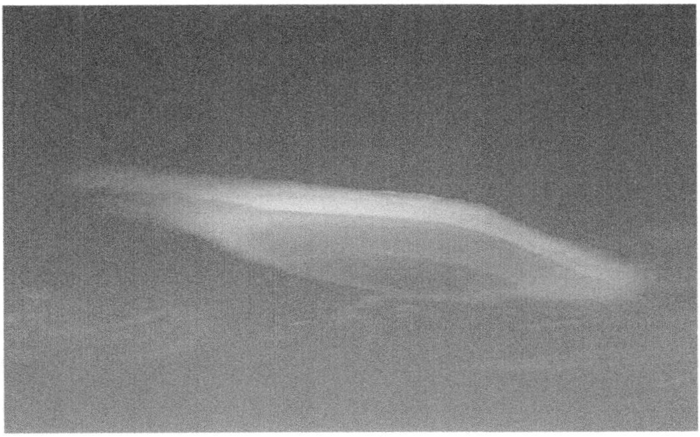

Fig. 3.58 What is it? A "flying saucer" or just a cloud?

use the broader term unidentified aerial phenomenon (or UAP), to avoid the confusion and speculative associations that have become attached to UFO (Fig. 3.58).

Studies have established that the majority of UFOs are observations of some real but conventional object – most commonly aircraft, balloons, or astronomical objects such as meteors or bright planets – that have been misidentified by the observer as anomalies, while a small percentage of reported UFOs are hoaxes. Only between 5 to 20% of anomalous sightings can be classified as unidentified in the strictest sense. The possibility that all UFO sightings are misidentifications of known natural phenomena [145] inspired some debate in the scientific community about whether scientific investigation was warranted given the paucity of available empirical data [146–150]. Very little peer-reviewed literature has been published in which scientists have proposed, studied, or supported nonprosaic explanations for UFOs [151]. Nevertheless, UFOs as a cultural phenomenon continues to be the subject of serious academic research [151] and amateur investigators continue to advocate that UFOs represent real and unexplained events, usually associated with alien encounters.

UFOs have been subject to investigations over the years that vary widely in scope and scientific rigor (Fig. 3.59). Governments or independent academics in the United States, Canada, the United Kingdom, Japan, Peru, France, Belgium, Sweden, Brazil, Chile, Uruguay, Mexico, Spain, and the Soviet Union are known to have investigated UFO reports at various times. Among the best known government studies are the ghost rockets investigation by the Swedish military (1946–1947), Project Blue Book, previously Project Sign and Project Grudge, conducted by the United States Air Force from 1947 until 1969, the secret U.S. Army/Air Force Project Twinkle investigation into green fireballs (1948–1951), the secret USAF Project Blue Book Special Report #14 [152] by the Battelle Memorial Institute, and Brazilian Air Force Operation Saucer (1977). France has had on

Fig. 3.59 One of the available Internet pictures of a UFO

ongoing investigation (GEPAN/SEPRA/GEIPAN) within its space agency CNES since 1977, as has Uruguay since 1989.

Jacques Vallée, a scientist and prominent UFO researcher, has argued that most UFO investigations are scientifically deficient, including many government studies such as Project Blue Book, and that mythology and cultism are frequently associated with the phenomenon. Vallée states that self-styled scientists often fill the vacuum left by the lack of attention paid to the UFO phenomenon by official science, but also notes that several hundred professional scientists continue to study UFOs in private, what he terms the "invisible college." He also argues that much could be learned from rigorous scientific study, but that little such work has been done [144].

No official government investigation has ever publicly concluded that UFOs are indisputably real, physical objects, extraterrestrial in origin, or of concern to national defense. These same negative conclusions also have been found in studies that were highly classified for many years, such as the UK's Flying Saucer Working Party, Project Condign, the U.S. CIA-sponsored Robertson Panel, the U.S. military investigation into the green fireballs from 1948 to 1951, and the Battelle Memorial Institute study for the USAF from 1952 to 1955 (Project Blue Book Special Report #14). However, the initially classified USAF Regulation 200–2, first issued in 1953 after the Robertson Panel, which first defined UFOs and how information was to be collected, stated explicitly that the two reasons for studying the unexplained cases were for national security reasons and for possible technical aspects involved, implying physical reality and concern about national defense, but without opinion as to origins. (For example, such information would also be considered important if UFOs had a foreign or domestic origin.) The first two known classified USAF studies in 1947 also concluded real physical aircraft were involved, but gave no opinion as to origins. (See American investigations immediately below) These early studies led to the creation of the USAF's Project Sign at the end of 1947, the first semi-public USAF study.

Project Sign in 1948 wrote a highly classified opinion (see Estimate of the Situation) that the best UFO reports probably had an extraterrestrial explanation, as did the private but high-level French COMETA study of 1999. A top secret Swedish military opinion given to the USAF in 1948 stated that some of their analysts believed the 1946 ghost rockets and later flying saucers had extraterrestrial origins. In 1954, German rocket scientist Hermann Oberth revealed an internal West German government investigation, which he headed, that arrived at an extraterrestrial conclusion, but this study was never made public. Classified, internal reports by the Canadian Project Magnet in 1952 and 1953 also assigned high probability to extraterrestrial origins. Publicly, however, Project Magnet, nor later Canadian defense studies, ever stated such a conclusion.

Another highly classified U.S. study was conducted by the CIA's Office of Scientific Investigation (OS/I) in the latter half of 1952 after being directed to do so by the National Security Council (NSC). They concluded UFOs were real physical objects of potential threat to national security. One OS/I memo to the CIA Director (DCI) in December read, "...the reports of incidents convince us that there is something going on that must have immediate attention... Sightings of unexplained objects at great altitudes and traveling at high speeds in the vicinity of major U.S. defense installations are of such a nature that they are not attributable to natural phenomena or any known types of aerial vehicles." The matter was considered so urgent, that OS/I drafted a memorandum from the DCI to the NSC proposing that the NSC establish an investigation of UFOs as a priority project throughout the intelligence and the defense research and development community. They also urged the DCI to establish an external research project of top-level scientists to study the problem of UFOs, now known as the Robertson Panel, to further analyze the matter. The OS/I investigation was called off after the Robertson Panel's negative conclusions in January 1953 [153].

Some public government conclusions have indicated physical reality but stopped short of concluding extraterrestrial origins, though not dismissing the possibility. Examples are the Belgian military investigation into large triangles over their airspace in 1989–1991 and the recent 2009 Uruguay Air Force study conclusion (see below). Some private studies have been neutral in their conclusions, but argued the inexplicable core cases called for continued scientific study. Examples are the Sturrock Panel study of 1998 and the 1970 AIAA review of the Condon Report.

In March 2007, the French Centre National d'Études Spatiales (CNES) published an archive of UFO sightings and other phenomena online [154]. French studies include GEPAN/SEPRA/GEIPAN (1977), within the French space agency CNES, the longest ongoing government-sponsored investigation. About 14% of some 6,000 cases studied remained unexplained. The official opinion of GEPAN/SEPRA/GEIPAN has been neutral or negative, but the three heads of the studies have gone on record in stating that UFOs were real physical flying machines beyond our knowledge or that the best explanation for the most inexplicable cases was an extraterrestrial one [155]. So, the French COMETA panel (1996–1999) was a private study undertaken mostly by aerospace scientists and engineers affiliated with CNES and high-level French Air Force military intelligence analysts, with ultimate distribution of their study intended for high government officials. The COMETA panel likewise concluded the best explanation for the inexplicable cases was the extraterrestrial hypothesis and went further in accusing the United States government of a massive cover-up [156].

The UK conducted various investigations into UFO sightings and related stories. The contents of some of these investigations have since been released to the public. Eight file collections on UFO sightings, dating from 1978 to 1987, were first released on May 14, 2008, to the UK National Archives by the Ministry of Defence (MoD) [157]. Although kept secret from the public for many years, most of the files have low levels of classification and none is classified Top Secret. Two hundred files are set to be made public by 2012. The files are correspondence from the public sent to government officials/Ministries, such as Margaret Thatcher and the Ministry of Defence. The MoD released the files under the Freedom of Information Act due to requests from researchers [158]. These files include, but are not limited to, UFOs over Liverpool and the Waterloo Bridge in London [159]. Moreover, on October 20, 2008 more UFO files were released. One case released detailed that in 1991 an Alitalia passenger aircraft was approaching Heathrow Airport when the pilots saw what they described as a "cruise missile" which flew extremely close to the cockpit. The pilots believed that a collision was imminent. UFO expert David Clarke says that this is one of the most convincing cases for a UFO he has come across [160].

British investigations also include the UK's Flying Saucer Working Party. Its final report, published in 1951, remained secret for over 50 years. The Working Party concluded that all UFO sightings could be explained as misidentifications of ordinary objects or phenomena, optical illusions, psychological delusions, or hoaxes. The report stated: "We accordingly recommend very strongly that no further investigation of reported mysterious aerial phenomena be undertaken, unless and until some material evidence becomes available" (Fig. 3.60).

A secret study of UFOs was undertaken for the UK's MoD between 1996 and 2000 and was code-named Project Condign. The resulting report, titled "Unidentified Aerial Phenomena in the UK Defense Region," was publicly released in 2006, but the identity and credentials of whomever constituted Project Condign remains classified. The report confirmed earlier findings that the main causes of UFO sightings are misidentification of man-made and natural objects. The report noted;

> No artifacts of unknown or unexplained origin have been reported or handed to the UK authorities, despite thousands of UAP reports. There are no SIGINT, ELINT or radiation measurements and little useful video or still IMINT." It concluded: "There is no evidence that any UAP, seen in the UKADR (UK Air Defense Region), are incursions by air-objects of any intelligent (extraterrestrial or foreign) origin, or that they represent any hostile intent. (Fig. 3.61)

The Uruguayan Air Force has had an ongoing UFO investigation since 1989 and analyzed 2,100 cases, of which they consider only 40 (about 2%) definitely lacking any conventional explanation. All files have recently been declassified. The unexplained cases include military jet interceptions, abductions, cattle mutilations, and physical landing trace evidence. Colonel Ariel Sanchez, who currently heads the

Passoria, New Jersey, 31 July 1952

Fig. 3.60 Amateur photo of an alleged UFO

investigation, summarized their findings as follows: "The commission managed to determine modifications to the chemical composition of the soil where landings are reported. The phenomenon exists. It could be a phenomenon that occurs in the lower sectors of the atmosphere, the landing of aircraft from a foreign air force, up to the extraterrestrial hypothesis. It could be a monitoring probe from outer

Fig. 3.61 What is that in the sky?

space, much in the same way that we send probes to explore distant worlds. The UFO phenomenon exists in the country. I must stress that the Air Force does not dismiss an extraterrestrial hypothesis based on our scientific analysis." [161].

The U.S. Air Force's Project Blue Book files indicate that approximately 1% [162] of all unknown reports came from amateur and professional astronomers or other users of telescopes (such as missile trackers or surveyors). In 1952, astronomer J. Allen Hynek, then a consultant to Blue Book, conducted a small survey of 45 fellow professional astronomers. Five reported UFO sightings (about 11%). In the 1970s, astrophysicist Peter A. Sturrock conducted two large surveys of the American Institute of Aeronautics and Astronautics and American Astronomical Society. About 5% of the members polled indicated that they had UFO sightings. For example astronomer Clyde Tombaugh, who admitted to six UFO sightings, including three green fireballs, supported the Extraterrestrial hypothesis (ETH) for UFOs and stated he thought scientists who dismissed it without study were being "unscientific." Another astronomer was Lincoln LaPaz, who had headed the Air Force's investigation into the green fireballs and other UFO phenomena in New Mexico. LaPaz reported two personal sightings, one of a green fireball, the other of an anomalous disc-like object. Hynek himself took two photos through the window of a commercial airliner of a disc-like object that seemed to pace his aircraft [163]. Even later UFO debunker Donald Menzel filed a UFO report in 1949. In 1980, a survey of 1,800 members of various amateur astronomer associations by Gert Helb and Hynek for the Center for UFO Studies (CUFOS) found that 24% responded "yes" to the question "Have you ever observed an object which resisted your most exhaustive efforts at identification?" [164].

A 1952–1955 study by the Battelle Memorial Institute for the U.S. Air Force included these categories as well as a "psychological" one. However, the scientific analysts were unable to come up with prosaic explanations for 21.5% of the 3,200 cases they examined and 33% of what were considered the best cases remained unexplained, double the number of the worst cases. (See full statistical breakdown in Identification studies of UFOs). Of the 69% identifiers, 38% were deemed definitely explained while 31% were thought to be "questionable." About 9% of the cases were considered to have insufficient information to make a determination. The official French government UFO investigation (GEPAN/SEPRA/GEIPAN), run within the French space agency CNES between 1977 and 2004, scientifically investigated about 6,000 cases and found that 13.5% defied any rational explanation, 46% were deemed definitely or likely identifiable, while 41% lacked sufficient information for classification.

An individual 1979 study by CUFOS researcher Allan Hendry found, as did other investigations, that only a small percentage of cases he investigated were hoaxes (<1%) and that most sightings were actually honest misidentifications of prosaic phenomena. Hendry attributed most of these to inexperience or misperception [165]. However, Hendry's figure for unidentified cases was considerably lower than many other UFO studies such as Project Blue Book or the Condon Report that have found rates of unidentified cases ranging from 6 to 30%. Hendry found that 88.6% of the cases he studied had a clear prosaic explanation, and he discarded a further 2.8% due to unreliable or contradictory witnesses or insufficient information. The remaining 8.6% of reports could not definitively be explained by prosaic phenomena; although he felt that a further 7.1% could possibly be explained, leaving only at the very best 1.5% without plausible explanation.

There are some reports of UFO phenomena of "Russian origin." Unfortunately the authors are unable to present in this book such sensitive information due to the specifics of their professional activities in the past and present. However, they present here some profile photos, connected with a UFO appearance in Russia (Fig. 3.62). This UFO photo (Soviet times) was taken in a forest in Russia after a reported UFO crash. We don't know who took the picture or who leaked the picture into the public realm. One of the latest reports is dated December 9, 2009 in relation to two separate amateur videos, one taken during the day, and one at night, which show the same unusual object in the sky above the Kremlin (Moscow). The UFO made its appearance on the same night as mysterious lights were spotted in Norwegian skies. The lights were later identified as a failed Russian rocket launch. Russian television has been showing the videos continuously and a former Ministry of Defence and UFO analyst Nick Pope called the footage "one of the most extraordinary UFO clips he'd ever seen."

To account for unsolved UFO cases, several hypotheses have been proposed.

1. ETH is defined by Edward U. Condon in the 1968 Condon Report as "The idea that *some UFOs may be spacecraft sent to Earth from another civilization, or from a planet associated with a more distant star*," further attributing the popularity of the idea to Donald Keyhoe's UFO book from 1950, [166] though the idea clearly predated Keyhoe, appearing in newspapers and various government documents (see immediately below). This is probably the most popular

Fig. 3.62 Russian UFO crash events

theory among ufologists. Some private or governmental studies, some secret, have concluded in favor of the ETH, or have had members who disagreed with official conclusions against the conclusion by committees and agencies to which they belonged [167–172].

2. The interdimensional hypothesis (IDH or IH) says that *UFOs are objects crossing over from other dimensions or a parallel universe*, popularly proposed by Jacques Vallée [173] though also predating him.
3. The paranormal/occult hypothesis is a variant of the interdimensional hypothesis, invoked to explain so-called *paranormal aspects sometimes associated with UFO reports.*
4. The psychosocial hypothesis, that what people report as UFO experiences is the *result of psychological misperception mechanisms* and is strongly influenced by popular culture.
5. That *UFOs represent poorly understood or still unknown natural phenomena*, such as ball lightning or sprites [174]
6. The *earthquake lights/tectonic strain hypothesis: UFOs are caused by strains in the Earth's crust near earthquake faults*, which can also supposedly induce hallucinations.
7. That *UFOs are military flying saucers, top secret or experimental aircraft* unfamiliar to most people. [175]

A comprehensive scientific review of cases where physical evidence was available was carried out by the 1998 Sturrock UFO panel, with specific examples of many of the categories listed below [176].

- Radar contact and tracking, sometimes from multiple sites. These have included military personnel and control tower operators, simultaneous visual sightings, and aircraft intercepts. One such recent example were the mass sightings of large, silent, low-flying black triangles in 1989 and 1990 over Belgium, tracked by NATO radar and jet interceptors, and investigated by Belgium's military (including photographic evidence) [177]. Another famous case from 1986 was the JAL 1628 case over Alaska investigated by the FAA.
- Photographic evidence, including still photos, movie film, and video (Fig. 3.63).
- Claims of physical traces of landing UFOs, including ground impressions, burned and/or desiccated soil, burned and broken foliage, magnetic anomalies (specify), increased radiation levels, and metallic traces.
- Physiological effects on people and animals including temporary paralysis, skin burns and rashes, corneal burns, and symptoms superficially resembling radiation poisoning, such as the Cash-Landrum incident in 1980.
- Animal/cattle mutilation cases that some feel are also part of the UFO phenomenon.
- Biological effects on plants such as increased or decreased growth, germination effects on seeds, and blown-out stem nodes (usually associated with physical trace cases or crop circles).
- Electromagnetic interference (EM) effects. A famous 1976 military case over Tehran, recorded in CIA and DIA classified documents, was associated with communication losses in multiple aircraft and weapons system failure in an F-4 Phantom II jet interceptor as it was about to fire a missile at one of the UFOs [178].
- Apparent remote radiation detection, some noted in FBI and CIA documents occurring over government nuclear installations at Los Alamos National Laboratory and Oak Ridge National Laboratory in 1950, also reported by Project Blue Book director Ed Ruppelt in his book.

Fig. 3.63 Something unexplained over the city...

Fig. 3.64 UFO in the mountains

- Claimed artifacts of UFOs themselves, such as the 1957, Ubatuba, Brazil, magnesium fragments analyzed by the Brazilian government and in the Condon Report and by others. The 1964 Socorro/Lonnie Zamora incident also left metal traces, analyzed by NASA. A more recent example involves "the Bob White object" a tear drop-shaped object recovered by Bob White and which was featured in the TV show UFO hunters.
- Angel hair and angel grass, possibly explained in some cases as nests from ballooning spiders or chaff.

Some attempts have been made to reverse engineer the possible physics behind UFOs through analysis of both eyewitness reports and the physical evidence, on the assumption that they are powered vehicles. Examples are former NASA and nuclear engineer James McCampbell in his book Ufology, NACA/NASA engineer Paul R. Hill in his book Unconventional Flying Objects, and German rocketry pioneer Hermann Oberth. Among subjects tackled by McCampbell, Hill, and Oberth was the question of how UFOs can fly at supersonic speeds without creating a sonic boom. McCampbell's proposed solution is microwave plasma parting the air in front of the craft. In contrast, Hill and Oberth believed UFOs utilize an as yet unknown antigravity field to accomplish the same thing as well as provide propulsion and protection of occupants from the effects of high acceleration [179] (Fig. 3.64).

Therefore, the UFO as a phenomenon already exists and nobody knows quite how to deal with it. But if we have a lot of as yet unknown transphenomenal events at the scale of the Globe we have to think about the security of civilization and the safety of the huge and very complicated infrastructure built over generations which needs to be protected from destruction by external forces as this would be disastrous. That is why among the prospective tasks of the global aerospace monitoring system, jointly created by all states of the World, the issue of abnormal events subsequently needs to be given attention.

References

1. Dealing with the threat to earth from asteroids and comets, In: Ivan Bekey, IAA. (ed) p. 140. (2009), http://iaaweb.org/iaa/Communication/announcement_asteroids.pdf
2. Gehrels, T. (ed.): Hazards due to Comets & Asteroids, p. 1300. University Arizona Press, Tucson (1994)
3. Chesley, S.R., Ward, S.N.: A quantitative assessment of the human and economic hazard from impact-generated tsunami. Nat. Hazards **38**, 355–374 (2006)
4. Ceplecha, Z.: Meteoroids: an item in the inventory. In: Rettig, T.W., Hahn, J.M., (eds.), Completing the Inventory of the Solar System, Astron. Soc. Pacific Conf. Ser. 107, pp. 75–84 (1996)
5. Stuart, J.S.: A Near-Earth asteroid population estimate from the LINEAR survey. Science **294**, 1691–1693 (2001)
6. Bottke, W.F., Morbidelli, A., Jedicke, R., Petit, J.-M., Levison, H.F., Michel, P., Metcalfe, T.S.: Debased orbital and absolute magnitude distribution of the Near-Earth objects. Icarus **156**, 399–433 (2002)
7. Wiegert, P., Tremaine, S.: The evolution of long-period Comets. Icarus **137**, 84–121 (1999)
8. Binzel, R.P., Lupishko, D.F., DiMartino, M., Whiteley, R.J., Hahn, G.J.: Physical properties of Near-Earth objects. In: Bottke Jr., W.F., Cellino, A., Paolicchi, P., Binzel, R.P. (eds.) Asteroids III, pp. 255–271. University Arizona Press, Tucson (2002)
9. Weidenschilling, S.J.: Formation of planetesimals and accretion of the terrestrial planets. Space Sci. Rev. **92**, 295–310 (2000)
10. Chambers, J.E., Wetherill, G.W.: Making the terrestrial planets: N-body integrations of planetary embryos in three dimensions. Icarus **136**, 304–327 (1998)
11. Morbidelli, A., Petit, J.M., Gladman, B., Chambers, J.: A plausible cause of the late heavy bombardment. Meteorit. Planet. Sci. **36**, 371–380 (2001)
12. Ryder, G.: Mass flux in the ancient Earth-Moon system and benign implications for the origin of life on Earth. J. Geophys. Res. **107**(E4), 5022 (2002), doi: 10.1029/2001JE001583
13. Gomes, R., Tsiganis, K., Morbidelli, A., Levison, H.F.: The cataclysmic late heavy bombardment triggered by a delayed start of planet migration. Nature, submitted (2005)
14. Hartmann, W.K.: Megaregolith evolution and cratering cataclysm models–Lunar cataclysm as a misconception (28 years later). Meteorit. Planet. Sci. **38**, 579–593 (2003)
15. Bottke Jr, W.F., Vokrouhlick, D., Rubincam, D.P., Broz, M.:The effect of Yarkovsky thermal forces on the dynamical evolution of asteroids and meteoroids. In: Bottke Jr, W.F., Cellino, A., Paolicchi, P., Binzel, R.P. (Eds.), Asteroids III, pp. 395–408. University Arizona Press, Tucson (2002)
16. Culler, T.S., Becker, T.A., Muller, R.A., Renne, P.R.: Lunar impact history from 40Ar/39Ar dating of glass spherules. Science **287**, 1785–1788 (2000)
17. Watson, F.: Between the Planets, p. 222. The Blakiston Co, Philadelphia (1941)
18. Baldwin, R.B.: The Face of the Moon, p. 239. University Chicago Press, Chicago (1949)
19. Atkinson, H., Tickell, C., Williams, D. (eds.): Report of the Task Force on Potentially Hazardous Near Earth Objects, p. 54. British National Space Centre, London (2000)
20. Öpik, E.J.: On the catastrophic effect of collisions with celestial bodies. Irish Astron. J. **5**(36), 128 (1958)
21. Morbidelli, A., Jedicke, R., Bottke, W.F., Michel, P., Tedesco, E.F.: From magnitudes to diameters: the albedo distribution of near earth objects and the earth collision hazard. Icarus **158**, 329–342 (2002)
22. Shoemaker, E.M.: Report of the Near-Earth Objects Survey Working Group, NASA Office of Space Science, Solar System Exploration Office (1995)
23. Shoemaker, E.M. (ed.): Report of the Near-Earth Objects Survey Working Group, NASA, Washington D.C., (1995)
24. Koeberl, C., MacLeod, K.G. (eds.): Catastrophic Events and Mass Extinctions: Impacts and Beyond. Geological Soc. America, Boulder CO, Special Paper 356, 2002, 746pp

25. Bottke Jr., W.F., Cellino, A., Paolicchi, P., Binzel, R.P. (eds.): Asteroids III. University Arizona Press, Tucson (2002). 785pp
26. Chapman, C.R.: The evolution of asteroids as meteorite parent-bodies. In: Delsemme, A.H. (ed.) Comets Asteroids Meteorites: Interrelations, Evolution and Origins. University of Toledo Press, Toledo OH (1977). 265.130
27. Brown, P., Spalding, R.E., ReVelle, D.O., Tagliaferri, E., Worden, S.P.: The flux of small near-earth objects colliding with the Earth. Nature **420**, 294–296 (2002)
28. Levison, H.F., Morbidelli, A., Dones, L., Jedicke, R., Wiegert, P.A., Bottke, W.F.: The mass disruption of Oort cloud comets. Science **296**, 2212–2215 (2002). 129
29. Jedicke, R., Morbidelli, A., Spahr, T., Petit, J., Bottke, W.F.: Earth and space-based NEO survey simulations: prospects for achieving the spaceguard goal. Icarus **161**, 17–33 (2003)
30. Chapman, C.R.: Space weathering of asteroid surfaces. Ann. Rev. Earth & Planet. Sci. **32**, 539–567 (2004)
31. Bus, S.J., Vilas, F., Barucci, M.A.: Visible-wavelength spectroscopy of asteroids. In: Bottke Jr., W.F., Cellino, A., Paolicchi, P., Binzel, R.P. (eds.) Asteroids III, pp. 169–182. University Arizona Press, Tucson (2002)
32. "MIT Course precept for movie", The Tech, MIT, October 30, 1979
33. Satellite observations of total solar irradiance
34. Willson, R.C., Hudson, H.S.: The sun's luminosity over a complete solar cycle. Nature **351**, 42–44 (1991)
35. "Solar Forcing of Climate". Climate Change 2001: Working Group I: The Scientific Basis. http://www.grida.no/climate/ipcc_tar/wg1/244.htm. Retrieved 10 March 2005
36. Weart, S.: (2006). "The Discovery of Global Warming". In: Weart, S. American Institute of Physics, Retrieved 14 April 2007.
37. http://acrim.com/TSI%20Monitoring.htm
38. Willson, R.C., Mordvinov, A.V.: Secular total solar irradiance trend during solar cycles 21–23. Geophys. Res. Lett. **30**(5), 1199 (2003). doi http://www.agu.org/journals/gl/gl0905/2008GL036307.
39. "Construction of a Composite Total Solar Irradiance (TSI) Time Series from 1978 to present". Physikalisch-Meteorologisches Observatorium Davos (PMOD).
40. Committee on Surface Temperature Reconstructions for the Last 2,000 Years, Board on Atmospheric Sciences and Climate, Division on Earth and Life Studies, National Research Council of the National Academies: climate forcings and climate models. In North, G.R., Biondi, F., Bloomfield, P., et al. (eds) Surface Temperature Reconstructions for the Last 2,000 Years. National Academies Press, Washington, DC (2006). ISBN 0-309-10225-1. Retrieved 19 April 2007
41. Lean, J.: Evolution of the Sun's spectral irradiance since the maunder minimum. Geophys. Res. Lett. **27**(16), 2425–2428 (2000)
42. Scafetta, N., West, B.J.: Phenomenological solar signature in 400 years of reconstructed northern hemisphere temperature record. Geophys. Res. Lett. **33**, L17718 (2006)
43. Arctowski, H.: On solar faculae and solar constant variations(PDF). Proc. Natl. Acad. Sci. U.S.A. **26**(6), 406–411 (1940)
44. Fritts, H.C.: Tree Rings and Climate. Academic Press, Boston (1976). ISBN 0-12-268450
45. "William Herschel (1738–1822)". High Altitude Observatory. Retrieved 27 Feb 2008
46. Camp, C.D., Tung, Ka-Kit., Quinif, Y., Kaufman, O., Van Ruymbeke, M., Vandiepenbeeck, M., Camelbeeck, T.: (2006). The Influence of the Solar Cycle and QBO on the Late Winter Stratospheric Polar Vortex" (PDF). EOS Trans. AGU **87** (52): Fall Meet. Suppl., Abstract #A11B–0862. Retrieved 28 April 2009.
47. http://www.pmodwrc.ch/pmod.php?topic = tsi/composite/SolarConstant
48. Willson, R.C., Mordvinov, A.V.: Secular total solar irradiance trend during solar cycles 21–23. Geophys. Res. Lett. **30**(5), 1199 (2003)
49. DeWitte, S., Crommelynck, D., Mekaoui, S., Joukoff, A.: Measurement and uncertainty of the long-term total solar irradiance trend. Sol. Phys. **224**, 209–216 (2004)

50. Fröhlich, C., Lean, J.: Solar radiative output and its variability: evidence and mechanisms. Astronomy and Astrophysical Reviews **12**, 273–320 (2004)
51. Eddy, J.A.: "Samuel P. Langley (1834–1906)", J. Hist. Astron. **21**: 111–20 (1990)
52. Foukal, P.V., Mack, P.E., Vernazza, J.E.: The effect of sunspots and faculae on the solar constant. Astrophys. J. **215**, 952 (1977)
53. Willson, R.C., Gulkis, S., Janssen, M., Hudson, H.S., Chapman, G.A.: Observations of solar irradiance variability. Science **211**(4483), 700–702 (1981)
54. Usoskin, I.G., Solanki, S.K., Schüssler, M., Mursula, K.,; Alanko, K.: A millennium scale sunspot number reconstruction: evidence for an unusually active sun since the 1940's (PDF). Physical Review Letters. (2003)
55. Solanki, S.K., Usoskin, I.G., Kromer, B., Schüssler, M., Beer, J.: Unusual activity of the sun during recent decades compared to the previous 11,000 years (PDF). Nature **431**(7012), 1084–1087 (2004). Retrieved 17 April 2007.,"11,000 Year Sunspot Number Reconstruction". Global Change Master Directory
56. Usoskin, I.G., Solanki, S.K., Kovaltsov, G.A.: Grand minima and maxima of solar activity: new observational constraints (PDF). Astron. Astrophys. **471**, 301–309 (2007)
57. Lean, J.: Contribution of ultraviolet irradiance variations to changes in the sun's total irradiance science. Science **244**(4901), 197 (1989). 14 April
58. Damon, P.E., Laut, P.: Pattern of strange errors plagues solar activity and terrestrial climate data (PDF). Eos **85**(39), 370–374 (2004). Retrieved October 5, 2005
59. Tinsley, B.A., Yu, F.: Atmospheric ionization and clouds as links between solar activity and climate. In: Pap, J.M., Fox, P.A. (eds.) Solar Variability and its Effects on Climate, vol. 141, pp. 321–339. American Geophysical Union, Washington, DC (2004). Retrieved 19 April 2007
60. Svensmark, H.: Influence of cosmic rays on earth's climate (PDF). Phys. Rev. Lett. **81**, 5027–5030 (1998). Retrieved 19 April 2007
61. Pallé, E., Butler, C.J., O'Brien, K.: The possible connection between ionization in the atmosphere by cosmic rays and low level clouds (PDF). J. Atmos. Sol-Terr. Phy **66**(18), 1779 (2004)
62. Pallé, E.: Possible satellite perspective effects on the reported correlations between solar activity and clouds (PDF). Geophysical Research Letters **32** (3): L03802.1–4 (2005)
63. "Astronomy: On the Sunspot Cycle". Retrieved 27 Feb 2008
64. Houghton, J.T., Ding, Y., Griggs, D.J., et al. (eds) (2001). "6.11 Total Solar Irradiance—Figure 6.6: Global, annual mean radiative forcings (1750 to present)". Climate Change 2001: Working Group I: The Scientific Basis. Intergovernmental Panel on Climate Change. Retrieved 15 April 2007
65. Kessler, D.: Upper limit on the spatial density of asteroidal debris. AIAA Journal **6**(12), 2450–2450 (1968)
66. Weart, S (2003). "Changing Sun, Changing Climate?". The Discovery of Global Warming. Harvard University Press. ISBN 0674011570. Retrieved 17 April, 2008
67. Scafetta, N., Willson, R.: ACRIM-gap and Total Solar Irradiance (TSI) trend issue resolved using a surface magnetic flux TSI proxy model. Geophysical Research Letter **36**: L05701 (2009)
68. Oswald, M., Stabroth, S., Wiedemann, C., Wegener, P., and C. Martin: "Upgrade of the MASTER Model", Final Report of ESA Contract No. 18014/03/D/HK(SC), Braunschweig (2006).
69. Hoots, F., Schumacher Jr, P.,.Glover, R.: History of analytical orbit modeling in the U. S. space surveillance system J. Guid. Control Dynam. **27**(2), 174–185 (2004)
70. Kessler, 1981, 1991, 2001, 2009
71. Schefter 1982, p. 48.
72. Hoffman, M.:"It's getting crowded up there." Space News, 3 April 2009.
73. Orbital Debris: A Technical Assessment 1995, http://www.nap.edu/openbook.php?isbn=0309051258

74. Liou, J.-C., Johnson, N.L.: Risks in space from orbiting debris. Science **311**(5759), 340–341 (2006)
75. Lovgren, S.: "Space Junk Cleanup Needed, NASA Experts Warn." National Geographic News, 19 January 2006.
76. Milne, A.: Sky Static: The Space Debris Crisis. Greenwood Publishing Group, Santa Barbara, CA (2002). 86
77. Wang, Y-.M., Lean, J.L., Sheeley Jr., N.R.: Modeling the sun's magnetic field and irradiance since 1713. Astrophys. J. **625**, 522–538 (May 2005)
78. "Orbital Debris: A Technical Assessment" 1995, p. 7.
79. Carroll, J.: Space Transport Development Using Orbital Debris. niac.usra.edu, 2 December 2002, p. 3.
80. McKie, R., Day, M.: Warning of catastrophe from mass of 'space junk'. The Observer, 24 February 2008.
81. Ford, M.: Orbiting space junk heightens risk of satellite catastrophes. Ars Technica, 27 February 2009.
82. What are hypervelocity impacts? ESA, 19 February 2009.
83. Collocation Strategy and Collision Avoidance for the Geostationary Satellites at 19 Degrees West. CNES Symposium on Space Dynamics, 6–10 November 1989.
84. Hanada, T.: Developing a low-velocity collision model based on the NASA standard breakup model. Space Debris **2**(4), 233–247 (2000)
85. Van der Ha, J.C., Hechler, M.: The Collision Probability of Geostationary Satellites. 32nd International Astronautical Congress, 1981, p. 23.
86. Project Icarus
87. Notification for Express-AM11 satellite users in connection with the spacecraft failure. Russian Satellite Communications Company, 19 April 2006.
88. Iannotta B., Malik, T.: U.S. Satellite Destroyed in Space Collision, space.com, 11 February 2009
89. Marks, P.: Satellite collision 'more powerful than China's ASAT test. New Scientist, 13 February 2009. Note: The collision speed was 42,120 kilometers per hour (11.7 km/s).
90. Two big satellites collide 500 miles over Siberia. yahoo.com, 11 February 2009. Retrieved 11 Feb 2009.
91. UCS Satellite Database. Union of Concerned Scientists, 16 July 2009.
92. Wright, D.: Debris in Brief: Space Debris from Anti-Satellite Weapons. Union of Concerned Scientists, December 2007.
93. David, L.,. China's Anti-Satellite Test: Worrisome Debris Cloud Circles Earth. space.com, 2 February 2007.
94. Fengyun 1 C - Orbit Data. Heavens Above.
95. Burger, B.: NASA's Terra Satellite Moved to Avoid Chinese ASAT Debris. space.com. Retrieved: 6 July 2007.
96. Pentagon: Missile Scored Direct Hit on Satellite. npr.org, 21 February 2008.
97. Wolf, J.: US satellite shootdown debris said gone from space. uk.reuters.com, 27 February 2008.
98. Anz-Meador, P., Matney, M.: An assessment of the NASA explosion fragmentation model to 1 mm characteristic sizes. Adv. Space Res. **34**(5), 987–992 (2004)
99. Debris from explosion of Chinese rocket detected by University of Chicago satellite instrument. University of Chicago press release, 10 August 2000.
100. Rocket Explosion. Spaceweather.com, 22 February 2007. Retrieved: 21 Feb 2007.
101. Than, K.: Rocket Explodes Over Australia, Showers Space with Debris. Space.com, 21 February 2007. Retrieved 21 Feb 2007.
102. Recent Debris Events. celestrak.com, 16 March 2007.
103. Spate of rocket breakups creates new space junk. NewScientist.com, 17 January 2007. Retrieved 16 March 2007.

104. Smirnov, V.M., et al.: Study of micrometeoroid and orbital debris effects on the solar panels retrieved from the space station 'MIR'. Space Debris **2**(1), 1–7 (2000)
105. Orbital Debris FAQ: How did the Mir space station fare during its 15-year stay in Earth orbit? NASA, July 2009.
106. Matson, R.: Satellite Encounters. Visual Satellite Observer's Home Page.
107. Christiansen, E.L., Hydeb, J.L., Bernhard, R.P.: Space shuttle debris and meteoroid impacts. Adv. Space Res. **34**(5), 1097–1103 (2004)
108. Kelly, J.: Debris is Shuttle's Biggest Threat. space.com, 5 March 2005.
109. Debris Danger. Aviation Week & Space Technology, Volume 169, Number 10, 15 September 2008, p. 18.
110. Harwood, W.: Improved odds ease NASA's concerns about space debris. CBS News, 16 April 2009.
111. Investigation of Shuttle Radiator Micro-Meteoroid & Orbital Debris Damage.
112. Lear, D., et al.: STS-118 Radiator Impact Damage, NASA
113. Thoma, K., Wicklein, M., Schneider, E., Danesy, D. (eds): Abstract New Protection Concepts for Meteoroid/Debris Shields. Proceedings of the 4th European Conference on Space Debris (ESA SP-587), 18–20 April 2005 in Darmstadt, Germany, ESA/ESOC, August 2005, p. 445
114. Nahra, H.: Effect of Micrometeoroid and Space Debris Impacts on the Space Station Freedom Solar Array Surfaces. Presented at the 1989 Spring Meeting of the Materials Research Society, NASA TR-102287. 24–29 April 1989
115. Junk alert for space station crew. BBC News, 12 March 2009.
116. Haines, L.: ISS spared space junk avoidance manoeuvre. The Register, 17 March 2009.
117. NASA – Part I – The History of Skylab. NASA's Marshall Space Flight Center and Kennedy Space Center, 16 March 2009.
118. NASA - John F. Kennedy Space Center Story. NASA Kennedy Space Center, 16 March 2009.
119. PAM-D Debris Falls in Saudi Arabia. The Orbital Debris Quarterly News, **6**(2)
120. Debris Photos. NASA.
121. Debris Warning. NASA.
122. Gibson, J.: Jet's flaming space junk scare. The Sydney Morning Herald, 28 March 2007.
123. Today in Science History. todayinsci.com. Retrieve 8 March 2006.
124. Orbital Debris FAQ: How much orbital debris is currently in Earth orbit? NASA, July 2009.
125. Orbital debris: Optical Measurements. NASA Orbital Debris Program Office
126. Stokes, GH., von Braun, C., Sridharan, R., Harrison, D., Sharma, J.: The Space-Based Visible Program. MIT Lincoln Laboratory. Retrieved 8 March 2006.
127. Klinkrad, H.: Monitoring Space – Efforts Made by European Countries. fas.org. Retrieved 8 March 2006.
128. MIT Haystack Observatory. haystack.mit.edu. Retrieved 8 March 2006.
129. AN/FPS-108 COBRA DANE. fas.org. Retrieved 8 March 2006.
130. UN Space Debris Mitigation Guidelines.
131. Anselmo, L., Pardini, C.: Collision risk mitigation in geostationary orbit. Space Debris **2**(2), 67–82 (2000)
132. de Selding, P.B.: CNES Begins Deorbiting Spot-1 Earth Observation Satellite. Space News, 20 February 2009
133. Christensen, B.: The Terminator Tether Aims to Clean Up Low Earth Orbit. space.com. Retrieved 8 March 2006.
134. Amos, J.: How satellites could 'sail' home. BBC News, 3 May 2009.
135. Campbell, J.: Using Lasers in Space: Laser Orbital Debris Removal and Asteroid Deflection. Occasional Paper No. 20, Air University, Maxwell Air Force Base, December 2000.
136. Bekey, I.: Project Orion: Orbital Debris Removal Using Ground-Based Sensors and Lasers. Second European Conference on Space Debris, ESA-SP 393, p. 699 (1997)
137. Mullins, J.: A clean sweep: NASA plans to carry out a spot of housework. New Scientist, 16 August 2000.

138. Carlson, E., et al.: Final design of a space debris removal system. NASA/CR-189976, 1990.
139. Michaels, D.: A cosmic question: how to get rid of all that orbiting space junk? Wall Street J., 11 March 2009.
140. Hitchens, T.: COPUOS Wades into the Next Great Space Debate. The Bulletin of the Atomic Scientists, 26 June 2008.
141. Orbital Debris – Important Reference Documents.
142. Air Force Regulation 200–2
143. Armando Simon, A.: The Zeitgeist of the UFO Phenomenon. UFO phenomena and the behavioral scientist (Scarecrow Press) (1979)
144. Vallée, J.: Alien Contact by Human Deception. Anomalist Books, New York (1990). ISBN 1-933665-30-0
145. Menzel, D.H., Taves, E.H.: The UFO Enigma. Doubleday, Garden City (NY, USA) (1977)
146. Sagan, C., Page, T.: UFOs: a scientific debate. Barnes & Noble. pp. 310. ISBN 978076070916. (1995)
147. McDonald, J.E.: Statement on Unidentified Flying Objects submitted to the House Committee on Science and Astronautics at July 29, 1968, Symposium on Unidentified Flying Objects, Rayburn Bldg, Washington, D.C (1968)
148. COMETA Report: http://www.ufoevidence.org/topics/Cometa.htm
149. Politicking and Paradigm Shifting: James E. McDonald and the UFO Case Study http://www.project1947.com/shg/mccarthy/shgintro.html
150. http://news-service.stanford.edu/news/1998/july1/ufostudy71.html
151. Cross, A.: The flexibility of scientific rhetoric: a case study of UFO researchers. Qualitative Sociology **27**(1), 3–34 (2004)
152. Project Blue Book Special Report #14
153. (CIA history of their involvement in UFOs)
154. Site du GEIPAN
155. Interview with GEIPAN director Yves Sillard; public statements of SEPRA director Jean-Jacques Velasco; 1978 GEPAN report by director Claude Poher.
156. COMETA Report (English), part1; COMETA Report, part2; COMETA Report summary by Gildas Bourdais; Summary by Mark Rodeghier, director of CUFOS
157. UK National Archives
158. news.bbc.co.uk Files released on UFO sightings
159. AFP Article: Britons 'spotted' UFOs, records say
160. BBC News Airliner had near miss with UFO
161. 'El Pais', Montevideo,Uruguay, June 6, 2009; English translation by Scott Corrales
162. Catalog of Project Blue Book unknowns
163. Hynek's photos in Hynek, The UFO Experience, 1972, p. 52
164. Herb/Hynek amateur astronomer poll results reprinted in International UFO Reporter (CUFOS), May 2006, pp. 14–16
165. Hendry, A.: UFO the Handbook: A Guide to Investigating, Evaluating, and Reporting UFO Sightings. Doubleday & Co., Garden City, N.Y (1979). ISBN 0-385-14348-6
166. Scientific Study of Unidentified Flying Objects, Section II Summary of the Study, Edward U. Condon, University of Colorado
167. Document quoted and published in Timothy Good (2007), 106–107, 115; USAFE Item 14, TT 1524, (Top Secret), 4 November 1948, declassified in 1997, National Archives, Washington D.C.
168. Schuessler, J.L.: Statements about Flying Saucers and Extraterrestrial Life Made by Prof. Hermann Oberth, German Rocket Scientist" 2002; Oberth's American Weekly article appeared in a number of newspaper Sunday supplements, e.g., Washington Post and Times Herald
169. Copy of FBI FOIA document; Text quotation in essay by Bruce Maccabee on military/CIA ETH opinions circa 1952
170. Dolan, 189; Good, 287, 337; Ruppelt, Chapt. 16

171. David Saunders, UFOs? Yes
172. Velasco quoted in La Dépêche du Midi, Toulouse, France, April 18, 2004
173. Vallée, J.: Dimensions: A Casebook of Alien Contact. Ballantine Books, New York (1989). ISBN 0345360028
174. Peter, F.: Coleman has advanced a theory that some UFOs may be instances of visible combustion of a fuel (e.g., natural gas) inside an atmospheric vortex. See Weather, p. 31, 1993; J. Sci. Explor. 20, 215–238 (2006), and his book Great balls of Fire–a unified theory of ball lightning, UFOs, Tunguska and other anomalous lights, Fireshine Press
175. Cook, N. (Narrator and Writer): An Alien History of Planet Earth. History Channel (2006)
176. Sturrock Panel abstract & summary; Sturrock Panel report on physical evidence; Other links to Sturrock Panel
177. Investigation and explanations of Belgium case
178. Fawcett & Greenwood, 81–89
179. Oberth's UFO antigravity opinion as to propulsion and atmospheric air flow control also quoted by Donald Keyhoe in his 1955 book Flying Saucer Conspiracy
180. Зайцев А.В. Международная система планетарной защиты « Цитадель-1». (in Russian).
181. Report of the Task Force on potentially hazardous Near Earth Objects. British National Space Center. http://www.spacecentre.co.uk/neo/report.html (2008). Retrieved 21 October 2008p. 12.
182. Morrison, D.: The Spaceguard Survey: Report of the NASA International Near-Earth-Object Detection Workshop. NASA, Washington, D.C (1992)
183. National Academy of Sciences. 2010. Defending Planet Earth: Near-Earth Object Surveys and Hazard Mitigation Strategies: Final Report. Washington, DC: The National Academies Press. Available at: http://books.nap.edu/catalog.php?record_id = 12842.
184. Stokes, G., Evans, J. (18–25 July 2004): Detection and discovery of near-earth asteroids by the linear program. 35th COSPAR Scientific Assembly. Paris, France. p. 4338. http://adsabs.harvard.edu/abs/2004cosp. . .35.4338. Retrieved 23 October 2007
185. Lincoln Near-Earth Asteroid Research (LINEAR). National Aeronautics and Space Administration. http://neo.jpl.nasa.gov/programs/linear.html. Retrieved 23 October 2007
186. The Spacewatch Project. http://spacewatch.lpl.arizona.edu/index.html. Retrieved 23 October 2007
187. Near-Earth Objects Search Program. National Aeronautics and Space Administration. . http://neo.jpl.nasa.gov/programs/. Retrieved 23 October 2007
188. NASA Releases Near-Earth Object Search Report. National Aeronautics and Space Administration. http://neo.jpl.nasa.gov/neo/report.html. Retrieved 23 October 2007
189. Morrison, D.: NASA NEO Workshop. National Aeronautics and Space Administration.
190. Charter, H.: Near-Earth Objects: Status of the Survey Program and Review of NASA's 2007 Report to Congress | SpaceRef Canada - Your Daily Source of Canadian Space News
191. We Saw It Coming: Asteroid Monitored from Outer Space to Ground Impact. Newswise, Retrieved 26 March 2009
192. Predicting Apophis' Earth Encounters in 2029 and 2036
193. Why we have Asteroid "Scares". Spaceguard UK. http://www.spaceguarduk.com/scares.htm.
194. Cole, D.M., Cox, D.W.: Islands in Space, pp. 126–127. Philadelphia, Chilton Books (1964)
195. Kleiman Louis, A.: Project Icarus: an MIT Student Project in Systems Engineering. MIT Press, Cambridge, Massachusetts (1968)
196. Systems Engineering: Avoiding an Asteroid, Time Magazine, June 16, 1967.
197. Day, D.A.: Giant bombs on giant rockets: Project Icarus, The Space Review (2004), Monday, July 5,
198. UCAR (September 13, 2006). Changes in Solar Brightness Too Weak to Explain Global Warming. Press release, Retrieved 18 April 2007.
199. Great Moments in the History of Solar Physics 1. Great Moments in the History of Solar Physics. Retrieved 19 March 2006

200. Perry, C.A., Hsu, K.J.: Geophysical, archaeological, and historical evidence support a solar-output model for climate change (PDF). Proc. Natl. Acad. Sci. U.S.A. **97**(23), 12433–12438 (2000)
201. Wang, Y.-M., Lean, J.L., Sheeley, N.R.: Modeling the sun's magnetic field and irradiance since 1713. Astrophys. J. **625**(1), 522–538 (2005)
202. Krivova, N.A., Balmaceda, L., Solanki, S.K.: Reconstruction of solar total irradiance since 1700 from the surface magnetic flux. A&A **467**, 335–346 (2007)
203. Steinhilber, F., Beer, J., Fröhlich, C.: Total solar irradiance during the holocene. Geophys. Res. Lett. **36**, L19704 (2009)
204. Lean, J.L., Wang, Y-M., Sheeley Jr, N.R.: The effect of increasing solar activity on the sun's total and open magnetic flux during multiple cycles: implications for solar forcing of climate. Geophys. Res. Lett. **29** (24), 77–1 ~ 77–4 (2002)
205. Foukal, P., Fröhlich, C., Spruit, H., Wigley, T.M.L.: Variations in solar luminosity and their effect on the earth's climate (PDF). Nature **443**(7108), 161 (2006)
206. Lockwood, M., Fröhlich, C.: Recent oppositely directed trends in solar climate forcings and the global mean surface air temperature (PDF). Proceedings of the Royal Society A **463**, 2447–2460 (2007)
207. Benestad, R.E., Schmidt, G.A.: Solar trends and global warming. J. Geophys. Res., 114, D14101, doi:10.1029/2008JD011639. (2009), http://pubs.giss.nasa.gov/abs/be02100q.html
208. Scafetta, N., West, B.J.:Phenomenological reconstructions of the solar signature in the Northern Hemisphere, surface temperature records since 1600. J. Geophys. Res. **112**: D24S03 (2007)
209. Moberg, A., Sonechkin, D.M., Holmgren, K., Datsenko, N.M., Karlén, W., Lauritzen, S.E.: Highly variable northern hemisphere temperatures reconstructed from low- and high-resolution proxy data. Nature **433**(7026), 613–617 (2005)
210. Clark, P.: Space Debris Incidents Involving Soviet/Russian Launches. friends-partners.org.
211. Iannotta, B.: U.S. Satellite Destroyed in Space Collision. Space.com, 11 February 2009. Retrieved 11 Feb 2009.
212. Reichhardt, T.: Satellite Smashers. Air & Space Magazine (2008), 1 March

Further Reading

1. Меньшиков, В.А., Перминов, А.Н., Ю.М. Урличич Глобальные проблемы человечества и космоса – М.: НИИ КС имени А.А. Максимова, 2010. (Russian) – 570с.
2. А.Н. Перминов, В.А. Меньшиков "Космос и беэопасность человечества.". – М: НИИ КС имени А.А. Максимова, 2009. (Russia) – 48 с.
3. Ranganath Navalgund, Valery Menshikov, Joseph Akinyede Space-Based disaster Managment: the Weed for international cooperation IAA, Bengalor-Paris, 2010, 80 p. (Heads of space Agencies Summit on November 17, 2010 in Washington)
4. Degtyar V. G., Volkov V. A., Kalashnikov S. T. "Development of principles fro adaptation of space and rocket complexes to be used for missions of mitigation of asteroids and comets threat." by book "Space for Security and Prosperity of the Peoples.", editors Contant J.-M., Menshikov V., IAA,- 2010, p. 189–208

The Conception of the International Global Monitoring Aerospace System (IGMASS) 4

These original materials – the conception of the International Global Monitoring Aerospace System[1] (IGMASS) as a system for forecasting destructive natural phenomena and man-made disasters in order to guarantee social, economic, seismic, environmental and geophysical safety, the prevention of other global space threats, as well as the development of information-navigation and telecommunication resources for the planet for the benefit of all humanity – were developed by an innovative team of specialists of the International Academy of Astronautics (IAA) and Russian Academy of Cosmonautics n.a. K.E. Tsiolkovsky (RAKTS). The proposal to create IGMASS was first openly expressed at the International Conference "Modern Space Technologies for the Prosperity of Humanity" (Dnepropetrovsk, Ukraine, 2007). Later it was reported at the International Scientific Forums "Space for Humanity" (Korolyov, Russia, 2008), "Advanced Space Systems and Applications" (Shanghai, China, 2008), at the Mediterranean Conference on Astronautics (Tunisia, 2008), and also at the Academic Day of the International Academy of Astronautics (Glasgow, Scotland, 2008). In 2009 a special international working group of experts (from the USA, France, Germany, Russia, Japan, Italy, India, China, Ukraine, Belarus, Bulgaria, and Tunisia) was formed within the IAA to study the possibility and prospects of the creation of IGMASS. The results of the work of this group were discussed and strongly supported by the heads of several national space institutions, the managers of leading enterprises of the rocket–space industry, outstanding scientists and administrators from more than two dozen countries during the First International Specialized Symposium on "Space and Global Security of Humanity," which was held in November 2009 in Limassol (Cyprus).

The conception includes the goals of the creation of IGMASS, tasks that are expected to be solved with its help, profile of the system, description of its functionality including receiving, processing, and dissemination of forecast data

[1] Monitoring is a process of systematic or continuous collection of information on the parameters of a complex object or process.

V.A. Menshikov et al., *Global Aerospace Monitoring and Disaster Management*, DOI 10.1007/978-3-7091-0810-9_4, © Springer-Verlag/Wien 2012

from aerospace monitoring, use of the information resource provided by IGMASS in order to solve urgent problems of humanity (eradication of illiteracy, distance education, disaster management, technological, human, and environmental disasters), and prediction of threats in and from outer space. Also presented herein are the organizational and economic aspects of establishing, developing, and full-scale use of the system, as well as the role of the UN and the International Academy of Astronautics in solving the problems inherent in realizing this ambitious project.

As already mentioned, the sustainable development of modern civilization is prevented by threats of natural and man-made origin, requiring the adoption of effective preventive measures to predict and protect against such threats. The most common sources of natural disasters are meteorological, climatic, and tectonic phenomena: floods, typhoons, hurricanes, droughts, forest and grass fires, earthquakes, volcanic eruption, tsunamis, landslides, mudflows, and avalanches. To predict their beginning, to warn about such phenomena and the disasters they cause, including accidents and (or) man-made emergencies, is in all respects more beneficial than to respond to the following destructive consequences.[2] Since a third of man-made emergencies are caused by natural phenomena, the effective monitoring and forecasting of the geophysical situation in the vicinity of complex technical systems would avoid many accidents and disasters (Fig. 4.1).

In addition to Earth-based disasters the planet is also threatened by dangers of solar, lunar, and cosmic origin. The former two are generated by solar activity and the movement of the Earth around the Sun and the Moon around the Earth, the last – by comets and asteroids. For example, the periodic increase of seismic activity of the Earth's crust and effects on the atmosphere, ionosphere, and magnetosphere were observed in accordance with the 11-year solar activity cycle.[3] The Moon has 2.2 times more powerful gravitational influence on the Earth than the Sun, creating daily cyclic gravity loads on the Earth.[4]

Over the past century our planet has been bombarded by meteorites, comets, and asteroids, including dangerous events such as Tunguska, Sikhote-Alin, or the

[2] Thus, in 2008 alone there were 137 natural and 174 man-made disasters that had taken nearly a quarter of a million lives on the planet. According to international organizations from 1970 to 2000 the amount of damage, i.e., the cost to humanity of natural and technological disasters, was about one and a half trillion dollars.

[3] The Sun is a magnetically active star with a strong electromagnetic field, the intensity and direction of which periodically varies. The Sun affects the Earth and in accordance with the rotation around its axis (27 days), the annual circulation of the Earth and its daily rotation. Variations of solar activity and the solar magnetic field exert an influence on the structure of the magnetosphere, the ionosphere, and the Earth's atmosphere, causing a variety of effects which altogether cause the Earth auroras and geomagnetic storms, disrupting the operation of communications, electricity, as well as having negative impacts on living organisms, including human beings. Solar and lunar gravitational effects on the Earth's crust are the "trigger" for the onset of earthquakes and volcanic eruptions.

[4] The influence of gravitation from our natural satellite pulls water on the surface of the Earth up to 0.5 m towards the Moon and causes tides.

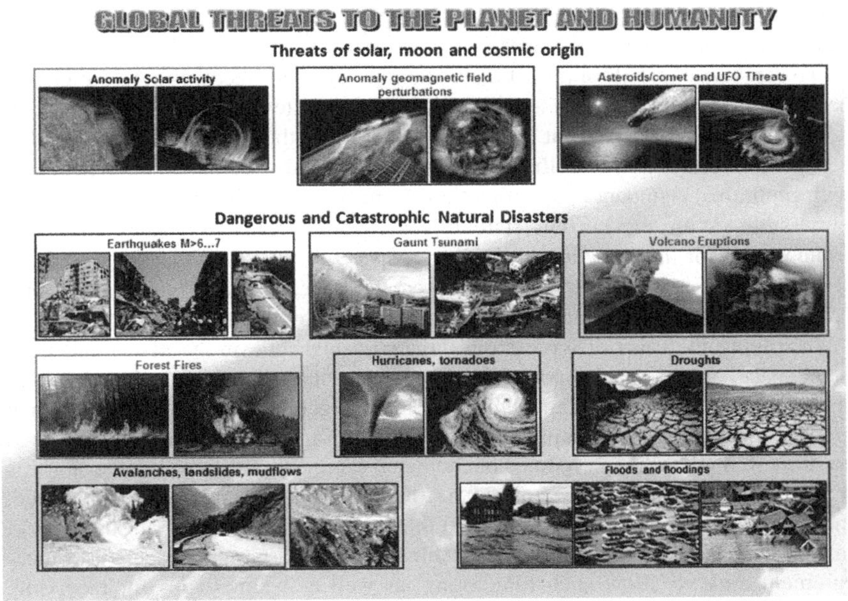

Fig. 4.1 Contemporary natural threats to the planet and humanity

meteorite that struck Brazil, warning us of a possible global or regional disaster to come, which could surpass in scale all known natural disasters. The danger comes especially from the increasingly more frequent instances of the trajectories of large but often unknown asteroids crossing the Earth's orbit,[5] the collision with which could have fatal consequences for the planet.

Work on monitoring of solar activity and the comet/asteroid danger is still at the stage of scientific experiments carried out with the help of ground-based optical and radar equipment. A number of satellites are used for basic research of solar–terrestrial relations, but observations of the Sun from space are conducted only periodically. The monitoring of outer space for detection of Earth-threatening

[5] An asteroid that approached the Earth and that was not detected by any telescope exploded on the 8th of October, 2009 in the atmosphere of the Earth at an altitude of about 15–20 km over the province of Southern Sulawesi (Indonesia). According to NASA, the destruction of this stone space object with a diameter of about 10 m and that entered the atmosphere at a speed of 20 km/s caused an energy release equal to 50 thousand tons of TNT, i.e., three times the power of the explosion of the atomic bomb over Hiroshima and it was detected by the Western Ontaric university observatory in Canada, which was 16 thousand km from the epicenter of the event.

comets and asteroids is limited by the number and physical abilities of ground-based optical and radio telescopes.[6]

To date less than half of the estimated number of 1-km sized cosmic "wayfarers" has been catalogued. Even the most powerful space telescope Hubble is able to detect 1-km sized asteroids at a distance of not more than 40 million km (or little more than 20 days before collision with the Earth) and even then only if provided with preliminary guidance to a potentially dangerous area in space which is quite impossible. With the current state of development of space technology and technologies for the detection of dangerous space objects, for a warning period of at least 5 days it would be necessary to create a special space system of several "patrol satellites" at a distance of millions of kilometers from the Earth.

In solving the tasks of forecasting threats arising both on the planet and from outer space, of particular importance is the continuous monitoring and comprehensive analysis of various parameters of anomalous geophysical phenomena, which precede the occurrence of natural disasters and man-made emergencies. It has been confirmed that such anomalous phenomena (precursors) occur in the Earth's magnetosphere, ionosphere, atmosphere, and lithosphere, and can be identified, measured, and used to predict the location, time, and effect of a catastrophe.[7] In many countries work is ongoing to establish ground-based and space-based measurement tools to allow such forecasting, as well as technologies for receiving, processing, and transfer of the necessary information that could form the basis for future integrated warning systems of natural disasters and emergencies. However, the well-timed detection of signs and the forecasting of emergencies of space, natural, and technogenic origin is possible only through the realization of large-scale international projects involving the complex use of both existing and prospective ground-based, air, and space facilities.

Concerted international efforts in this regard have been undertaken from the very beginning of the space age. The International Global Space System for hydrometeorological support (the *Global Climate Observation System*) has been created and is currently operational. It is able to forecast forthcoming disasters of meteorological origin and dangerous climatic anomalies (floods, typhoons, hurricanes, and storms). Today, there are space systems, which are capable together of solving up to 300 basic and applied issues of Earth Remote Sensing (ERS), including problems of the consequences and evaluation of natural and man-made disasters. However, the instrumental composition, size, and structure of ERS satellite constellations does

[6] The maximum range of detection by ground-based facilities of asteroids with a size no less than 1 km does not exceed 2–2.5 million km. This means that at an average closing rate of the asteroid with the Earth (20 km/s) a collision can occur in less than 1.5 days, which is insufficient for any effective security measures to be undertaken.

[7] The number of such precursors is more than 300. However, statistics confirming the reliability of forecasting according to precursors – is absent, only individual events are described. Therefore, problems related to forecasting can only be effectively solved by registration of precursors in their totality in all environments.

not allow the solving of forecasting problems and early warning of coming danger.[8]

The IGMASS project introduced herein is a large organizational and technical system, which should be created under UN guidance according to the principles of coordinated international cooperation and long-term partnership in the field of technical design, development, and exploitation of ground-based and aerospace resources for the solving of a wide range of forecasting issues.

Project realization in regard to its practical implementation would initiate a new, unified strategy of space exploration aimed at achieving environmentally sound and socially sustainable development of the world community based on common, lasting values to sustain life on our planet.

4.1 Analysis of the Development of the Emergency Monitoring Space Facilities

In recent years the world has paid a great deal of attention to the development of space systems for the monitoring of emergency situations. During more than five decades since the launch of the first ESV several generations of spacecraft as well as target and communication equipment, new multispectral and hyperspectral devices, multispectral radiometers and radars, lasers, heliogeophysical equipment, computers, communication facilities, and many other assets have been developed. New technical and technological solutions have been developed for the small and microsatellites.[9] As a result, modern observation satellites (ERS), which have a mass from 300 up to 800 kg, solve effectively issues related to the monitoring of the atmosphere and the Earth's surface. Because of mass and cost reductions for spacecraft it has become possible to create multisatellite systems, providing high efficiency, reliability, and integrity of monitoring of the various objects and processes.

Today, relevant projects and initiatives at various stages of realization are ongoing in the United States of America, Canada, the EU countries, and the states of South and South-East Asia. Both national and corporate space systems for monitoring and security are developing rapidly. They include multipurpose multisatellite space remote sensing, communication and data broadcasting, navigation, hydrometeorological and topogeodesic support and technological purposes as well. We can surely say that in recent years a world space industry and information

[8] There are also a number of international, regional, and national projects and programs (UN-SPIDER, "Global Earth Observation System of Systems" (GEOSS), "The European Global Monitoring of Environment and Security" (GMES), "The System of Monitoring Natural Disasters in Asia Pacific Region" (Sentinel Asia), "The International Charter on Space and Major Disasters" (Disaster Charter), "The monitoring system of natural and technological disasters" "Ionosat" (Ukraine), etc.), which focus more on providing mitigation of consequences than on their prevention, and still less – forecasting.

[9] According to the current classification space vehicles with a mass of 100–1,000 kg belong to the category of small spacecraft (SSV), about 100 kg – microspacecraft.

infrastructure for observation has been formed with the participation of almost all the leading nations of the world (the USA, Canada, France, Italy, Germany, UK, Israel, India, China, Russia, and Japan), international consortia, and about 20 countries from all the continents of the Earth.[10]

Space monitoring facilities are generally accepted to be divided into hydrometeorological and ERS systems, although during the solution of applied monitoring problems information from both systems is used jointly. *Hydrometeorological systems* are usually deployed in low polar geosynchronous[11] and geostationary[12] orbits. They provide meteorological monitoring and forecasting of dangerous meteo-phenomena and can only partially be used to solve monitoring problems related to geophysical processes taking place in the lithosphere. The geophysical facilities on newly launched low-orbit meteorological satellites can catalogue in the atmosphere and ionosphere only some of the precursors of large earthquakes and heliophysical anomalies.

Contemporary space remote-sensing facilities are represented by a very extensive nomenclature of satellites: American (Landsat-7, EO-1, Ikonos-2, Quick Bird-2, OrbView-3, Geo Eye-1, World View-2, World View-3, USA-200); Indian (IRS, Cartosat-2A, Risat, IMS-1); Israel (EROS-B, EROS-C, TECSAR); French (Spot-5 and Jason-2); Japanese (Adeos-1, Adeos-2, Alos); Canadian (Radarsat-1 and Radarsat-2); Chinese (HJ-1A,-1B, Yaogan-5); Italian (Cosmo-Skymed, Cosmo-3); European (ERS-2, Envisat-1); German small and microsatellites (TerraSar-X, Sar-Lupe, Rapid Eye); Russian (Resurs DK). Algeria, Brazil, Nigeria, Taiwan, Thailand, Turkey, South Korea, and other countries also have their own space observation satellites, created in cooperation with the leading space powers.

It is proposed that IGMASS be created as a supranational system according to the principles of using the whole potential of modern space, including international space projects in relation to disaster monitoring, and its realization will substantially contribute to the development of a global process for providing information on emergency situations in various regions of the Earth. Analysis of these projects shows that they are all mainly focused on the issues related to identifying the harmful effects of natural disasters and emergencies. Thus, the final result of the international project Global Earth Observation System of Systems (GEOSS), based on a 10-year plan (2005–2015), initiated by the United States "Group on Earth Observations" (GEO), will be a publicly available global infrastructure, which will

---

[10] In 2007–2008 the proportion of spacecraft for communication, broadcasting, navigation, and hydrometeorology exceeded 85% of the total number of all spacecraft launched by the world community into Earth orbit (92 of 113 spacecraft in 2007 and 87 of 97 spacecraft in 2008).

[11] Presently, there are about a dozen meteorological satellites belonging to the U.S. (NOAA-K, DMSP5D-3), ESA (Metop-A), China (FY-1D, FY-3), and Russia (Meteor-M) in the subpolar geosynchronous orbit.

[12] Spacecraft, created by the U.S. (GOES), the European Union (Meteosat, MGS), Japan (MTSAT-1R), India (Metsat-1, Insat-3A), China (FY-2 C, D, E), and Russia (Elektro-L in 2010) are placed in geostationary orbit.

provide a wide range of users with comprehensive, processed space monitoring information[13] at a near real-time scale. Though because of recent investment it has become possible to unite disparate monitoring tools and software to measure the physical, chemical, and biological parameters that characterize the integrated potentially dangerous processes occurring on the Earth within GEOSS, this project is not intended to create its own orbital constellation which significantly limits its ability to solve the declared tasks of prediction of dangerous natural and man-made phenomena.

The international system of space monitoring of natural disasters (Disaster Monitoring Constellation – DMC), for the realization of which an international consortium was created in 2002 (Algeria, UK, Nigeria, China, Thailand, and Turkey), includes a low-orbit constellation of seven British-developed 80–130 kg microsatellites in polar orbits, equipped with a multispectral optoelectronic complex of medium resolution of 20–30 m. Microsatellites in the DMC are owned and operated by the United Kingdom, Algeria, Nigeria, Turkey, China, Thailand, and other countries, exchanging, if necessary, space data. The possibilities of such a system are very limited – it can catalogue only past major seismic or man-made phenomena, it focuses on obtaining information only in the visible spectral range, and is designed to provide operational information to competent organizations and professionals only in those countries on whose territory an emergency situation arises.

The European initiative "Global Monitoring for Environment and Security" (GMES), aimed at the establishment of a European monitoring potential (the project includes France, Italy, Germany, Canada, Israel, and a number of specialized aerospace companies in other countries), represents the EU contribution to GEOSS. This system is to include space ERS, navigation, and communication systems. It will be applied, according to its framework, to create a global environmental monitoring system of the planet, which will consist of analytical centers, ground stations, and a space constellation. Although some parts of the system are already in operation, it is still under development and completion of formation of the orbital constellation is planned for 2012.[14]

The orbital constellation GMES includes 13 observation spacecraft, as well as satellites Gelios-2, Pleiades, Cosmo-Skymed, SAR-Lupe, Spot-5, Rapid Eye, DMC2 (Topsat 2), and TerraSAR-X.[15] In the future, ESA is planning to create

[13] At the same time in GEOSS it is expected that various ground-sensor equipment, weather stations, weather sensors, sonars and radars, a set of 60 satellites, including the navigation constellation "NAVSTAR," a powerful package for modeling, simulation and forecasting, as well as facilities for early warning of the population in countries and regions at risk, will be integrated.

[14] The program budget has been approved in the amount of 2.2 billion Euros.

[15] In relation to the fact that in 2008 ESA initiated the deployment of the global space navigation system Galileo, it has its own space hydrometeorology systems (9 SV), communication and broadcasting (16 SV), and as a part of the constellation GMES in some periods can operate with over 70 satellite vehicles.

the satellite constellation (among them – SV Sentinel, ERS, ENVISAT, GOCE, SMOS, CryoSat-2, Swarm, ADM-Aeolus, Earth CARE, MSG, MetOp, JASON-2, and PLEIADES), which are expected to be equipped with C-band radar (for interferometric shooting), optical cameras with medium spatial resolution (for mapping and hyperspectral shooting), and optical equipment and a radar altimeter (for detailed monitoring of ocean waters, the Earth's atmosphere from low and geostationary orbits). Although the GMES project has its own orbital constellation, development and acquisition of the satellites, as well as the coordination of space-based assets of European national satellite operators is realized by ESA, but it does not include issues related to the identification of precursors and forecasting of natural and man-made disasters. In addition, a number of satellites in GMES are designed to meet the challenges of defense departments and its resources are unlikely to be used on a regular basis for the purposes of international global monitoring.

Initiated in 2000 by ESA and the French Space Agency the International Charter "Space and Major Disasters," which space agencies and organizations of Argentina, India, Canada, the USA, and Japan subsequently joined for the purposes of its realization, aims at creating a unified space data system, designed to provide necessary information to victims of natural or man-made disasters. Although the orbital segment of the project includes national ERS satellites of member states – ERS, ENVISAT (ESA), SPOT (France), RADARSAT (Canada), IRS (India), GOES (USA), SAC-C (Argentina), ALOS (Japan), because of its specific objective focus (coordinated use of space facilities in case of natural or man-made disasters and providing free space monitoring data to the affected countries), the Charter does not solve a wide range of forecasting issues in relation to occurring natural disasters on the planet.

The project "Sentinel Asia," proposed in 2004, includes 51 organizations, as well as 44 agencies from 18 countries, and provides for the creation in the Asia-Pacific region (APR) of a control system for the management of the consequences of natural disasters using space ERS technologies in quasi-real time, in conjunction with GIS-mapping technologies and the modern global network, the "Internet."[16] However, considering the limited size of the on-board equipment used in the SV-project and the specifics of the orbital constellation's construction, it is unlikely to be possible to solve issues related to the forecasting of natural and man-made phenomena on a global scale within the project.

In concluding our analysis of the status and prospects of the development of space facilities and monitoring systems for emergency situations and their objective focus, the complete absence of issues related to global planetary threats (for example related to the danger from meteoroids and asteroids, solar activity, etc.) is apparent.

[16] The architecture of the project is being developed with the possibility of receiving and processing imagery and textual information voluntarily submitted by Asia-Pacific countries from satellite remote sensing systems, including geostationary platforms.

4.2 Purpose of the Creation of IGMASS and Its Missions

The objective behind the creation of IGMASS is the provision of timely warning to the international community of upcoming emergencies as well as natural and man-made disasters via global monitoring and forecasting using the scientific and technical potential of Earth-based, air and space-based monitoring systems all over the world and the further development and gradual integration of navigation, telecommunication, and information resources of the planet with the purpose of solving global human issues (Fig. 4.2).

The purpose of IGMASS is global and effective forecasting of potentially dangerous natural and man-made disasters on the Earth and in outer space, based on integrated global aerospace monitoring resources.

The priority missions of the system, which will be solved by using ground, air, and space-based facilities, are:

- continuous and uninterrupted space monitoring of the lithosphere, atmosphere, and ionosphere of the Earth, and the near-Earth space, to identify early signs of dangerous natural and man-made disasters (Fig. 4.3);
- obtaining, on-board preprocessing, and transfer of monitoring data from satellite to the ground receiving stations;
- compilation and complex use of the global monitoring data, received from space, air, and Earth assets, processing with the use of national, regional, and international situation centers, and its interpretation, storage, and display;
- operating operational delivery of information on identified natural and man-made threats to the relevant organizations of the countries at risk and specialized UN structures;

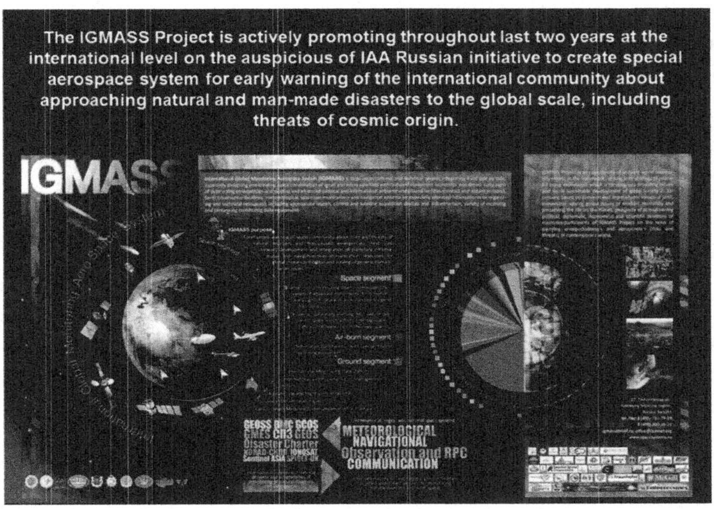

Fig. 4.2 Purpose of IGMASS creation and its applicability

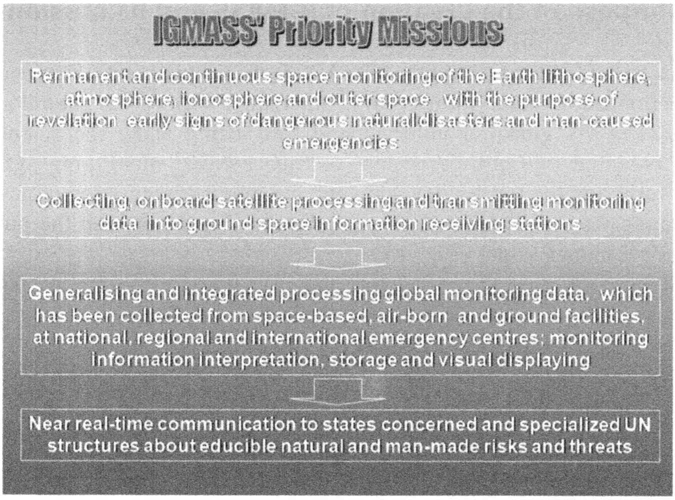

Fig. 4.3 Priority missions of the system

- guaranteed navigational and telecommunication support for customers all over the world in the interest of performing emergency response, disaster medicine, human operations; creating a system of transport corridors, optimizing the movement of people and goods; eradication of illiteracy, development of distance learning for preservation of cultural values;
- warning about global threats in and from outer space: the threat from asteroids and meteoroids and abnormal phenomena of different nature;
- gradual formation of a united, planetary "information security space" in order to reduce global risks and emerging threats.

Concerning the objective focus of IGMASS, the primary tasks of the system should be: identification of earthquake-prone areas, detecting and documenting the precursors of dangerous geological phenomena for real-time warning of their coming, their evolution in time and space, and the subsequent permanent control of environmental danger (seismic, aggressiveness, variability, etc.) to man-made systems and its components (Fig. 4.4).[17]

Appreciating the necessity of optimizing the terms of creation of IGMASS, other issues, assigned to the system, will be solved in two stages. The first stage – telecommunication and navigation support activities to decrease the negative consequences of disasters, support humanitarian operations; development of distance learning and specialist training in various fields. The second stage – long-range

[17] Some dangerous man-made disasters occur as a result of the gradual merging and interaction of complex technical systems with the natural environment (geotechnical processes and systems).

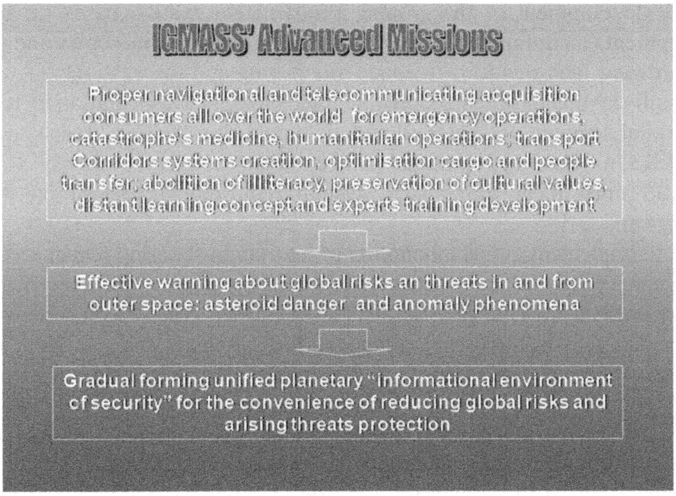

Fig. 4.4 The system's advanced missions

objectives of preventing global threats in and from outer space and the gradual formation of a united "information security space."

4.3 The Main Principles of the Creation of IGMASS and General Requirements for the System

To perform short-term forecasting of natural disasters and man-made disasters it is necessary to ensure efficient delivery, specialized processing and transmission to decision-makers, of specialized information concerning the evolution of parameter changes in the Earth's lithosphere, atmosphere, and ionosphere and near-Earth space which can be achieved by the use of a low-orbit and GEO satellite constellation complex, equipped with specialized on-board facilities, combined with the aircraft as well as land-based radar control tools and effective ground infrastructure for receiving, processing, and analysis of information.

IGMASS is based on the following basic principles:

- Absolute observance of the norms and principles of international space law, as well as the relevant single and multistate responsibilities in the field of space activity.
- The widest possible use and ensuring the continuity of results of relevant research and development programs, conducted in the framework of the international space programs for aerospace monitoring.
- A stage-by-stage approach to the creation of system parts taking into account the priority tasks of forecasting global natural and man-made phenomena, technological progress in terms of developments in the field of aerospace monitoring and resource use.

- Priority development of the ground infrastructure system based on full-scale development (including testing and practice) of basic technologies and software and hardware facilities for aerospace monitoring and forecasting.
- Broad informational, organizational, and technological cooperation of IGMASS' own orbital segment with ground and air monitoring facilities and with existing space ERS, navigation, communication, and data broadcast systems.
- IGMASS must meet three basic system requirements, based on the main purposes and tasks.

Firstly, to implement global monitoring of the current state and dynamics of potentially dangerous processes, early detection of their expressions, updating coordinates of the disaster origin areas with prior assessment of impacts on basic ecosystems and the human population in order to develop adequate mitigation measures for prevention and protection.[18] And in relation to this the following should be provided:

- Search, identification, and cataloging of earthquake-prone regions, active faults of the Earth's crust, updating of the global maps of seismic hazard, including the precise mapping and identification of signs of such activation[19].
- Receiving and transmission of information by ground-based sensors from meteorological, seismic, hydrological, geophysical stations and observation points in earthquake-prone regions.
- Cataloging precursors of dangerous geological phenomena, notification of their manifestations, positioning of seismic objects on the Earth's surface.
- Regular monitoring of the development of seismic phenomena, determination of their devastating effects in real-time.
- Mapping areas for construction of objects related to potentially dangerous industries, monitoring the construction progress of these objects.
- Monitoring the impact of geotechnical processes on the largest and most dangerous man-made objects and systems, as well as their environment.
- Warning of forthcoming meteorological, seismic, hydrological, geomagnetic, and other dangerous phenomena, which provide a threat to technical objects, and may cause unauthorized interference in their work.

[18] Criteria for forecast requirements for IGMASS can be divided into four groups, which are long-term (years, decades), medium-term (1 year), short-term (up to 10 days), and operational (day-hour) types of forecast. Short-term and operational forecasts apply to dangerous meteo-phenomena, medium – to the prevention of meteoroid/asteroid dangers and natural disasters of geological origin, and long-term – global natural disasters of geological origin.

[19] Today a number of anomalous phenomena in the atmosphere, ionosphere, and on the Earth's surface are known that can potentially be considered as signs of an upcoming seismic phenomenon. These include a sudden change in the concentration of electronic components and arising of large-scale irregularities in the F2 layer of the ionosphere, and ultra-high-and-low frequency electromagnetic waves, abnormal changes in the quasi-steady electric field and magnetic field; variations in the composition, concentration, flow rate, and temperature of the ionospheric plasma; intense glow of the atmosphere at frequencies corresponding to the vibrational spectra of atomic oxygen and hydroxyl; emissions of radon and metalized surface of aerosols in the atmosphere; raising the Earth's surface temperature, forming of aerosol clouds above active faults, etc.

- Monitoring of solar weather and gravitational action of the Moon and the Sun, in order to forecast geomagnetic effects.
- Monitoring of the near-Earth space to provide warning of meteoroid and asteroid threats.[20]

Secondly, to provide an opportunity to inform in-time the competent authorities of the concerned countries and the international community about dangerous natural and man-made disasters that are forthcoming in the short term.

Thirdly, to provide a wide range of consumers with high-precision navigational and telecommunication services to monitor progress in emergency response, conduct evacuation efforts, optimize conveyance of persons and goods, and solve other social and economic issues (distance education with using advanced space and information technologies, disaster medicine, training of relevant IGMASS experts and specialists in other areas of science and engineering).

4.4 Structure and Formation of IGMASS

IGMASS as a large organizational and technical system is intended to integrate in its structure, along with its own specially created space segment – a microsatellite constellation with onboard detection equipment to detect early signs of destructive disasters, both existing and future national and international air and ground-based facilities, including contact and remote sensors, the ERS space systems, communication and broadcasting, meteorological and navigational support (or allocated information, organizational, and technical resources), together with appropriate ground infrastructure for the insertion, deployment, and maintenance of spacecraft, as well as for the receiving, processing, and distribution of monitoring information. This will provide for the monitoring of emergencies and disasters developing and occurring on the Earth and in the near-Earth space with worldwide coverage, and enable reliable forecasting of their occurrence with the purpose of taking the necessary measures in the world community to prevent or mitigate their devastating effects, including well-timed evacuation of people, preserve material resources and cultural values, along with the use of navigation, telecommunication, and information resources of the world community to solve an entire range of contemporary issues facing Humanity (Fig.4.5).

The combination of IGMASS' own orbital constellation with the information resources of existing space systems that possess the ability to monitor the Earth's surface, atmosphere, and near-Earth space will provide for global warning of dangerous geophysical and meteorological phenomena and efficient data transfer to monitor their precursors to almost anywhere in the world (Fig. 4.6).

Space-based and air-borne IGMASS facilities should also be used to obtain data on the situation in areas of large-scale destruction (state of the power grid,

[20] Space warning subsystem on the asteroid and meteoroid risk within IGMASS in conjunction with existing and pro-spective ground-based facilities should ensure high reliability of objects detection larger than 50 m at distances of at least 15 mln km.

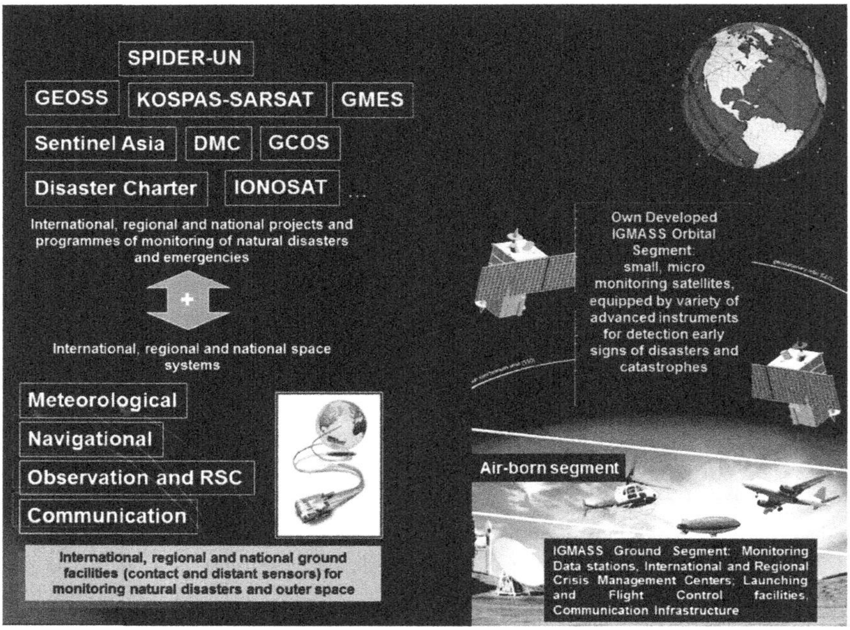

Fig. 4.5 The system's structure

pipelines, roads, etc.) for future forecasting of adverse climatic and meteorological conditions and seismic disturbances (crustal movements, landslides and rockfalls, caves-ins, mudflows, avalanches, etc.), that could threaten the integrity of technical facilities. Of particular importance is the cataloguing of all kinds of anomalies both during the construction and operation of objects (emissions to air or water of poisonous or radioactive materials, flammable gases, dust, aerosols, etc., unauthorized access to pipelines, disruptions in transport infrastructure, etc.)

The ground-based component of IGMASS must provide for the collection of objective, telemetry, and navigational information from space-based and air-borne facilities, and the deployment and replenishment of the space echelon of IGMASS using space-rocket complexes at ground, sea, and air bases.

IGMASS' own specially created space segment will consist of a low- and high-orbit constellation of small and microsatellites, placed in low, sun-synchronous (polar) orbits and GSO, respectively. In low orbits MSV (Modular Space Vehicles) would be deployed, equipped with consistent monitoring facilities and specialized geophysical equipment (side-looking radar and interferometry, multifrequency, polarimetric and multistation radar with antennas with synthesized aperture). Consistent microsatellite platforms to solve heliophysical observation issues as well as experiments with advanced scientific equipment and communication and

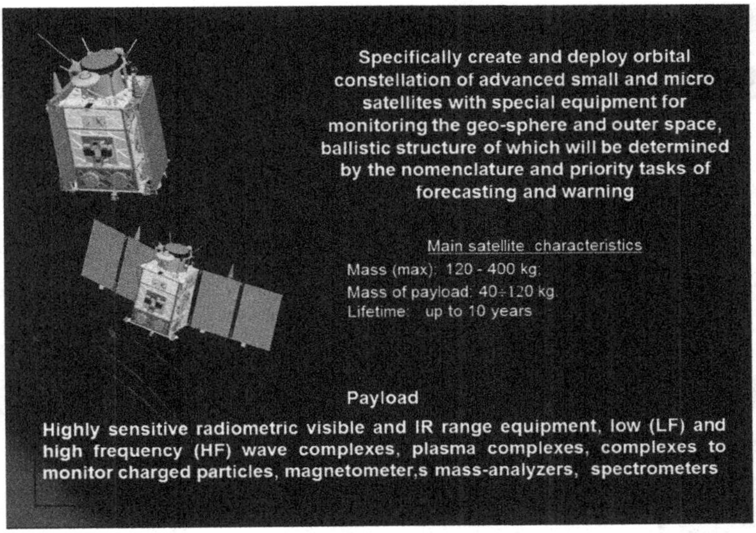

Fig. 4.6 IGMASS' own (specialized) orbital segment

broadcasting, will be available in the area of the geostationary orbit. Developments, such as those that would be required to create low-orbit observations via MSV are already underway in many countries: Russia in cooperation with companies from the UK, Israel, Germany, France, Italy, and the USA; Canada in cooperation with companies from the USA and several European countries (Fig. 4.7).

The integration into IGMASS' own specially created space segment of other national and international facilities – geosynchronous and low-orbit space complexes and systems of hydrometeorological support, ERS, communication and broadcasting ground-based complexes for the receiving, recording, and processing of space monitoring information – will provide an integrated picture of precursors of natural and man-made disasters.

With the possibility of integration of foreign and international systems the space segment of IGMASS would include the following special equipment:

- Geophysical monitoring facilities for solar activity and the identification of physical abnormalities of the Earth's magnetosphere, ionosphere, and atmosphere.
- Radar S and X-range with multiple polarizations.
- Microwave radiometers (from 10 to 200 GHz and more) for the registration of small gaseous components in terms of temperature, humidity, and other atmospheric parameters.
- Optoelectron devices with high and medium spatial and radiometric resolution for registration of the Earth's surface temperature.
- Radio-tomography facilities for the ionosphere using signals from low-orbit navigation satellites and ground-receiving stations.

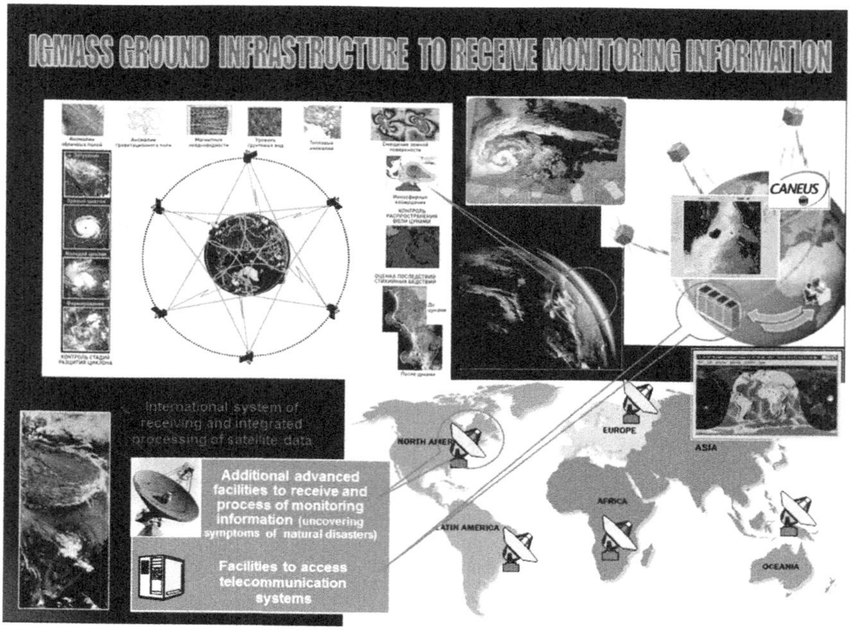

Fig. 4.7 Monitoring data ground infrastructure for the IGMASS project

- Heliophysical equipment for solar activity anomaly registration.
- Powerful optical telescopes to monitor asteroid and meteoroid threats[21] and for efficient warning about the dangers of "space debris."

The air-borne segment of IGMASS will include national air-borne facilities (airplanes, helicopters, dirigibles, meteo-sounding balloons, pilotless aircraft), used by the project member states. Pilotless aircraft systems for remote sensing have been intensively investigated in recent years and will assume an important place in the air-borne segment of IGMASS, especially for solving forecasting issues in relation to large-scale man-made disasters.

The ground segment of IGMASS will include ground facilities to support orbital position of satellites and their control subsystem that provides consumers with

[21] One of the projects in deep space is expected to deploy three satellite-telescopes: two of which are placed in the orbit of revolution of the Earth around the Sun, providing the detection of large asteroids at distances of up to 10 million km, and the third – with a long-focus telescope (17 m), located in the Lagrangian libration point between the Earth and the Sun, providing accurate determination of motion parameters of identified objects threatening the world and forecasts of objects closing to dangerous proximity to the Earth (with a warning period of at least 3 days). See Chapter III of the book.

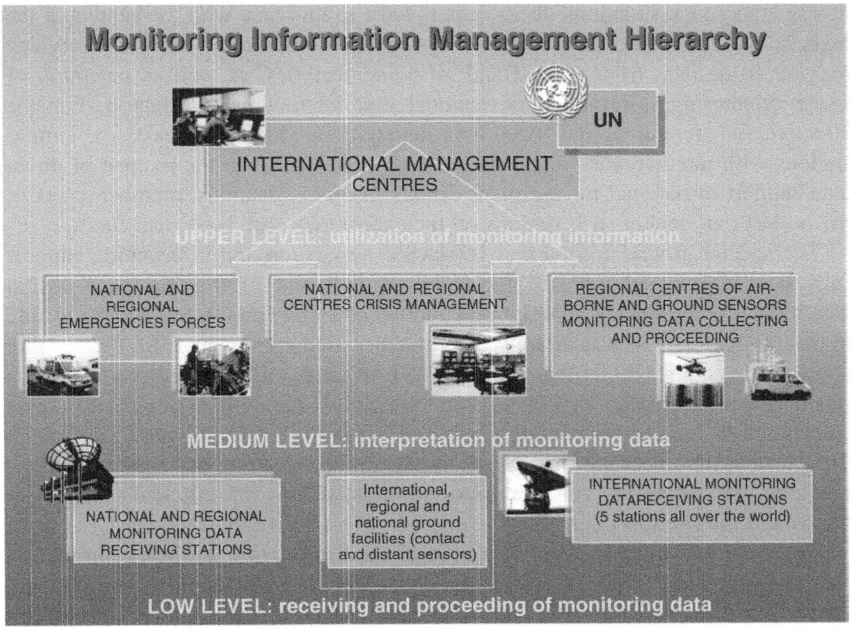

Fig. 4.8 IGMASS monitoring information management

monitoring information, and a special complex of navigation and information support subsystems.

IGMASS launch facilities shall provide necessary capabilities for autonomous, complementary, cluster injection of small or micro spacecraft in order to deploy and maintain system's space segment augmented by ground, marine and air assets.

The ground control complex of IGMASS should guarantee the collection of telemetry data from its own orbital segment, conduct flight control missions, develop long- and short-term plans for the application of orbital constellations, work programs and corrective commands of onboard equipment, etc.

The ground-based global subsystem provides consumers with acquisition, structural recovery, processing, storage, and circulation of all types of monitoring data as well as the planning of special applications for the system. The system is represented by a complex of interconnected and topologically distributed ground-based receiving facilities, for multilevel processing, storage, and transmission of the monitoring and forecast data, obtained from space and ground-based information sources. The main subsystem within IGMASS will have a hierarchical, three-level structure with a radial topology, combining international and national monitoring and crisis management centers as well as ground stations for receiving the monitoring information (Fig. 4.8).

The highest level of the subsystem will include international centers for control during emergency situations, situated in Russia, Asia, Europe, and America. The medium level of the hierarchy of the subsystem includes national centers for control during emergency situations; these centers have to interface with the centers at the highest level. The lower level of the subsystem will consist of ground stations which provide space data from ground and air-borne facilities, as well as reception of space monitoring information of national and international affiliation. For the effective functioning of IGMASS it is necessary to deploy at least five similar stations with international status, spread across the Globe. In the interest of direct data support of national management centers in crisis situations, member states of the project can deploy such stations on their territories with their own funds.

The special ground complex of IGMASS' navigation and information support subsystem is designed to create a unified navigation and information space, within which the whereabouts of an unlimited number of mobile and fixed objects are able to be automatically and accurately determined, based on signals from the satellite navigation systems GLONASS, GPS NAVSTAR, and Galileo. The structure of this special ground complex will include an "integrated telematic system of transport corridors," designed for evacuation activities in case of natural disasters, enhancing the transportation network throughput, ensuring traffic safety, protecting the environment, and increasing cargo transportation as well as the efficiency of people movement via transport corridors.

An important independent component of the information and telecommunication resources of IGMASS will be systems for distance learning and disaster medicine, which will provide a qualitative expansion of opportunities for citizens to obtain access to various types of distance education programs supplied by the projects member states, as well as emergency medicine in case of natural or man-made disasters.

4.5 Organizational and Technical Background of IGMASS Project Implementation

The practical realization of IGMASS requires implementation of the following interdependent measures:

1. Research carried out in international cooperation. Along with the definition of system shape and the basic components of data elaboration (the Academy of Cosmonautics n.a. K.E. Tsiolkovsky in 2009 was instructed to conduct research on the shape of IGMASS and its functional organization, the results of which are presented in five volumes, totaling over a thousand pages), system projects related to hardware implementation issues for the revealment of primary signs of disasters of seismic origin were launched. Furthermore, this phase is to include development of the complex of technologies and scientific and technical solutions for the conduction of experiments in outer space (a material-friction, atmosphere limbic sensing, measurement of space radiation flows), creation of radiation-hard base elements of microsatellite equipment, braked airbag elements with lowered workload, etc. (Fig. 4.9).

   Since February, 2010 the IGMASS project management (ICPI) has signed dozens of Memorandums of Understanding for project promotion at the

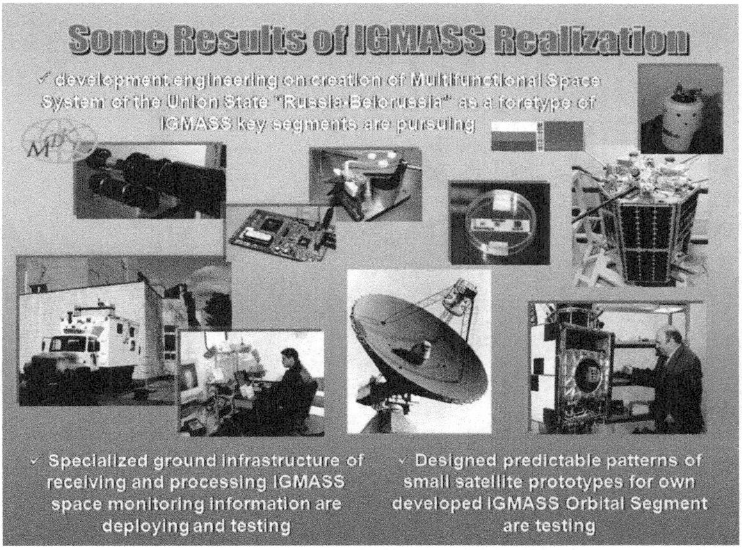

Fig. 4.9 Technical and technological background of IGMASS

international level relating to cooperation with relevant organizations and projects (SPIDER-UN, GEOSS, DMC, the Charter "Space and Major Disasters," etc.), definitions of a rational shape of the IGMASS ground infrastructure (including collection, processing, storage, and distribution of monitoring information), creation of IGMASS' own orbital segment, based on small and microsatellites, including cooperative development financed through international investment mechanisms.

2. Development activity in the framework of creation of the "Multipurpose System of the Union State Russia-Belarus" as a prototype of IGMASS' main segments. Base elements of Russian (Jubileiny, Moscow region) and Belarusian (Minsk) segments of two multinational information systems – providing Russian and Belarusian consumers with monitoring information as well as an integrated navigational and information system providing highway traffic control and management – are to go through system and structural design to the closing stage. In the Moscow region navigational and informational (telematic) technologies have passed practical approbation (Fig. 4.10).
3. Development and determination of specialized ground infrastructure elements for reception and processing of satellite monitoring information. This should include an information reception complex, including a reflector with a diameter of 9 m, providing diplex reception of information from the Russian and Belarusian Earth remote sensing satellites in the frequency range 8.025–8.4 GHz with speeds up to 123 Mbit/s.

Fig. 4.10 Group of Russian and Belorussian scientists and designers of the Multipurpose System of the Union State Russia-Belarus at the hand-over of the vehicular complex "Sadko"

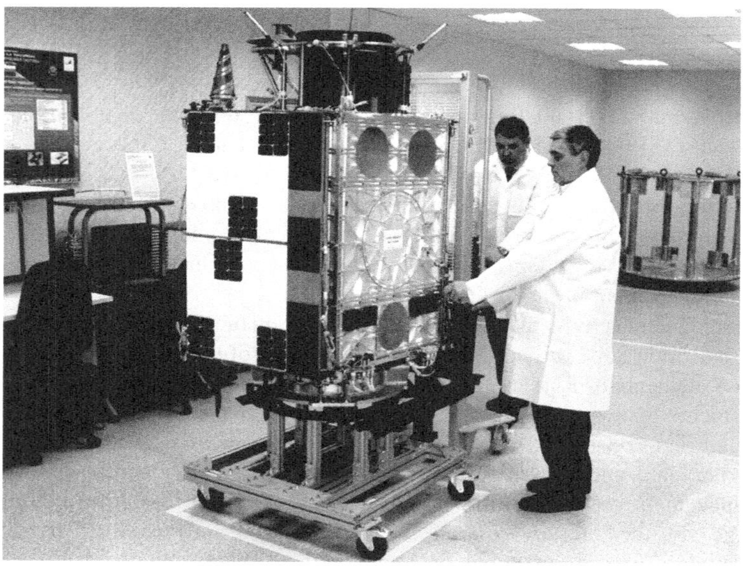

Fig. 4.11 The advanced Russian small satellite "SOYUZ-SAT-O"

The "Sadko" vehicular complex has entered into experimental operation, allowing for provision to the fire and medical services and the Emergencies Ministry of specialized space monitoring data on emergency situations in separate regions of the Earth, as well as data received from its own on-board sensor systems (Fig. 4.11).

The vehicular complex is equipped with several subsystems for reception and data distribution in real time, first developed and introduced by soft mechanisms of efficient complex processing of monitoring information, obtained from space and ground-based sources that provides the possibility of a complex batch operation over not less than 2 weeks.

Development of next-generation telemetry equipment providing highly efficient reception of information, obtained from MSVs in unequipped regions including mobile telemetry tracking station equipment (transportable in hand luggage), hardware and software systems for the processing and display of space data using GIS-technologies, and a high-resolution positioning subsystem, equipped with data security features within information exchange allowing for definition of a base line with a margin of error from one to tens of centimeters are all required.

4. Designing and prototype testing of small and microsatellites for the IGMASS projects own orbital segment. In this regard, particularly, the ERS small satellite "SOYUZ-SAT-O" has been finished including development of a dimensional model for dynamic tests. Development of on-board facilities to include both specialized and secure systems with resolutions to 2.5 m with lower mass and dimension parameters than existing analogues, providing an operational lifespan of 10–12 years. Rather significant scientific and engineering groundwork have been conducted in the field of microsatellite propulsion design. Advanced engineering solutions include ablation-, laser plasmatic-, oxygen hydrogen- class propulsion systems to be used for microsatellite attitude control and stabilization, reboost, interorbital transfer. These systems may be operated as parts of upper stages and space tugs.
5. Active promotion of IGMASS at the international level and attraction of foreign participants (in the framework of the United Nations and the International Academy of Astronautics forums).

4.6 Stages of IGMASS Project Realization

Taking into consideration the fact that the problem of forecasting natural and man-made disasters is of an obviously international character as well as the necessity to solve a number of scientific and applied tasks dealing with the development, testing, and utilization of up-to-date facilities as part of the important efforts towards promotion of the project, the following organizational steps should be taken:

1. To provide the IGMASS project with adequate organizational, political, and perhaps financial support from the UN. This will be necessary for its future practical realization.
2. To promote the IGMASS project at the interstate level.
3. The ultimate goals of the IAA working group on the IGMASS project should present concrete proposals concerning the creation of IGMASS on the basis of efficient development and joint use of aerospace capability and advanced technologies, based on wide-ranging international cooperation.

The potential member states of the project are Russia, the USA, Canada, the EU, Japan, China, India, Indonesia, Australia and other countries in the Asia-Pacific region, Africa, and South and Central America. Countries most prone to major natural disasters (earthquakes, tsunamis, floods) and thus more interested in the well-timed prediction of these events should be involved from the very beginning of IGMASS project implementation.

Organizational forms for IGMASS project management during its realization may include an International Coordinating Council, a Management Company, or an International IGMASS Consortium.[22]

The prototype of the International Coordinating Council was created at the "Symposium on Space and Global Security of Humanity" in Cyprus and named "The International Committee on the Project IGMASS Implementation" (ICPI), which includes heads of several national space agencies, rocket industry and space industry leaders, outstanding scientists and administrators, as well as dedicated specialists and politicians of more than 20 countries of the world.

The Committee is to become a governing body for the project and function in the framework of the IAA. The main purpose of the Committee is: to draw wide public attention to the IGMASS project both in the participating countries and at International level, profile scientists and experts from all over the world and their integration within the project, search for new ideas and make technical decisions in regard to the field of forecasting natural and man-made emergency situations, and engage in administrative and financial roles for IGMASS project implementation.

The main tasks of the Committee are:

- System-wide project management
- Political and juridical support of the IGMASS project
- Coordination of the collaboration of state participants of the project
- Managing scientific research and engineering at the international level
- Scientific/technical, financial and organizing supporting and promoting of the project at all steps of IGMASS "Life cycle"

Members of the Committee: Organizing Committee of the Cyprus Symposium (according to its Resolution) + Members of the IGMASS Study Group + new members from NGOs and private persons.

The proposed Committee structure (Fig. 4.12) includes: Managing Board (Chair, Co-chairs, vice-chair, plus members), five profile subcommittees, headed by Managing Body members (vice-chairs) and the Executive Secretariat – the Committee's Working Body (3–4 project managers responsible for realization of the decision-making process).

There are five subcommittees and a description of each is presented below.

[22] The management company or an International consortium legalized in a special agreement, a temporary association of independent business and government agencies for the realization of a major international project, co-location of industrial orders, conducting industrial large-scale credit, financial and marketing operations, and coordination of activities to obtain beneficial contracts and their joint performance.

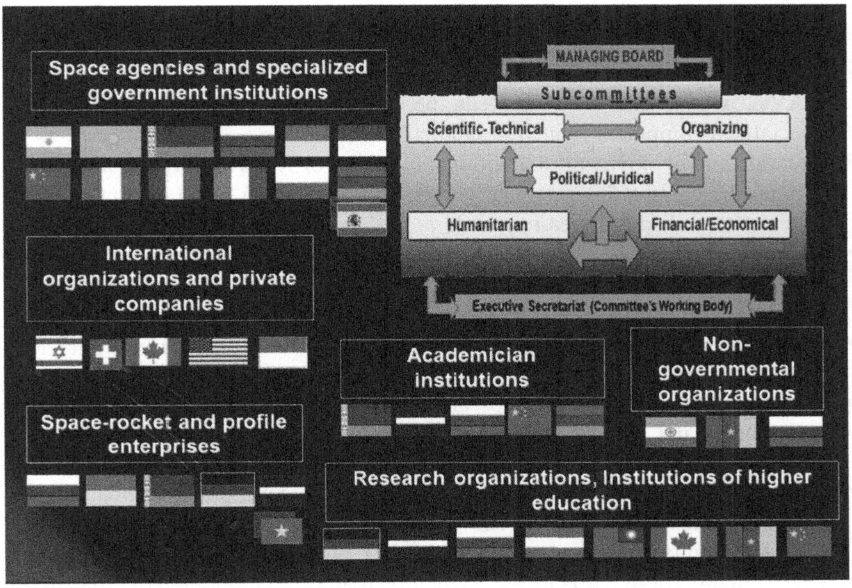

Fig. 4.12 The International Committee for IGMASS project implementation

The Scientific-Technical Subcommittee missions (scientific and technical support of the process of IGMASS creation at all stages of the life cycle of the system):

- Collection, compilation, systematization, and evaluation of hardware feasibility of methods, techniques, and technologies relating to the forecasting of natural disasters and man-made catastrophes (ND&MC) and its practical (technical) realization
- To undertake research looking for new scientific information relating to early signs (foreshocks) of ND&MC in the ionosphere, atmosphere, and the Earth's lithosphere, and techniques and methods for their effective identification and registration
- To undertake research in relation to collection and systematization of data and information about mathematical models that could be used for (ND&MC) prediction and forecasting
- Searching for up-to-date (including nontraditional) scientific ideas and technical solutions for ND&MC early warning and forecasting
- Creation and development of microsatellites and their on-board specialized equipment in the framework of international cooperation for predictive monitoring (including research engineering, production, and procurement)
- Deploying and maintaining IGMASS' own orbital segment constellation
- Control of IGMASS' own orbital segment constellation
- Systematization of data on existing terrestrial infrastructure, which can be used in the frame of IGMASS (sensor equipment, stations and facilities for receiving

and processing space information, international information and analysis centers, international centers for crisis management, etc.)

- Systematization of data on existing air assets (including balloons, airships, UAVs), whose characteristics can be used as IGMASS air platforms for specialized equipment related to predictive monitoring
- Creation and deployment of additional elements of IGMASS ground infrastructure on the basis of international cooperation (including R&D, manufacturing, procurement, disposal area required)
- Creation of advanced aircraft carrier-platforms for specialized predictive monitoring equipment in the framework of international cooperation (including research engineering, production, and procurement)
- Using and managing IGMASS predictive monitoring information (collection, processing, accumulation, storage, and distribution)
- Estimating necessary procedures and conditions for attracting into IGMASS information and navigational and telecommunication resources from existing and advanced space systems
- Development and establishment of international cooperation in relation to hardware and software for pairing and integration of informational resources from IGMASS and other systems (including R&D, production, and procurement)
- Determination of how to engage navigational and telecommunication resources of existing and advancing aerospace systems into the project
- Research relating to collection and systematization of available data on paths to a common planetary "information security space" (including international R&D cooperation programs)
- Systematization of data on existing infrastructure relating to asteroid hazard warning, which can be used by IGMASS (ground and orbital optical and radar facilities, databases, GPS directories, locations for receiving and processing information, etc.)
- To undertake research looking for new scientific information relating to potentially hazardous early signs of global anomalies in the ionosphere, atmosphere, and lithosphere of the Earth, in outer space, and techniques and methods for their effective identification and registration
- Establishing and developing international cooperation programs relating to hardware and software to prevent and ward off the asteroid hazard and global anomalies (including R&D, production, and procurement)
- Estimation of procedures and conditions for attracting into IGMASS information resources of existing advantageous space control systems to prevent and ward off the asteroid hazard and global anomalies
- Searching for, collecting, and systematization of data on technical and technological innovations that can be used during the practical implementation of IGMASS
- Determination of the mode and terms of engagement into IGMASS of advantageous technical and technological innovations (the observance of intellectual property, international patent law, compensation, etc.)

Political/Juridical Subcommittee missions (comprehensive solution to political and legal issues arising in the process of IGMASS creation at all stages of its life cycle)

- Consideration of political issues, related to project implementation
- Analysis of international legal aspects of project implementation
- Analysis of other juridical aspects of project implementation

Organizing Subcommittee missions (interaction with states and international organizations, system integration of the project segments during its implementation):

- Organization and maintenance of international cooperation of the project participants
- Coordinate the efforts of project implementation at the international level
- Establishing and maintaining cooperation with the institutions of the UN
- Establishing and maintaining cooperation with relevant international and national NGOs
- System linkage of the project's elements in the frame of wide international cooperation at all stages of IGMASS implementation
- The project "system of system" integration

Financial/Economic Subcommittee missions (financial and economic support of project creation at all stages of the life cycle of IGMASS, marketing, advertizing, and investments)

- Cost/efficiency estimates of technical and economic decisions of the project at various stages of its life cycle
- Searching for project funding across the entire spectrum of possible approaches to IGMASS implementation
- Engaging state, interstate and private investments for the project
- Estimating the international market for monitoring data for the prevention of natural disasters and technological events. Sales assessment of the information obtained from the space capabilities of IGMASS
- Identification of IGMASS funding mechanisms

Humanitarian Subcommittee missions (humanitarian aspects of the project: prospects of using informational and communicational IGMASS resources for distance learning and education, disaster medicine, and other humanitarian problems of the world community)

- Using Utilization of IGMASS navigational, informational, and telecommunication resources for emergency and disaster medicine activities
- Developing IGMASS telematics functions including creation of transport corridors, optimizing movement of people and goods
- Estimating the prospects of using IGMASS resources for implementation of humanitarian operations, literacy, preservation, and dissemination of cultural values, etc.
- Development of the use of navigational information and IGMASS telecommunication resources for the purposes of distance education and training in various fields

Therefore, on the basis of the Committee in the future the Project Management Company or the IGMASS Consortium could be built. Among its founders may be governments of its member states (represented by the relevant ministries and departments), the international and national academies of astronautics and cosmonautics, various international and national financial funds, space agencies and departments of the project member states, specialized Russian and foreign companies with state, joint and private capital, corporations, and individuals. Thus, the IGMASS Consortium could effectively meet the challenges of organizational and technical, legal and financial aspects of the project, including the effective coordination of the system creation.[23]

At the initial stage of project realization a preliminary technical and economic system study should be carried out, also it will be necessary to take concrete steps to create a legal project and its governing body, to acquire international patents on IGMASS and its constituent parts, to acquire the cooperation of relevant enterprises and organizations – the participants of the project, and to determine the UN institutions interested in supporting and promoting IGMASS at the United Nations level.

Integrated scientific trial study is a high-priority work to be done on initial phase of the project. This would cover studies of IGMASS problematic issues, including development of principles and methods for parametric control of geophysical hazardous phenomena precursors, dedicated situational models which provide the opportunity to asses spatial and time evolution of the phenomena, development of mathematical, logical and SW models to be used for monitoring data processing. IAA can become a customer for this study. It would be reasonable to involve leading scientific organizations, R&D companies, ventures from IGMASS participating states. This study shall result in technical specification (TS) for IGMASS design and development. TS is the first IGMASS implementation stage, to be followed by draft design, development of test models for IGMASS key items, and well as working documents for test products of the system (2011–2013); production of the test products, standalone tests, and updates of the working documents (2012–2013); integrated and interagency tests, and updates of the working documents (2015); start of flight tests, preparation of the documents for IGMASS items and serial production, system commissioning (2016); start of full-scale system deployment (2017).

Today's changes in the state management linked with transition from accident recovery to accident caution and warning, justify economical and "managing" effectiveness of IGMASS implementation by the world community.

[23] The basis of the IGMASS Consortium consists of: suppliers of space products and services, corporate users, research institutes, universities and research labs, and interested government agencies of various countries. Consortium members send technical experts and representatives to attend its various groups: these groups will conduct major technical work for the Consortium – the result of their activities are technical reports on various aspects of creating and using the system, subprojects, software tools with open access to various monitoring products and services.

4.7 Some Cost Estimates of IGMASS Project Implementation

The sources of funds for the IGMASS project are share capital of the Consortium authority (should it be created), loans from international and domestic commercial banks (should the system be created), and then – funding for services to provide monitoring and forecasting information to state and commercial organizations, with the start of the system and operation of its components. The financial resources of the project will be formed from its own (share) capital, borrowed funds, and retained earnings from transactions. Shares of the Consortium can be distributed among the states wishing to participate in the IGMASS project. On the part of the project member states half of the investment is public in the form of the system's ground infrastructure (spread of equipment to provide and operate aerospace facilities, etc.), and the remaining part is private investment in this project. Subscription of the interested countries for Consortium shares will be realized in accordance with their economic power following the example of the World Bank (Fig. 4.13).

The following estimates for IGMASS project realization have been derived from the prefeasibility development and creation cost study of megasystems that required large investments at national and international levels. Inasmuch as the technical configuration of IGMASS is at the early stage of development, it is difficult to give comprehensive figures for the realization cost of the system lifecycle, starting with research and advanced development and ending with tests and deployment of complex technical and organizational components of ground, air, and space segments. Meanwhile, the estimates indicate a figure of up to 6–9 billion US$ including the cost of integrating information resources and infrastructure

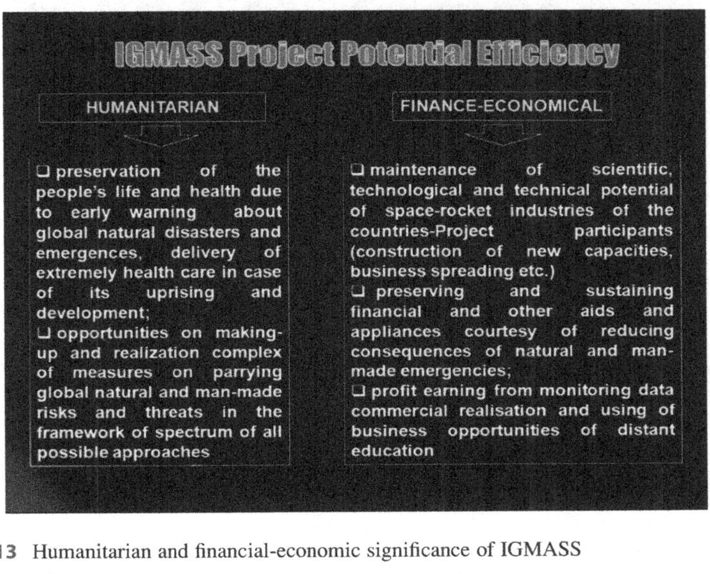

Fig. 4.13 Humanitarian and financial-economic significance of IGMASS

components of potential participating countries. The share of the specially created first stage of the IGMASS space segment (without the warning system of asteroid danger), made up of 16–20 spacecraft will be about 10% of this amount.

Certainly, these figures which cover seven-year period of IGMASS implementation will be distributed differently. In terms of potential returns for the International Consortium project, as well as expanding the number of participants, one may talk about its profitability both in terms of direct financial investment and long-term investments for the development of the member states' rocket and space industries. Moreover, the initial creation of IGMASS with wide international cooperation and under the guidance of the UN, as well as the subsequent operation of the system will be characterized by a pronounced effect on the socio-political, humanitarian, and economic characteristics of the member states.

The socio-political significance of the IGMASS project will be understanding by the international community of the necessity for peaceful uses of outer space and integrating efforts to solve global problems of the XXI century, the strengthening of foreign policies of the member states leading to improvements in the prevention of scientific, technical, and political issues related to third-party threats and risks in today's multipolar world (Fig. 4.14).

The humanitarian effect of the practical realization of this international project includes both health and life preservation of hundreds of thousands of people with the help of permanent control and forecasting of natural and man-made disasters, early warning of the population about impending natural disasters and global calamities, rendering timely medical assistance in case of their occurrence and adverse development and finally – the possibility of developing and realization of

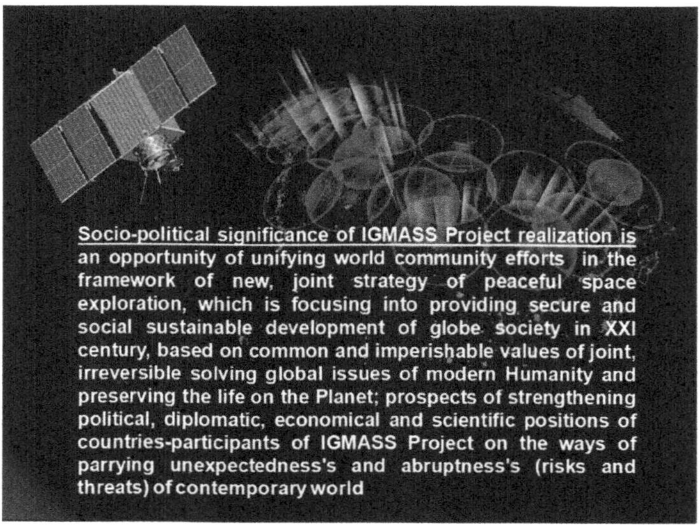

Fig. 4.14 Significance of IGMASS project implementation

effective measures to parry natural and man-made threats across the spectrum of possible approaches by the world community.

The economic aspect of IGMASS project realization is directly or indirectly positioned as retention and buildup of scientific, engineering, and technological capability of participant states (the possibility of creating thousands of new jobs in the rocket and space industry), and annual savings in terms of financial and material resources to the amount of several millions of dollars in the USA through mitigation of the negative impact caused by natural and man-made disasters. A direct economic effect of the use of IGMASS also consists of the profits from the sale of monitoring information and services to consumers and commercial resource management of distant education and telemedicine. A commercial sale of packages of distant education programs, increased investment through the expansion of public funding and attracting private investors, etc., can be considered as examples.

So, taking into account the challenges, threats and risks, with which Humanity enters the postindustrial phase of civilization development, it is hard to overestimate the importance of global international projects focused on receiving and distribution of information which defines all types of national and planetar resources on political and economical levels.

References

1. Меньшиков В.А. и др. Комплексное использование космических средств наблюдения за поверхностью Земли, метеорологического и навигационного обеспечения в интересах оперативного контроля природных ресурсов, стационарных и подвижных объектов. 57 Международный космический конгресс. – Валенсия, Испания, 2006. (Russian).
2. Материалы по Глобальной системе систем наблюдения Земли (GEOSS). www.epa.gov/GEOSS (Russian).
3. Материалы по Системе глобального мониторинга в интересах окружающей среды и обеспечения безопасности (GMES). www.gmes.info/ (Russian).
4. Материалы по Системе предупреждения о катастрофах и стихийных бедствиях («Sentinel Asia»). www.asiasentinel.com/index.php (Russian).
5. Материалы по Системе мониторинга стихийных бедствий (DMS). www.pryroda.gov.ua/index.php.newsid. (Russian).
6. Материалы по Проекту « Международная хартия « Космос и крупные катастрофы». www.oosa.unvienna.org/pdf/reports/ac105_841R.pdf. (Russian).

Further Reading

1. Меньшиков, В.А., Перминов, А.Н., Ю.М. Урличич Глобальные проблемы человечества и космоса – М.: НИИ КС имени А.А. Максимова, 2010. (Russian) – 570с.
2. Ranganath Navalgund, Valery Menshikov, Joseph Akinyede Space-Based disaster Managment: the Weed for international cooperation IAA, Bengalor-Paris, 2010, 80 p. (Heads of space Agencies Summit on November 17, 2010 in Washington)

Space Exploration and Formation of the "Information Field of Global Security": A New Paradigm of Sustainable Development

5

The main aim of this chapter is to show the absence of an alternative to space facilities as a source of data required to attend to the problems and risks burdening mankind as it enters the postindustrial phase of civilization development – the informational society. In this new phase of our development information has become an economic asset and defines all types of national and global resources. At the same time an intellectual basis is being formed to solve many contemporary issues affecting the World including the constant struggle against global risks of acts of nature and technogenic accidents, threats from abnormal phenomena in space and from space, easing of social frictions and religious intolerance related to humanitarian problems.

The technological preconditions for the shift to the information society have already been created. Practically unlimited information volumes including new knowledge are gained, processed, and distributed globally by means of the well-developed communications networks which have appeared as a result of the convergence of the initially independent information, satellite, and component technologies. Such communication types form a special informational infrastructure, a so-called "infosphere" [1] represented by the material and spiritual components of the scheme (Fig. 5.1). The former is a combination of different informational communications in the form of the global webs for obtaining, transferring, processing, and accumulating information. The latter is represented by the humanistic outlook of people, their psychology and mentality defining national and civilized value scales. Only on achieving a certain harmony between these two components can an actual shift from the postindustrial society, which brought mankind into conflict with the environment, to the information society, on which great hopes are pinned in terms of the ecological and social crises of the planet, become possible. If some courtiers are if not socially but technologically ready to enter the information society, the others have the well-balanced infosphere components but are much lower than level which defines not even the information society but its postindustrial and industrial stages.

V.A. Menshikov et al., *Global Aerospace Monitoring and Disaster Management*,
DOI 10.1007/978-3-7091-0810-9_5, © Springer-Verlag/Wien 2012

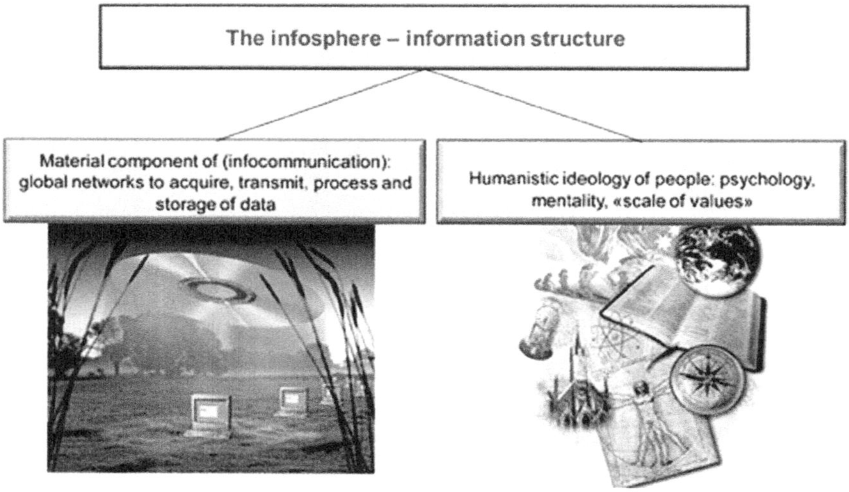

Fig. 5.1 "Infosphere" components

5.1 Space Facilities and the Formation of the Information Society of the Twenty-First Century

The fast development of telecommunications for transference of any data and revolutionary changes in information technologies in the twenty-first century greatly influence people's lives and provide greater opportunities. As a result of globalization the information society – changing the lives of the whole of mankind – dawned.

The term "information society," which practically substituted the term "postindustrial society" at the end of the last century was first used by the American economist F. Mashloop in his work

"The knowledge production and spread in the USA" (1962). This work was one of the first devoted to the informational sector of the economy. The introduction of this term in scientific practice is attributed to Prof. U. Khajashi of the Tokyo Technological Institute who presented to the Japan Government the following reports: "Japanese information society: theme and approaches," "The outline of the cooperation policy of the Japanese society informatization" (1969), and "The information society plan" (1971).

In modern philosophy and some other sciences the term "information society" exists as a conception of a new social order, which differs greatly by its characteristics from its predecessors. At first the concept "postindustrial society" was postulated (Darrendorf, 1958). In the frames of this society knowledge production and dissemination start to prevail in the economy and thus a new economic branch – an information economy appears. The fast development of this branch defines its control over the business and governmental sphere (Gelbreit, 1967).

The organizational base of the control is defined in the application to social structure (Boldwin, 1953, White, 1956); that means the appearance of the new class, the so-called meritocracy (Young, 1958, Goldner, 1979).

The production of information and the creation of communications soon became a centralized process (the theory of the "global village" Mac-Luen, 1964). In the end information is defined as a main resource of a new postindustrial order (Bell, 1973). One of the most interesting and well-worked philosophical conceptions of the "information society" belongs to the well-known Japanese scientist E. Masuda who is eager to understand the coming social evolution. Its main principles were presented in the book "Information society as a postindustrial one" (1983). The main theme is that computer technologies as a substitution or as an accelerator of people's mental activity have become the basis of the new social order. The informational revolution will turn into a new production force and make mass production of cognitive information, technology, and knowledge possible. "The border of known" will become a potential market and as a result the possibility of solving some of the current problems of mankind and the development of international cooperation will grow. Intellectual production will be the leading branch of the economy. Information will be accumulated and then disseminated via synergic production and fractional usage. The main subject of the social activity will be "a free community" and the political system will turn into a "participatory democracy." The main aim in this society will be the realization of the "value of time" thus it acquires the possibility to harmonize societal relations as it will be functioning on the principle of synergic innovation as a substitute for the free competition principle.

From the viewpoint of understanding the processes that really happen in the modern postindustrial society the works of J. Benninger, T. Stounier, and J. Nisbet are also valuable. These scientists consider the integration of the existing system of values with the newest means of mass communications as the most plausible result of societal development in the nearest future. The development of the new informational order doesn't mean the disappearance of the postindustrial society. In addition there is the real possibility of a setting of total control over information, its production and dissemination. Having become the main product, information becomes a powerful resource and its concentration in one place may lead to the appearance of a new variant of the totalitarian state. This possibility is not excluded even by those western futurologists (E. Masuda, O. Toffler) who are optimistic about the future changes in social order.

Since 1992 immediately after Japan other countries started to use the term "information society" (in the USA the term "national global informational infrastructure" came into usage after a well-known publication of the National Science Foundation and a well-known report of Clinton–Gore). At the same time this term appeared in the works of the expert group of the European Commission on information society programs under M. Bangemann and such terms as "an information highway" and "an information superhighway" appeared in Canadian, British, and American publications. In Russia this term was also widely used at this time.

Today the terms "information society" and "informatization" are well known and widely used not only by specialists in the area of information but also by politicians, economists, teachers, and scientists. They are mostly associated with the development of information technologies and some means of telecommunication, which help in making a new evolutionary leap and allow states to enter the present century at least at the first stage of the information society. It's remarkable that on the 27th of March, 2006 the UNO General Assembly accepted a resolution (A/RES/60/252) which set the 17th of May as an Annual International Information Society Day.

Some politologists and economists consider it necessary to share the conceptions of the information society and "postindustrialism." But in spite of the fact that the former was created as a substitution of the latter, the supporters of "postindustrialism" further develop the most important statements of technocratism and traditional futurology. It is interesting that those who earlier formed the postindustrial society theory (D. Bell, for example) now stand for the information society conception. For D. Bell this conception became a new development stage in his postindustrial society theory. "In the modern century the setting of the new way of life based on telecommunications gains great meaning for economic and social life, the methods of knowledge production, and the characteristics of people's labor. The revolution in the organization and cultivation of information and knowledge where the computer plays the leading role occurs at the same time as the formation of the postindustrial society."

Another well-known specialist, Prof. W. Martin made an attempt to find and formulate five main characteristics of the information society by the following criteria: technological (the key factor of development is information technologies which are widely used in production, municipals, systems of education, and everyday life), social (information is used as an important accelerator for lifestyle changes, an "information mentality" is formed), economic (information becomes the key factor of the economy as a resource, service, goods, a source of added cost and employment), political (the freedom of information leads to democratization of the political processes which are characterized by growing participation and consensus between different classes and social levels), and cultural (Recognition of data cultural importance through facilitating consolidation of informational value for evolution of a single human or the whole society). W. Martin remarked in relation to the information society that it must not be understood literarily but as a model of the changes in modern society. Among these changes are structural changes in the economy, the growing understanding of the importance of information and information technologies, the governmental support of the development of computer and microelectronic techniques, and social and educational computerization and informatization. Taking these changes into consideration the "information society may be defined as a society where the quality of life as well as the perspectives of social changes and economic development mostly depend on information and its usage, where life standards, forms of labor and rest, educational systems, and markets are under the significant influence of the need to achieve in the sphere of information and knowledge." The possibility to govern large organizations which demand the coordination of millions of people became a new challenge. Fast development was and is occurring in scientific branches connected to the problems of organization of millions of people, such as informational theory, informatics, cybernetics, the theory of making decisions, game theory, and so on.

Speaking of the changes which accumulate to shift modern society into a new stage the supporters of this conception have as a base the objective processes of the scientific, energetic and labor branches of the economy, the processes of production automatization, computerization, and informatization of the main spheres of social and political life. At the present time decisions related to such important questions as economic growth, employment, and growth in the quality of life depend on new scientific and energy-safe components of the technique. They touch upon all the main principles of the modern society revealing questions related to social and political changes which the insertion of the information technology brings. This fact influences the perspective of the social-historical development of mankind, people's destiny and their place and role in this process. At the same time the main difference between the information society and the industrial one is the fact that machines have begun to accumulate and use information and knowledge without man's participance. If in the process of industrialization the automatization of physical and manual labor was used and as a result the quality of life had risen, the information society aims at the automatization of mental activity and this may result in unpredictable positive and negative events.

One of the most unpleasant aspects of the present situation is the tendency of the information society to lose its social steadiness, when because of the growth of the role of information different small groups can influence practically everyone (for example, via terrorism which is the result of the loss of society steadiness as a result of information growth). The steadiness may be reacquired with the help of biometry, for example. After the 11th of September, 2001 through an initiative introduced by the USA international passports with automatic biometric identification came into use. But the most unpleasant situation may happen soon. In the information society the speed of a car will be controlled by a camera, hanging over the road, its door will be unlocked by a biometric lock, and one's hand-written text and voice will be known by a personal electronic secretary. Thus, the unsanctioned usage of personal information that should be known only by governmental services becomes acceptable. Russia became the first developed country to formulate a national technical policy on this problem and create (since 2004) national standards providing safety, confidentiality, and anonymity of personal citizen data.

Several stages may be defined in the Russian authorities' activity on the question of the development and realization of governmental policy in the field of information society development. The first stage (1991–1994) forms the basis of country informatization. The second (1994–1998) may be characterized by changes in priorities from informatization to determination of information policy. The third stage is still ongoing. In this stage the conception of the information society is realized. The most important step of this stage is the acceptance of the Russian Federal Program "Digital Russia 2002–2010." This program strongly enhances information society development in the regions. (A new program with the same name will be launched in Russia for the period 2011–2018).

The occurrence of the numerous and various communication networks "entangling" our planet, demands a new view on the growth of information. Here we mean the consequences of prompt expansion of the World Net (Internet), expressed by the leveling of firm moral values, and the transformation of such concepts as "information," "good" and "bad," "myth" and "reality." So, if before the communication facility met our needs for transfer, search, ordering, and collection of helpful information with time the growth of the Internet became an end in itself, and network services became the goods in demand. Spam which has become the symbol of our defenselessness before possible information terrorism has rushed into the Web and filled it with substitutes of a reality which have caused the destruction of a world picture. The information (bits, gigabytes) became a fetish regardless of its content and the balance between "good" and "bad," "truth" and "lie" became secondary. The most favorable aspect encouraging the impetuous growth of entropy became the free character of relations in the Network. "If something is not forbidden, possible to have fun, mocking everything", sadly says the well-known Polish science fiction writer and futurologist Stanislav Lem, giving examples of falsifications of socially important events, and "computer revival" of well-known people for advertising purposes. Most dangerous, in this opinion, is that the "network" ideology is transferred to other rather sensitive areas of human life.

The present stage of the information revolution would be impossible without prompt development of applied astronautics. The score of constantly operating space systems of military, dual and civil purpose (status and national identity, under the present conditions of globalization, integration, and wide international cooperation, can hardly be defined) provide decision-making capability for the broadest spectrum of problems in the interest of safety and steady development both of separate states and regions, and the planet as a whole.

So, in particular, the forecasting of external global threats to our planet and its inhabitants is carried out on the basis of observations of space from the Earth, and global internal threats on the observation of the Earth from space. The information on external threats is gained both by ground and space facilities. However, the opportunities for data collection by ground systems are limited by the masking properties of the atmosphere whilst space facilities, placed in circumterraneous, circumsolar, and more removed orbits are free of this problem.

For the control of internal threats modern space has been turned into a global integrated information field turned towards the Earth, representing a data carrier that is adequate for global threats and risks. The basic sources of information in this field are space vehicles for remote sounding of the Earth of various classes and purposes: global navigation, communication, relaying and data transmission, meteorology, geodesy, search and rescuing.

Today, this "integrated information field" already provides unique universal opportunities for operational provision of any kind of information services. So, the opportunity by means of virtual hardware-software "filters," to create private information files is given to users of any scale. In Fig. 5.2 the types, objectives, and functions of the global monitoring which is carried out by space information systems are presented. The most informative and beneficial space systems for these purposes are the monitoring which is carried out with the use of systems for geodetic navigation and for the remote sounding of the Earth (Figs. 5.3 and 5.4 schematically show the space global information field with its real and potential users).

Information acquired from space is demanded practically in all activity fields of mankind. However, the space geodetic navigation systems and remote sensing systems represent an adequate toolkit for the creation and actualization of a specialized global infrastructure of geospatial data used in appliances of planet, regional, national, corporate, and even individual scale.

Today the environmental problem is one of the most important for our planet (Fig. 5.5). Thousands of elements that are absent in nature enter the atmosphere. On the territory of Russia nearly 20 million tons of chemical substances are released annually into the air. The world ocean is already polluted with nearly 20 billion tons of dust and annually ten million tons of oil [2] is thrown into the ocean. At the present time the number of global extreme situations of technogenic character, leading to aggravation of environmental problems is growing. Thus, the primary goal of the monitoring of potentially dangerous objects consists of using information in a preventative fashion to maintain a potentially dangerous process within a "holding zone" of admissible risk, and in the case of its escape from this holding zone, in blocking further dangerous developments until its return to the safety zone.

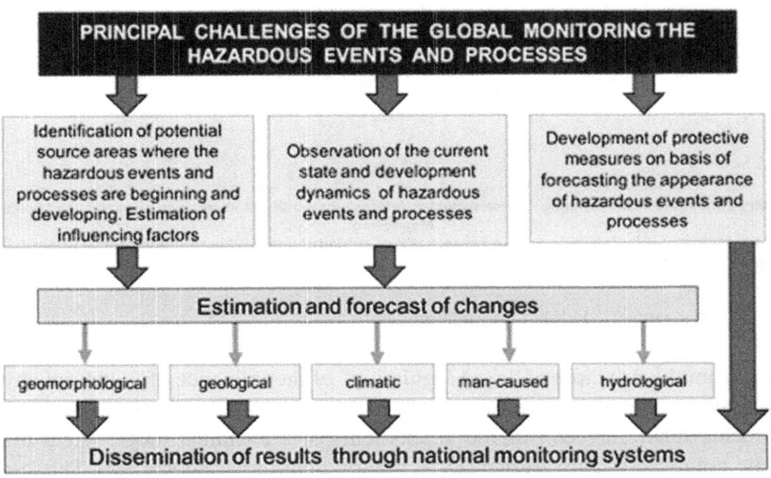

Fig. 5.2 Role and place of global monitoring in the contemporary world

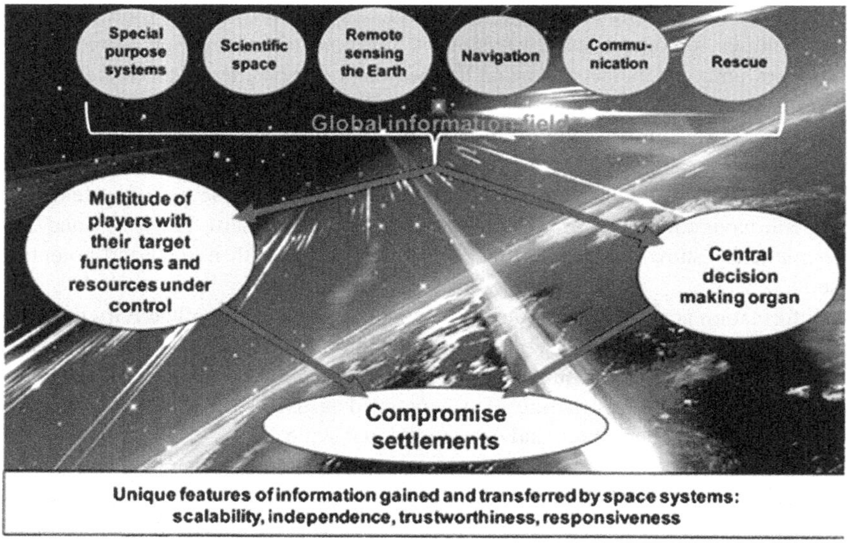

Fig. 5.3 Global informational field of the planet

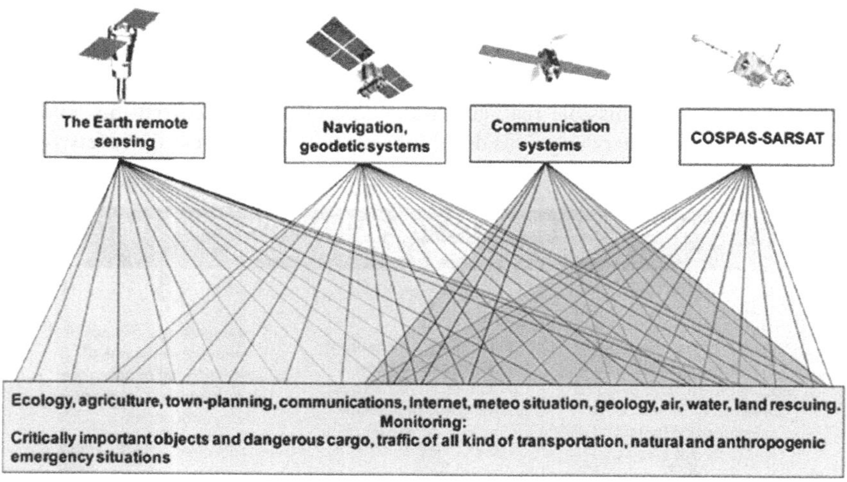

Fig. 5.4 Space-supported global information field

A shift from the principle of "rigid regulation" of the influence of industrial objects on the environment to a system approach is close. This approach means that objects of a technosphere are considered as elements of a technical megasystem that is closely connected with Vernadsky's theory about the noosphere. The global

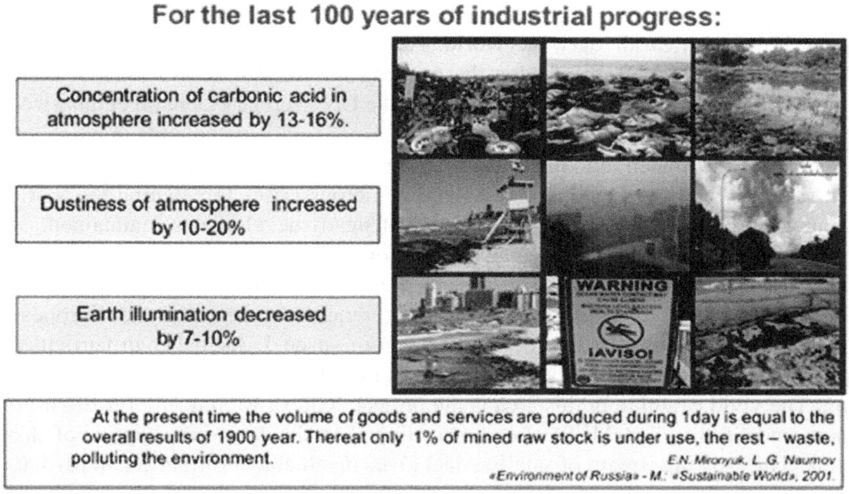

Fig. 5.5 Environmental issues of the postindustrial society

monitoring of such a "megasystem" can be carried out with the use of space facilities only.

The modern development of communication facilities and communications goes in two opposite directions – globalization and personification which leads to the creation of a global general telecommunication environment with gradual integration into it of the national coherent networks. The personification of communication consists of granting to each user a wide spectrum of multimedia services. As a result, in spite of location the user has an opportunity to use the world information resource created with his/her participation. The globalization of communication creates technological preconditions for information supply for the development of different ecological, social and, probably, military decisions, on the basis of the relevant information coming from all regions of the planet in real time. The satellite communication systems play the main role here.

The intellectual basis of a modern information society is made with fundamental and applied scientific knowledge of a planet and its current physical condition. The modern space infrastructure allows the study of the Earth, on the one hand, as a complete system cooperating with space, specifying and receiving new knowledge on the structure of the Universe, and on another – as a morphological structure with processes proceeding in it. Study of the Earth from space is carried out with the widest international cooperation which is proven by the fact that over 50 years of the space era in practically all the launched Russian scientific satellites foreign equipment has been installed.

At the end of 2003 in Geneva (Switzerland) the World Summit on Information Society (WSIS) was held under the auspices of the United Nations. A thousand participants representing the various national and international organizations gathered. They included scientific institutes, private enterprise, and mass media from 176 countries all over the world. An overall objective of the forum was discussion of the question of what the ongoing information revolution can bring to the world community. Opening the forum, the UN Secretary General emphasized that the advantages of an information society promising huge benefits to mankind, cannot extend by itself, and the efforts of the most developed countries are needed. Only so are the hopes of overcoming the numerous crises threatening the world community and for the promotion of a way of steady development maintained.

One of the most active participants of the forum was ESA, which also presented at the international conference that preceded the forum "The Role of science in the information society." Speaking at the conference, the head of the European programs of remote sounding of the Earth from space J. Ashbaher in particular, spoke about the contribution which ESA satellites bring to environmental monitoring. The vivid example in this area is the project "Global Monitoring for Environment and Security" (GMES) already mentioned in the previous chapters of this book. The growing stream of satellite data gives invaluable information, in particular, for the management of wildlife, an estimation of consequences of natural and technogenic accidents, and the distribution of humanitarian aid.

The space community does much to avoid division of mankind into categories based on availability of electronic technologies and aspires to making sure that satellite data can be received by those who really need it. GMES can become a powerful ESA contribution to the creation of "the International space system of global monitoring" which as we hope, will unite the satellite data with similar results received on land, in the sea, and in the air. The overall objective of the created system consists of making these data demanded and acceptable to everyone including those who are at the initial stage of the information revolution.

In 2003, at the World Information Forum and the conference connected with it, ESA presented three projects, aimed at the world-wide development of an information society. First of all this was the "International Charter of Space and Major Natural Disasters" – the large international agreement on the use of space data and equipment for the organization of first aid and regenerative works which various space agencies of the leading space powers have joined or are in the process of joining. It is necessary to mention also the joint project of UNESCO and ESA on the monitoring and rescue of cultural objects included in the United Nations World Heritage List of various architectural and natural monuments, national parks, bird reserves, and areas with rare and endangered animals. One of the organizers of the conference – the European Organization for Nuclear Research (CERN) even suggested ESA setup their presentation on its own stand to convince everyone of the exclusively important role of space research in the transition from an industrial society to an informational one.

According to the space experts opinion the WSIS (World Summit on Information Society) forum provided a deeper understanding of the role of science, including space research (and, in particular, the international programs), in the transition to the information society. There is no doubt that results acquired from outer space exploration, remain the major contribution of the scientific and technical revolution to this transition to a postindustrial society in which the most perfect space technologies become general property.

5.2 Humanitarian Aspects of IGMASS Project Implementation

The implementation of a large project like IGMASS involves worldwide information and telecommunications support of consumers addressing a wide range of humanitarian issues such as emergency medicine, distance learning, distribution of space education and personnel training in advanced technologies with the use of advanced space systems.

5.2.1 Distance Learning and Distribution of Space Education

Only an educated individual with a relevant level of knowledge which he constantly maintains is able to resist natural and man-made disasters. So knowledge is an inseparable component of global security. Today, the effective distribution of knowledge is possible only by using new education techniques based on common learning and research systems employing modern information and telecommunication technologies, including those in space.

It has become imperative, both in Russia and abroad, to create a distance learning system in the field of space exploration. This is so because society, science, and the space industry need skilled young specialists. Today's progress in Russian astronautics depends heavily on the availability of skilled personnel. Many young well-educated people capable of becoming specialists in the industry, prospective designers of space rocketry, as well as outstanding scientists and well-educated and trained space explorers, prefer to take up more prestigious and economically lucrative jobs, i.e., they become lawyers, doctors, managers, sportsmen, businessmen, software developers, and the like. Therefore, given that the average age of the Russian aerospace worker has of late grown considerably, the industry may face a generation gap in the next 10–15 years.

The space industry urgently needs replenishment by young specialists in such areas as fundamental space research, geo-information systems, space communications and navigation, Earth remote sensing, and thematic space photograph processing. In this context it is highly desirable to fully use all the achievements of space exploration and to distribute knowledge and specialized training in space technologies. Today, it is safe to say that astronautics can help itself in addressing issues related to the training of both young scientists and a wide range of specialists involved with rocket research and space monitoring system development.

The study of modern trends in economics, science, and culture shows that two processes are now underway across the globe. Those are the differentiation and integration of labor in all walks of life of the world community. Those processes arise from the fact that quite a few momentous issues have cropped up today, the tackling of which calls for large-scale economic and intellectual spending unaffordable to individual developed countries or even powerful transnational groups. The co-existence of differentiation and integration of global labor seems, at first glance, mutually exclusive and leads to globalization in all types of human activities.

This process has become objective and implies a free transfer of products, finance, and information. The products of intellectual work, new principles of physics, new technologies, new models etc. become not only sources of considerable profits but also a strategic reserve of the state. The export of knowledge, education, and training turn into serious self-sufficient businesses, whereas intellectual property becomes their costliest and most lucrative element. Intellectual work has become one of the most popular types of human activity, especially in applied astronautics, a motive force of progress in science and technology.

As a science embracing the achievements of various branches of research and engineering, astronautics has begun to markedly impact mankind's lifestyle. Space communications, navigation, monitoring, surveillance, etc. have become an inseparable component of people's economic and cultural life the world over. Therefore, it is easy to forecast that the number of people involved with space exploration will, over time, grow, which is why it is increasingly important to promote education in space affairs. Under such circumstances various forms of distance learning with the use of the newest telecommunication and information technologies gain importance.

Today, it is generally recognized that distance learning is a synthetic, integral, and humanistic form of education which must address many current problems. Under modern social and economic conditions these problems have become universal because the traditional forms of education and learning patterns cannot fill the need for education available mostly in large cities.

According to some experts [3], the meaning of "education" began to change in the mid-1990s. In a broad philosophical sense, education is "a means and a process of settling the fundamental controversy between the biological and social nature of man." In a narrower sense, this is "a blend of knowledge obtained in training." The terms accepted by the authors for training and education imply that the former is a goal-oriented, systematic, organized process of obtaining knowledge, skills, and habits, and the latter – the result of training, upbringing, and development of a person. Prepared by the Russian Universal High Education Academy, the "Concept of distance learning on the basis of telecommunication technologies" defines distance learning as "remote training in which a trainer and a trainee are separated in space" [4]. As for the use of space-based telecommunications and information technologies for humanitarian purposes, we shall further employ, with reference to distribution of knowledge and preservation of cultural assets, the "distance learning" term.

Distance learning has drawn the attention of both trainers and trainees because it can be practiced in various forms depending on the organization and technique of the training method. Before the era of information technology this manner of training was mainly the exchange of printed mails (along with printed matter, other widely used means were video records, training TV and radio programs) and sporadic interviews of trainers with their trainees during checks and examinations. This is the so-called extramural training. It is still widely used in all Russian educational institutions. Different from the two traditional types of training – full-time and extramural education – distance learning is, essentially, a hybrid. From the extramural form it inherits the "remoteness" of the trainee from his/her trainer. However, unlike the extra-mural distance learning it is individual in its nature. As in face-to-face tutorials, this is a personalized curriculum combined with constant interaction with a tutor whose functions may be performed by a computer. Its role is to monitor the training, to consult on complex matters, to check on tests, and to aid in preparation for exams. The traditional full- and part-time (extramural) training differs from distance learning in that it presupposes an independent choice of the subjects and a rate at which they will be tackled. For example, during one semester it is possible to undertake a course, which under the full-time arrangement, takes up to a year, or, on the contrary, it can be stretched over 2 years. Another feature of the distance learning is the elimination of formal boundaries between the high school, the college, and the postuniversity training.

Distance learning is, in the first place, an efficient tool for getting education in distant regions. People in the provinces, especially in Russia, don't have many opportunities for basic and higher education, or, for that matter, for advanced professional learning. So, the best university tutors give up work because the local educational institutions are unable to pay them adequately, or to somehow attract them. Under such circumstances it is obvious that higher schools, corporations, and commercial training centers must create, develop, and promote distance learning programs. Such programs are already being implemented in Russia. The distance learning scheme has become quite popular today all over the world. For instance, in the US more than 200 universities offer this kind of education [5].

Thus, distance learning is "a synthetic integral humanistic form of training based on the use of a wide range of conventional and new information technologies and technical means of delivery of the subject matter, its independent study, organization of an on-line communication between a trainer and a trainee when the teaching is not affected by their position in space and existence in time, or by their belonging to an educational institution" [5]. This definition also shows the regular trends in progress of science and technology, such as the automation of production of industrial and agricultural products, as well as the automation of distribution of education and knowledge in general. This makes it possible to characterize the new form of distance learning as a fast process of obtaining authentic, complete, and well-protected knowledge. Such a technocratic approach is particularly useful in space education which encompasses many specific areas of science.

The importance of distance learning in the field of applied astronautics and space education (*space distance learning*) is obvious. Here it is necessary to single out the priority learning techniques, and to organize and carry out research on the possible integration and mutual adaptation of national education systems of various countries. This will provide an attractive opportunity of cooperation in this area for all the interested parties. In addition, it would be reasonable to develop a dedicated project under which the universities of the space powers (Russia, leading members of the CIS, the US, France, Germany, China, India, and others) could use small satellites for implementation of bilateral and multinational linguistic, technical, and general education programs. Here it is essential that the space agencies, dedicated scientific and educational establishments, major UN institutions, and other organizations should provide their support.

It seems logical to give every possible support for higher schools and other research and public organizations developing light and small satellites and their payloads. The support could be in the form of easier availability of means of conveyance, reduced cost of delivery of payloads to space, and preparation of programs to aid such endeavors sponsored by the UN and UNESCO, space agencies, and other interested users. The elaboration of proposals to maintain such a transnational program, with regard for the consolidated efforts of space agencies, universities, high schools, and design organizations, could start with building a specialized small spacecraft database in a transnational education computing network.

Such a global network of information and education, which is accessible to all the interested parties and which brings together the major aerospace research centers, design and engineering organizations, and universities will dramatically ease communication between individuals and organizations wishing to receive space education, largely thanks to giving them free access to sources of information. It would make sense to ponder a possibility of a preferential, and then a free access to this network for a number of users so long as it is sponsored by international humanitarian or commercial organizations.

A space education system for youth can only operate if Transnational Youth Space Centers are created. The need in Russia for its own education and propaganda centers supporting science, engineering, and culture became obvious long ago. The work in this area began at Vorobyovy Gory Engineering Center for Youth, Moscow, with the aid of foreign partners. In the US, a similar center was opened in Huntsville, Alabama. Under the auspices of the UN (UNESCO, COPUOS) such centers could help create in the society a positive attitude to space exploration and promote the spread of knowledge about space. Fitted with an infrastructure of technical inventory, they could help in implementation of new projects of close international cooperation in space.

Worth noting in the field of transnational space education are pioneering technologies being developed by Russia for use in such roles as accident prevention, natural and man-made disaster response, and counter-terrorism missions. To this end, new techniques for rescue operations are being developed, advanced models of rescue equipment are being optimized, and modern means of reception and transmission of surveillance and forecast information are being designed.

Russia's achievements in civil defense technology are widely used for creating relevant educational systems. Excellent research, production, and laboratory facilities have been built and are now in operation. Those include modern computing centers for aerospace monitoring of the environment on Earth, forecasting natural and man-made disasters and evaluation of their economic and social consequences. The capabilities of those facilities enable development of a technology for training specialists with the aid of geo-information systems, space photographs of the Earth's surface, and other information. Thanks to them, it is possible to elaborate a disaster scenario and ways of responding to it using applied software for the evaluation of material damage and humanitarian aid needed by the population in distress.

Also, it should be noted that Russia's advanced information technologies are increasingly widely used in the training of civil defense specialists in Europe under deals for establishing integrated networks in the field of national and international security. Such networks, the agreements for which had been signed between Russia and the EU in May 2004, should support, among other things, the progress of science, culture, and education. The commencement of cooperation between Russia and Europe shows that the rapid introduction of advanced information technologies is becoming ever more important and must be regarded as a serious factor contributing to security of the world community, which is steadily transforming into an information society.

The attempts to spread space education have given birth, in France, to the International Space University. This was founded "for creating a peaceful, prosperous and free future through exploration and development of space for the good of mankind" [6]. The educational program of 2–3 years is conducted in English (French is accepted as the second language) in a full-time format by providing separate learning sessions lasting several weeks or months in different countries.

However, the progress in modern educational technologies and applied astronautics offers essentially new techniques for distribution of space education through use of the telecommunications resources of IGMASS (International Global Monitoring Aerospace System). It is possible to use in this way new methods both in organization of the training and in improvement of its techniques.

The concepts of the "Global Space University of Distance Learning" (hereinafter referred to as the University) are the basis for building a future world system of knowledge distribution and education, which will help address mankind's humanitarian issues.

Such a University (Fig. 5.6) must be created using the leading training centers and universities for training and retraining skilled specialists in the fields of space research, space rocketry, and applied astronautics. Also, this will be useful for individual distance learning of citizens of all countries with involvement of specialized Russian, foreign, and international educational institutions with the fullest possible employment of information and telecommunication resources of IGMASS. An international system of distance learning in the monitoring of natural and man-made disasters will function as an independent constituent unit of the University. Such distance learning provides the following opportunities for learners:

- The study of subjects and the mastering of the entire course independent of a timeframe.
- The formation of a package of independent and alternative subjects and courses; programs that meet the learner's individual or collective needs.

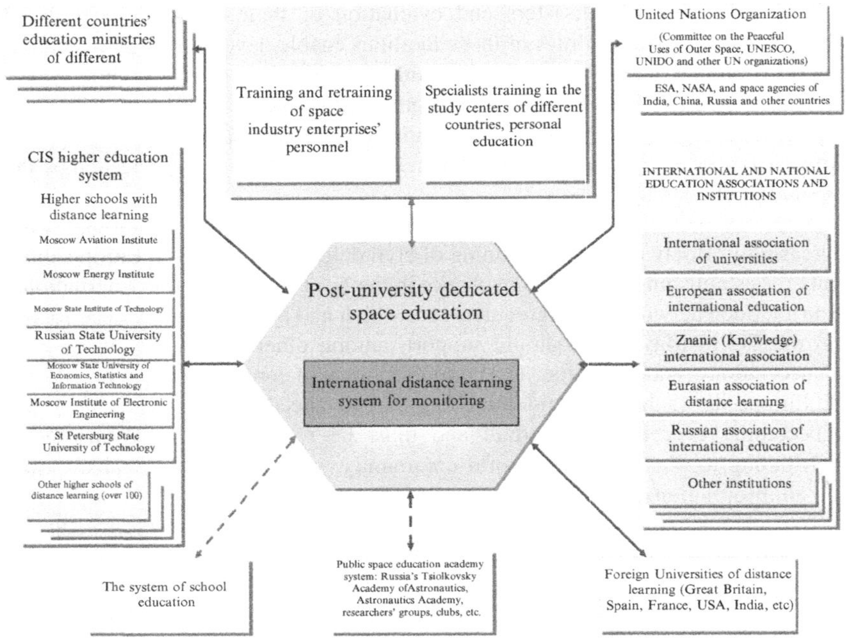

Fig. 5.6 The World University of Distance Learning in the international education system

- The acquisition, simultaneously with professional skills and habits, of theoretical knowledge, which is achieved by combining educational, manufacturing, and professional activities and the employment of virtual modeling software.
- The diversification of sources of study information (electronic libraries, information banks, databases etc.), communication via the Internet with one another and skilled trainers.
- The reduction of spending on personnel training through effective use of learning premises, technical aids, and vehicles and the concentrated and standardized presentation of learning information and multiple access to it.
- The use in the training process of the latest achievements in information and telecommunication technologies that help the learner enter the information network.
- Export and import of achievements in the world's educational services market.
- The enhancement and update of the role of the trainer in the training process.
- The buildup of creative and intellectual capability of trainees through self-organization, stimulation of the desire to learn, and the honing of skills in the operation of computer technologies.

According to the founders' concept, the University can provide three levels of education: secondary space education (based on learners with an incomplete secondary school education) with a duration of 1 year, secondary specialized space education (based on general or dedicated technical secondary education) lasting 2 years, and higher space education (based on general higher education or specialized higher education) lasting 4–5 years. Specialists of the above-mentioned levels will be trained, respectively, by the following departments of the University: lyceum, college, and the University proper. Specialization will be accomplished by training at a number of faculties, which train in various aspects of space exploration. Those faculties are:

- Space rocketry, spacecraft, and orbital stations.
- Spacecraft control systems.
- Spaceport equipment.
- Carrier rockets and spacecraft production technology.
- Distance learning and Earth remote sensing, forecasting, and emergency prevention and estimation.
- Space law.
- Space optics and radio-electronics.
- Space information technology.

The structure of the University comprises the executive organs, faculties, and departments. At each faculty the training process is supervised by a dean. The preparation of methodological materials and assignments in subjects as well as tests for checking theoretical knowledge are under the control of methods managers. The hardware-based distribution of methodological materials and assignments as well as their checking and supervision of interaction with trainers are controlled by methodology operators. The University chairs can be controlled by the entire University or by specialized faculties. That is where dedicated literature and assignments will be prepared and the training subject matter elaborated with the help of skilled trainers, most of whom may work part-time, residing in other towns and countries. In this context, much effort will have to be dedicated to elaboration of the curriculum, i.e., the determination of the learning subjects, the creation of digitized textbooks, assignments, and tests for distance learning. According to the founders' concepts, the structure of the University must also include divisions and representations (support stations) located, as a rule, in engineering higher schools in major cities of Russia, the CIS, and abroad.

A serious issue in establishing the University is overcoming the language barrier. Initially it is planned to use only five languages (Russian, English, Spanish, German, and French).

Obviously, before launching the training, it is necessary to undertake market research in order to determine the number of future trainees. This often necessitates the conduct of a large-scale advertising campaign in electronic and printed mass media. According to preliminary estimates, it is expected that in the first training year the number of students may be around 300 people. By the third year it may increase to several thousand. Considering a possible reserve of up to 20%, this number can be accepted as a basis for requirement in the output of the distance learning facility. The minimum number of the University's permanent staff may vary between 40 and 45 people.

After 2 years of study at the University and the passing of exams in major subjects, those wishing it, may take a bachelor's degree and, upon graduation from the University, study for a magister of astronautics. In the future, postgraduate and doctorate courses may be offered at the University for training certified experts to the highest standard.

It should be pointed out that during the setting-up of such a University it will be necessary to simultaneously make serious investments in the preparation of the training facilities, the creation of a technical training complex, and the outfitting of the training premises. Other necessary jobs are accreditation, licensing, and certification. In terms of business the investments in education are called long money, i.e., the payback period and profit-taking may reach 10 years and more.

Table 5.1 shows approximately the subjects of the World Space University of Distance Learning. Its curriculums, technology, and the level of training must meet the requirements of colleges and universities of the world's leading countries, to provide the mutual recognition of their graduates' degrees.

As the University's conception was elaborated, much attention was paid to the formation of a technical guide of a multifunctional system of distance learning distributed over territories. Among its functions are the dissemination of space knowledge, the processing, storage, and distribution of a large amount of training and research information. A version of the technical concept of such a system is shown in Fig. 5.7. This will include the following structural elements.

1. The engineering center where the information and computing facilities are located for storing and processing information, as well as equipment for communication and data transmission, lecture rooms, trainers' TV studio, and the conventional and digitized library.
2. The interacting institutions of distance learning institutions (regional and foreign engineering centers);
3. Remote terminal complexes of trainers (lecturers) involved in teaching specialized syllabuses of dedicated institutions, enterprises, and organizations having relevant knowledge and experience in other cities and countries;
4. Trainees' terminal stations for study information reception, storage and presentation systems divided into collective (located at educational institutions) and individual stations for, respectively, in-group and personal communication with trainers and for access to study information;
5. Telecommunication systems for transfer of study information from the engineering center and trainers' terminals (trainers' automated workstations). The tasks addressed with the help of the Technical complex of the University of Distance Learning are shown in Fig. 5.8.

The engineering center's role is to support the functioning of the University's distance learning system, which includes:

- Interaction during real-time instruction of trainers and trainees;
- Development and preparation of video and digitized courses in different subjects;
- Conduct of laboratory work and computing practice;
- Support of storage aids, compilation of catalogs and updating of information and education resources;

Table 5.1 Study contents of the World Space University of Distance Learning

Training contents	Themes studied
General scientific problems of astronautics	Space philosophy and noospherology
	Politology and humanitarian problems of astronautics
	Space law
	Space astrophysics
Space programs and systems	Integrated programs of space-based study of environment and ecology
	Space systems for the monitoring of global natural and man-made disasters
	Space-based remote Earth sensing systems
	Manned flight programs
	Space navigation systems
	Space communications and TV systems
	Space meteorological systems
	Space geodesics and cartography systems
	Space systems for fundamental studies and outer space exploration
Space systems design and application facilities	Spacecraft delivery systems: carrier rockets, booster units, and space tows
	Aerospace systems
	Piloted spacecraft and orbital stations
	Automatic spacecraft
	Lunar and planetary space bases
	Space power supply systems
	Space robotics
	Automatic space rocketry design systems
	Choice and optimization of in-orbit space systems
Space systems manufacturing technology	Organization of space rocket production
	Spacecraft materials study
	Space technologies
	Technology of operations in space
	Technology of assembling assets on-Earth and in-space
	Static and dynamic strength of space systems
	Aero-gas-dynamics and heat exchange in space
	Onboard power supply systems for space rockets
Spaceports	Spaceport technical facilities
	Launch sites
	On-Earth infrastructure of spaceports
	Measuring systems of spaceports
	Space system transportation facilities
	Security in transportation, preparation and launch of space rocketry

(continued)

Table 5.1 (continued)

Training contents	Themes studied
Control systems	Orientation and stabilization systems of carrier rockets, booster units, and spacecraft
	Rocket movement control systems. Software- and terminal-aided control
	Spacecraft automatic control systems. On-board and on-Earth control stations
	Information-supply and measuring devices for controlling spacecraft and their onboard systems
	Telecontrol of carrier rockets and spacecraft
	Measuring devices of current navigation parameters of carrier rockets and spacecraft
	Determination and forecast of parameters of spacecraft movement
	Onboard time scale formation and storage systems
	Systems for transfer of command and program information to spacecraft
Preparation, processing, and testing of space rockets	Space rocket production quality control
	Autonomous and all-inclusive tests of space rockets at the manufacturers' facilities
	Preparation and testing of space rocketry at spaceports' maintenance and launching sites
	In-flight and engineering trials of space rocketry systems
	Rocketry acceptance trials

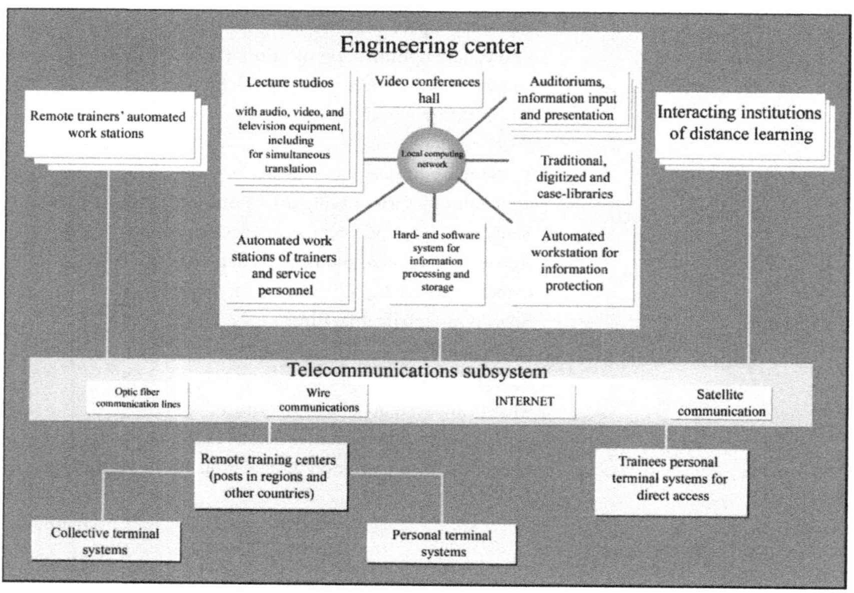

Fig. 5.7 The structure of the distributed distance learning system

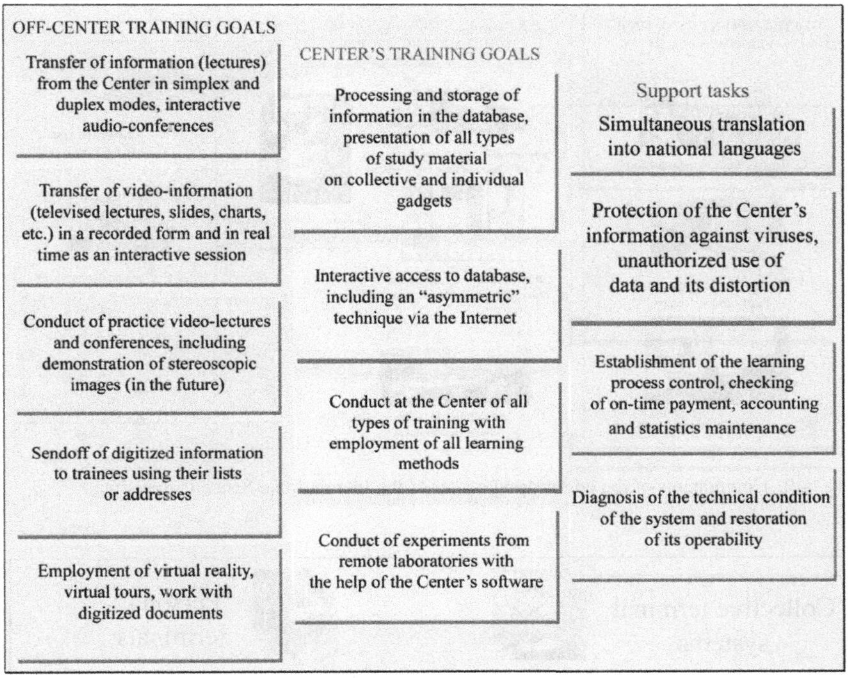

Fig. 5.8 Tasks of distance learning of the International Space University

- Production and distribution of study materials for distance learning;
- Adaptation of study materials for use in distance learning;
- Archiving, backup, and restoration of software;
- Periodic checking of software operability and database integrity, detection of errors resulting from incorrect finishing of work during training (unfinished testing or disruption of connection, etc.);
- Hard- and software-aided system protection against unauthorized entry.

The University's Engineering Center must comprise the following key components (Fig. 5.9): hard- and software systems of the Center's organizational and technical support, conduct of collective and individual video sessions in real time, development of electronic and video courses, information archiving, storage, and presentation, commutation unit with hard- and software systems for information protection, technical facilities of the local computing network.

The facility for organizing and conducting so-called "multiple-point" video sessions in real time is created using the currently available hard- and software systems of video conferencing. The main technical assets of the complex are video terminals fitted with technology for processing and reception/transmission of video images and sound in real time. A version of the required network infrastructure is

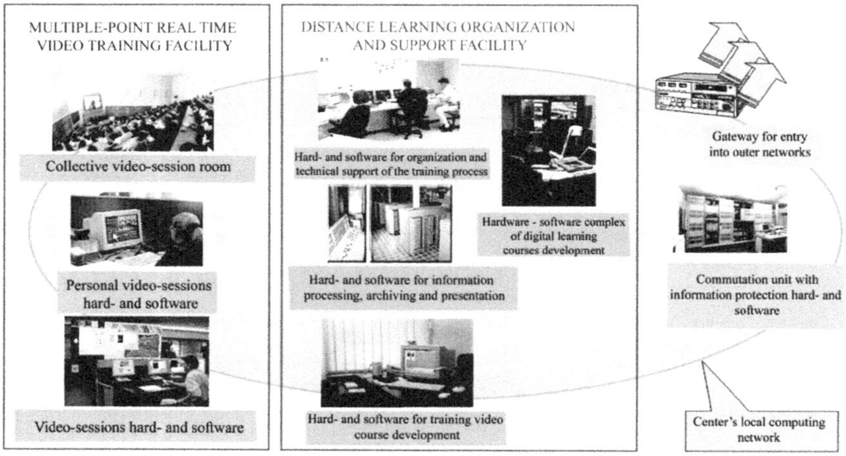

Fig. 5.9 Components of the engineering center of the International Space University

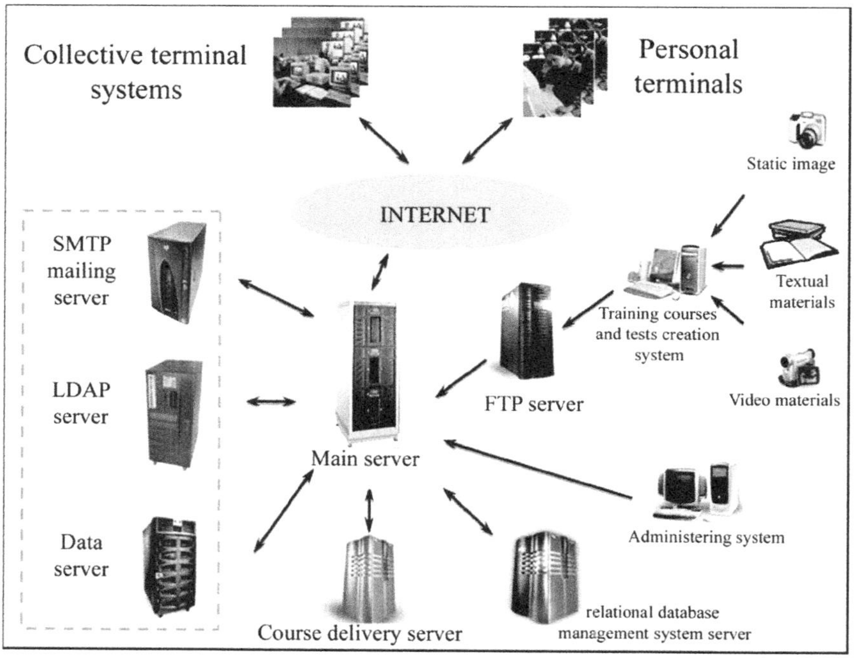

Fig. 5.10 The distance learning organization and support system

shown in Fig. 5.10. This includes the digital networks with integration of services (ISDN) and IP nets.

The distance learning organization and support system belongs to the learning content management systems maintained by a server through which the users access courses of the University's Engineering Center. The delivery server is part of the principal server which delivers the required study materials to the users, launches course materials, and performs management functions. As part of the launch procedure from the relevant catalog, the main server refers the trainee (user) to the delivery server which briefly describes the course. This enables the user to optionally choose separate sections. When the user intends to load the required course sections, the delivery server addresses the data server where the relevant file is stored and shows the data on the user's terminal. The data used in the University's distance learning system (related to the courses and other resources to which, if necessary, access is provided) are stored in a relational database management system.

For storing data related to organization and support of distance learning at the University, three databases are used: the main server, the delivery server, and the auditing server. The former two store information related to users, courses, and resources. This information can be used in various ways. The auditing database stores system information concerning the use of the system resources. It is used for system management.

Each course calls for availability of data (texts, images, multimedia data), which the trainee uses in studying a particular subject. The relevant data files are stored on data servers. Access to them is possible from the brief description of the course with the help of a URL link. For dispatching automatic messages to users and for sending requests for help to the user support service maintained by the distance learning organization and backend facility, recourse is made to the e-mailing system. This necessitates the availability of the SMTP server which relays messages in the system in the required mode. Before loading the courses onto the main server, they are stored on the FTP server.

The University's distance learning curriculum production systems are dedicated software attachments for developing

1. Digitized training courses, tests, exercises, and other kinds of teaching aids which enable the creation of course structure (the purpose of training, tests, theoretical and practical materials), interactive training sessions, situational games or interactive drill sessions protected against unauthorized entry; structuring of training material and its rapid transformation in compliance with demands; use of test results for further checking of knowledge; integration of additional external programs (editors) for processing and correcting multimedia objects; support of standards and specifications of digitized educational resources;
2. Multimedia and digitized catalogs, encyclopedias and textbooks, as well as supplements using Web-CD technology, saturated with multimedia effects and a large amount of textual, hypertextual, and structured information;
3. Video lectures and other teaching aids.

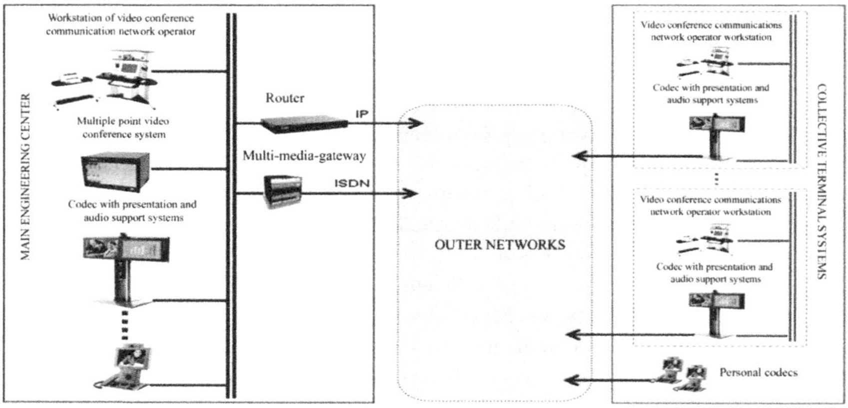

Fig. 5.11 Scheme for support of the distance learning nets

Collective terminal distance learning systems of the University are designed to obtain and present information from the engineering centers in a fashion convenient for learners, for interaction between trainers and trainees, as well as to protect the system against unauthorized entry into hard- and software. The collective terminals include (Fig. 5.11) hard- and software for conducting collective video sessions in real time and for individual distance learning; a commutation unit with hard- and software for protecting information, and include the technical facilities of the local computing network.

Personal terminals of individual users are designed to provide access to the distance learning space education engineering center resources for trainees in order to obtain and transmit the required study information. Personal terminals must be equipped with computers that have a sound card, Web camera, microphone, and that have access to IP and ISDN networks.

The telecommunication subsystem is designed for transmitting different study information and includes communication channels and terminal devices (multigateways, IP routers, modems), located in major engineering centers, collective terminal systems, and on personal terminals.

At present, there is a great variety of on-earth, satellite-based, and mixed telecommunication systems owned by various entities (state agencies, corporations, and even private individuals) differing in capabilities, performance, information exchange protocol, rates, etc. As operating costs go, the highest are usually those associated with payments for renting communications channels. The choice of satellite or on-earth major channels is not simple and depends on the speed of information transmission, communication distance, type of access to the network, the nature of transmission (symmetrical – speed of exchange in both directions is equal or asymmetrical – the speed in the direct (request) channel is lower (by several times) than in the reverse (respond) channel). So, at an asymmetry of

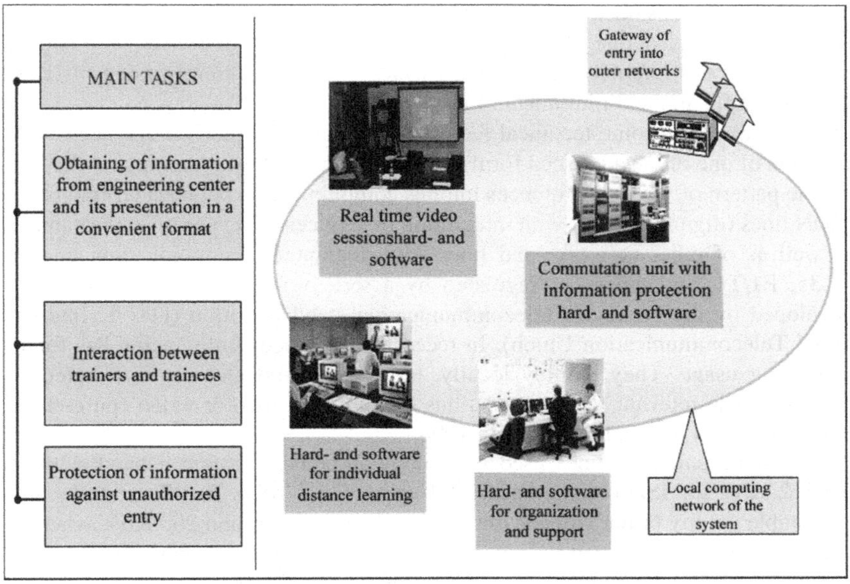

Fig. 5.12 Typical hard and software for distance learning

¼ it makes sense to order a satellite channel for a user remoteness of 700 km and more.

The choice of a means of communication depends on the types of information accepted in the system, which information must be transmitted; on the required speed of transmission; on the cost of equipment and the extent of development of the system in terms of its implementation stages. It is expected that various types of communication systems are used, such as wire-aided channels [conventional telephone networks, digital transmission networks (ISDN type, IP telephony)]; space communication channels, and optic-fiber communication lines. The regional networks may use cellular systems (GSM) Fig. 5.12.

Internet technologies will become the most widely used means of distance learning; the global network is an inseparable part of modern distance learning. So, with the help of the Internet it is possible to organize teleseminars and teleconferences, in which the dialog mode will enable hundreds of users to participate. In this case, the computer monitors will present statements and responses of all participants. There are instances of implementation of multiuser object-oriented mode that ensures real-time contact. Thus, the computer becomes a terminal of a remote host machine in the center where participants meet in virtual mode and simulation of virtual objects takes place, for instance by using slides and virtual boards on which questions are recorded for discussion. This technology does not need high-speed communication and data transmission lines. The well-known WWW (World Wide Web) service allows one to present textual and graphical information, to transmit sound and images, to reproduce simulation models, to carry out tests, etc. Also one should remember the most popular and very simple capability of the Internet called e-mail, which enables the transmission of texts, graphics, software, and multimedia files.

Let us dwell on some technical features of organizing video-conferences as an example of one of the advanced forms of training in a distance learning mode. The classic pattern of video-conferences implies communication between terminals via ISDN lines (digital network with integration of services). The use of such channels as well as of other networks and lines with guaranteed communication quality (V.35, E1/T1, and others) is regulated by a series of recommendations H.320, developed by the sector for telecommunication standardization (ITU-T, International Telecommunication Union). In recent years, video-conferencing has found increasing usage. They employ locally, territorially and globally distributed IP networks. The relevant recommendations (H.323 standard) for video-conferences via IP networks were adopted in late 1996.

The communication lines used for video-conferences feature a bandwidth of 64–512 Kb/s for ISDN conferences and up to 1–1.5 Mb/s for IP networks. An acceptable quality of the video is obtained at a speed of around 200 Kb/s, whereas high-quality images in good systems are obtained at a speed of around 300 Kb/s and higher. The IP systems need wider bands: because of the specific nature of information transmission in networks with package commutation (addition of headlines, RTCP protocol service packages) the bandwidth is increased by 20–30%. Practice shows that the quality of video-conferences is approximately the same with the use of three BRI (Basis Rate Interface) channels (384 Kb/s) or an IP channel with a speed of around 500 Kb/s.

The leading manufacturers of video-conferencing systems, such as VCON, produce multiprotocol (H.320/H.323) systems that operate well simultaneously in IP- and ISDN networks. There is a range of specialized gadgets, the use of which dramatically improves video-conference capabilities. For example, the arrangement of multipoint video-conference (MCU, Multi Conference Unit) systems, often referred to as video servers, is used for organizing video-conference sessions when they are attended by several (three and more) people at a time. The problem is that in case of multipoint communication, if no special solutions are employed, the load on each workstation grows in proportion to the number of video conference participants, whereas on the whole network it is proportionate to its square. That is precisely why the processing of a huge flow of information circulating in the network needs video servers. For instance, a computer serving a video conference with ten participants at a time, must process nine incoming data flows from interlocutors. If, however, the network has a video server, it receives all the data flows and sends only one, which is already processed.

Another type of dedicated system for video conferences is gateways that transmit information at the junction of heterogeneous networks. In addition to computer IP nets there are high-speed telephone networks. The transmission of audio and video information through them proceeds in their own formats. The IP nets are networks with package commutation, whereas telephone networks are nets with channel commutation. For solving the problem of compatibility and recoding of audio and video flows at the junction of networks a special gateway is fitted. For searching the stations and gateways and for linking to multipoint conferences the Gatekeeper software is used. This is a key device for organizing video conferences in the IP net.

For video sessions of several users at a time (similar to video selector conferences) the hardware includes multipoint conference MCU control. The operating principle of such devices is organization of a multipoint communication system that mixes audio flows and enables the participants to hear one another. Video flows are switched over in such a way that everyone can see only one participant of the session. This choice can be made by a chairman of the video conference or by an operator, or can occur automatically depending on the voice intensity. The MCU, as a video terminal, can be arranged by means of hard- or software. For example, the company White Pine offers software-aided realization of the Meeting Point conference server. To establish the communication it is necessary to obtain an isolated server running on the Windows NT 4.0 or Sun Solaris operating system. One of the most difficult jobs of the server is to recode video- and audio-signals. As in the case with software-aided realization of video terminals the general application processor output is insufficient for solving the task successfully.

Net broadcasting is used ever more widely today. This is transmission in the net with the help of group addressing of previously recorded or "live" video programs. To establish an adequate broadcasting server it is necessary to have a rather costly set of equipment. However, the receiving party can make do with simply viewing software (if no superior image quality is needed). The universally recognized leader in this area is the Cisco IP/TV project. But in many cases one can do without this costly broadcasting server, substituting it with a conventional hardware codec compatible with Cisco IP/TV. For example, even using the most economic codec, VCON Escort 25 and Cisco IP/TV Viewer, it is possible to equip a mini-studio for net-aided transmission of images received through the video camera or a video recorder. By combining these two products the terminals can operate as sources of video images transmitted to any personal computer using the Cisco IP/TV Viewer software. The shareware distribution of this product makes such a combination extremely attractive when it is necessary to transmit a video conference to an unlimited number of users, for instance while viewing or during training.

It is well known that digital television uses data transmission channels much more sparingly than the analog type. Therefore, it makes sense to implement distance learning that needs a bilateral communication, with the help of a compressed digital video (CDV) method. Digital video allows one to transmit audio and video information via special telephone lines known as ISDN. This ensures data transmission with a speed of 56 Kb/s to 3 Mb/s. The computer compresses the TV signal to the required degree, after which it is fed to the telephone network, the use of which provides a bilateral interactive communication. In this case no special studio or extra equipment is needed, whereas standard TV quality can be achieved on lines with a throughput capacity of 112–384 Kb/s. The thus-constructed distance learning systems allow one to hold seminars, organize training, communicate with colleagues, and to transmit documentation.

Another type of communication system used in distance learning is a two-way video method. This brings remotely spaced trainers and trainees together in a virtual classroom. For communication channels they often use digital telephone lines (ISDN), optic fiber networks, or satellite systems. Also, there is a similar concept, known as "one-way video," using a voice feedback communication method. This is a version of the "conventional" TV supplemented with the capability of communicating via voice reproduction systems (audio conferencing). However, the most economic method of delivering information to many points located thousands of kilometers apart is satellite TV. The information is transformed into a video signal, which is amplified and sent from a transmitting antenna to a geo-stationary satellite and relayed to Earth with a large effective radius. On Earth the signal is received by small antennas. The cost of equipment is steadily falling, which stimulates the massive use of such systems. However, the satellite systems have the essential drawback of a low interactive capability.

Thus, the University's telecommunications system can use narrow (transmission speed up to 30 Kb/s) and wide (tens to hundreds of Kb/s and even several Mb/s) band data transmission channels. Referred to as wide band lines are also optic cables and satellite communication channels. In recent years satellite communication systems have introduced ever more achievements of science and technology into the field of telecommunications such as multimedia communication systems based on MF-TDMA technologies (multifrequency access), asymmetric Internet access technologies, and VSAT (Very Small Aperture Terminal) satellite stations employing IP/TCP user protocol which is particularly effective for scalable corporate solutions. The new generation of multimedia stations will cost less than their predecessors and will be more suitable for interacting with on-Earth telecommunication networks, without duplicating the functions of transformation and routing the information flows.

To organize the main and regional communications of the World Space University of Distance Learning in the framework of IGMASS it is possible to rent the channels of Russian and foreign satellite systems. For instance, out of 13 geo-stationary orbital positions, ten (Long. 110 W to 1,450 E) are assigned to the Space Communication Federal Unitary Enterprise, which uses Gorizont, Ekran-M, Express-A, and Express-AM types of satellites. The latter are the most advanced systems that guarantee a high probability of trouble-free operation and a service life of not less than 15 years. The Express-AM satellites, in addition to TV broadcasting, are used for establishing the main lines of satellite communication and backing up the optic fiber lines. Also, they maintain the VSAT corporate networks. The Space Communication Federal Unitary Enterprise has on-Earth facilities in the space communication centers of Dubna, Vladimir, Medvezhyi Ozyora, and Shabalovka. Those are united by an own optic fiber communication network that has access to the main on-Earth lines. The Space Communications Enterprise has the Bonum-1 direct TV transmission satellite (orbital position Long. 560 E). Thus far this is the sole Russian satellite for direct TV broadcasting in Ku range, which is going to be used for data transmission in filling the needs of Russia's Education Ministry. It is possible to use the services of interactive television and asymmetric access to the Internet.

The Yamal satellite communication and broadcasting system belongs to OJSC Gazprom. Three artificial satellites are in orbit. One is Yamal-100 (Long. 90°E) and there are two Yamal-200 (Long. 90°E and Long. 49°E). The satellites at Long. 90°E service the territory of Russia, the CIS, and foreign countries. The artificial satellite at Long. 49°E services the European part of Russia and the CIS as well as some countries of Europe, the Middle East, and South and Southeastern Asia. The Yamal integrated communication center includes three teleports (No.1 in Moscow, No.2 in Korolyov, Moscow region, and No.3 in Medvezhyi Ozyora village, Moscow region), a digital satellite television center, and a trunk optic fiber communication line with a throughput capacity of 622 Mb/s. The digital television center is fitted with modern equipment that converts signals to DVB/MPEG-2 digital transport flow.

For the performance of the University's functions the Eutelsat and Intelsat fixed satellite communication systems are worth noting. Eutelsat involves 47 participant countries. The interests of Russia are represented by the Space Communication Enterprise. Over 20 geo-stationary artificial satellites of Eutelsat, Hot Bird, Seesat, and other types are in orbit. The orbital positions of those artificial satellites are distributed over the range Long. 7°W to 45°E. One satellite, Eutelsat-2 F1, is at Long. 70°E (previously Long. 48°E). It provides services for the Internet in a channel operating at up to 2 Mb/s. Eutelsat is actively expanding its VSAT network. On-Earth D-SAT stations operate in the DAMA (Demand Assigned Multiple Access) mode and are oriented towards conjunction with ATM (Asynchronous Transfer Mode) networks. The channels ensure an information transmission speed of up to 160 Mb/s. Under development is a project for interactive access to the Internet.

In addition to the above mentioned, there are many other systems of fixed space communication (FSC) as well as of mobile communication with a satellite on a geo-stationary orbit (Inmarsat) and with an artificial satellite in low orbits like Globalstar, Iridium, and others. However, for creating the University telecommunication system in regions with a well-developed communications network, it is sufficient to employ only FSC systems. The mobile satellite communication systems can be used in the case of a fairly large number of trainees in sparsely populated areas where the communication network is not adequately developed.

The FSC systems feature a fairly well-developed sector of VSAT technologies. In 2000, a DVB-RCS (Digital Video Broadcasting – Return Channel via Satellite) – ETSI EN 301 790 standard was adopted. The standard sets technical requirements for subscriber interactive terms oriented to an asymmetric traffic typical of Internet access. It should be noted that the C-range holds no promise for DVB-RCS systems since obtaining in the reverse channel a minimum speed of 144 Mb/s that provides adequate system efficiency, is only possible with an antenna measuring 2 m in diameter.

The review above of the various communication and data transition facilities allows the conclusion that the telecommunication system of the World Space

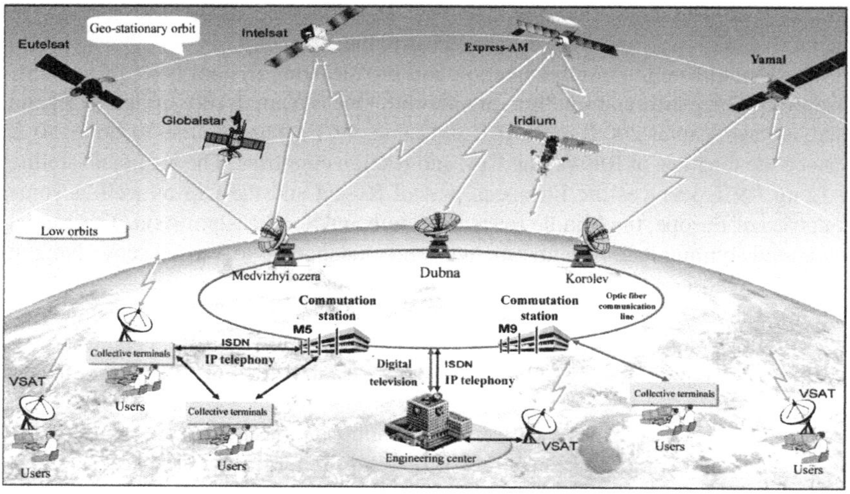

Fig. 5.13 Telecommunication system of the World Space University of Distance Learning

University of Distance Learning (Fig. 5.13) must use on-Earth wire- and optic fiber communication channels of ISDN type and IP telephony with data compression for digital television as well as fixed satellite communication systems based on VSAT technologies and using teleports that can operate both with existing and future spacecraft for communication and data transmission.

To complete the review of the concept of the University as one of the best thought-out humanitarian components of IGMASS, let us consider the types, forms, and peculiarities of distance learning and the technical facilities involved. First off, let us draw the readers' attention to Table 5.2 where the types and the required amount of study and methodological information are shown with reference to the main forms of training at the University. These well-known forms have been optimized in the course of development of telecommunications networks at the turn of the century. So, belonging to the main instruments of organization of communication between the trainers and the trainees are such widely used forms of information delivery as e-mail and the numerous tools for conducting virtual discussions (textual and graphic forums, telecommunications, etc.)

Virtual discussions make it possible to organize the collective work of remote trainees in an asynchronous mode. Virtual discussions (seminars) can proceed, for instance, under the supervision of a trainer (or a group of trainers) acting as a conductor. In this case, the participants can scan all incoming messages and send their own, thus participating in the discussion. The widely used and very popular textual forum (chat) is used for organizing a textual dialogue between two or more users in real time. A graphic forum (whiteboard) organizes a multiple user virtual many-page board on which it is possible to present a diagram, or drawing, or any picture that will be visible to participants of the discussion. The graphic forum can

Table 5.2 The required types and amounts of learning and methodological information for the World Space University of Distance Learning

Type of training	Types and required amounts of training and methodological information
Lecture course (including digitized versions)	A semester lecture course built on a modular layout concept in the form of semester topical files containing on average 17 lectures with a total amount of 9.0 Mb per semester
	Course of lectures in subjects under study arranged as semester topical files or video films are entered in the relevant structural module or video card of the University's study site. Topical files with lectures can also be delivered to learning centers or personally to students (by order) using e-mail. Video films with lectures are chosen from a video library of the University's study site, as a rule, by training centers
Course work (projects)	Course tasks, tasks for doing course work and projects, seminar and colloquium topics, tasks for lab work, and home tasks are entered in structural units of the study site and transmitted to training centers and students (by order) via e-mail
Lab work	Reports for lab work with tabulated and standard graphic materials takes up on average 60 Kb, a course work ~ 200 Kb, a course project ~ 400 Kb, a statement for a seminar ~ 30 Kb
Home course tasks	Fulfilled statements for seminars, course works (projects), reports for lab work and home tasks are submitted to the engineering center via e-mail and introduced into a databank
Seminars	Seminars and colloquiums (2–3 per semester) are conducted in the form of video conferences in an interactive mode at training centers and at the University
Tests	Digital tasks in 17 tests (home tasks), lab works, plans of seminars and colloquiums, initial data and tasks for course work (projects) total ~ 500 Kb
Colloquiums	A digital textbook of the semester course under study with lecture files, tasks, and initial data for tests, lab and course works (projects) take up 9.5–10 Mb of memory
Checks	Checks and exams are given at the University and (or) training centers. In the latter case the questions for the check and examination cards are prepared and transmitted with the help of computer and e-mail. The answers are

(continued)

Table 5.2 (continued)

Type of training	Types and required amounts of training and methodological information
	submitted via e-mail to the University's engineering center
Examinations	Accounts concerning a subject studied during the semester total around 0.7 Mb
Total in semester	The permanently stored amount of study material related to the subject studied during the semester is 9.5–10 Mb; temporarily stored accounting materials for one semester in one subject under study is 0.7 Mb per trainee
Postgraduate training and skill improvement	The amount of study and methodological material in after-graduate training and skill improvement courses for individuals with vocational education is designed for four learning days per week, 6 h in a session. Three hours a week are allocated to each subject. A total of eight subjects are studied during a semester. Thus, the annual amount of digital study and methodological materials in one specialty due for long-term storage is ~ 480 Mb
3-year training at eight faculties of postgraduate education and four levels of professional skill improvement courses for individuals with vocational training	The total amount of long-term stored digital study and methodological materials in each learning language is ~ 5,800 Mb. In case of teaching in five basic languages – Russian, English, French, German, and Spanish – the total of long-term stored information reaches ~ 30 Gb

support information transmission (texts, files) via a clipboard common to all participants of the virtual discussion.

Audio- and video-sessions organized in the framework of the telecommunication network make it possible to conduct in real time lectures, seminars, and collective and individual consultations. The facilities used for organizing joint work and collective interaction (groupware) incorporate various net instruments of communication for establishing remote cooperation in different forms (Table 5.2). Thus, the lectures can be both asynchronous (during use of conventional television broadcasting) and synchronous (during use of interactive television and video conference technologies). The laboratory work in distance learning is done using the so-called "software simulators" imitating the operation of equipment and laboratory benches and by means of remote access to the actual hardware. The first version allows work at a convenient time without dependence on a specific learning time (asynchronous mode). The second version implies a remote via-net access to the equipment at a scheduled point of time (a synchronous mode).

Seminars are probably the most intense form of study in a distance learning mode. Seminars can be held in asynchronous and synchronous modes. Virtual tours are used in the asynchronous mode. The advantage of asynchronous seminars is that their participants communicate at a time which is most convenient for them. At any moment any participant can study the history of the development of the discussion and join it. A trainer can evaluate the assimilation of material based on the activity of a participant in the discussion. Synchronous seminars can be conducted with the help of video conferences and computer forums, which (like synchronous technical forms of communication) call for simultaneous presence of all participants.

Consultations, both collective and individual, are one of the forms of managing the trainees' work and aiding them in studying the subject. The most frequently used technical aids for consultation during distance learning are telephone and e-mail. The facilities used less often are virtual discussions and video conferences, which are almost the same as consultations held, for instance, for training in conventional face-to-face mode.

Independent study can be both individual and in groups. Naturally, no communication facilities are needed for self-study. However, if independent work proceeds in groups, use is made of mostly asynchronous training aids. Those are textbooks, manuals, video lectures on digital carriers, and digital instructional devices. Also, it is very useful to employ synchronous aids, such as textual and graphic forums.

To complete the consideration of how distance learning is organized at the World Space University, let us examine the Center's terminal systems as well as users' and remote training centers. Being the final component of the distance learning system, technically such systems are divided into users' and trainers' systems (Fig. 5.14). Both are designed for delivering information in a convenient

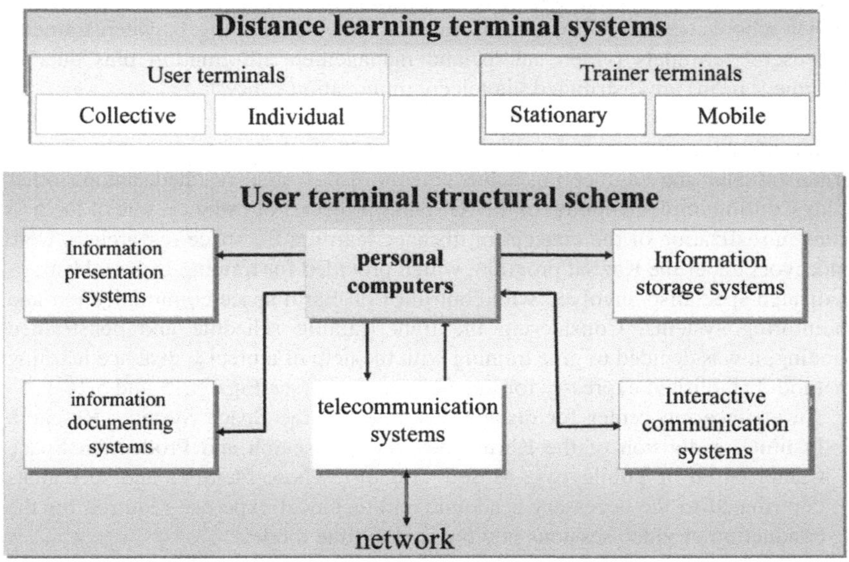

Fig. 5.14 Terminal systems of the World Space University of Distance Learning

and useful fashion and for providing an effective interactive dialogue between the user and the source of knowledge. Therefore, both users' and trainers' systems must include computers and devices for documentation, storage and presentation of information as well as for interactive and telecommunication sessions. In terms of usage pattern, users' terminals can be collective and individual.

The collective users' terminals belong to engineering and remote training centers. These terminals differ mainly in having powerful multimedia stereo systems. Depending on who possesses them, collective users' terminals may differ in their design solutions. For instance, terminals in the engineering center for conjunction with the center's hardware don't need extra telecommunication subsystems. However, for activating the collective users' terminals of remote training centers it is necessary to have extensive telecommunication subsystems providing a bilateral high-speed or asymmetrical, in terms of speed, communication with the engineering center.

The terminals of some users are fitted with traditional information presentation systems and devices for interactive feedback with the trainer. Such systems can be united functionally.

In terms of engineering decisions there is practically no difference between users' and trainers' terminals. However, the latter, by virtue of their application, are used more for information reception, which is why they must be fitted with more advanced systems for data introduction and presentation (this applies in the first place to video cameras). Trainers' terminals can be used both in stationary and mobile versions. The stationary terminals are installed in engineering and remote training centers. Mobile ones, designed for outdoor training events, are installed directly at dedicated facilities (spaceports, research centers, factory workshops, higher schools, etc.) Multimedia information flows circulating between trainers' and users' terminals (video, audio, and management information plus data in document form) are distributed via telecommunication subsystems.

To conclude our consideration of issues related to distance learning and distribution of space education, let us turn to some facts related to Russia's accrued organizational and engineering achievements in this area, reached, among other things, during implementation of the IGMASS project. For instance, one of the first steps in realization of the concept of distance learning for space researchers were endeavors under the KazSat program, which provided for training of Kazakhstan's dedicated specialists involved with control of on-Earth space communication and monitoring systems. Considering the tight training schedule and constrained funding, it was decided to give training with the help of a precise distance learning method. Established expressly for this purpose were (see Figs. 5.15 and 5.16):

- An engineering center for distance learning (at the Space Systems Research Institute, a division of the Khrunichev State Research and Production Space Center based in Yubileynyy, Moscow region), whose facilities and specialists contributed to the necessary academic and technical expertise required for the conduction of video sessions in close to real time mode;

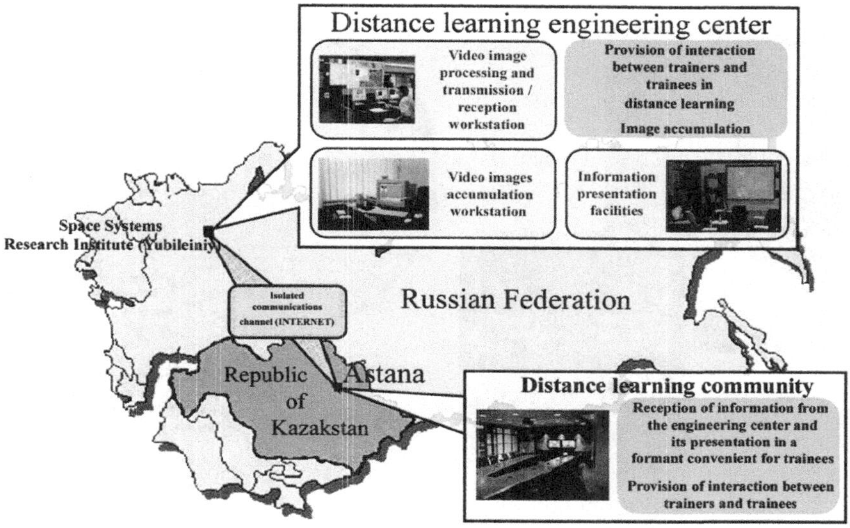

Fig. 5.15 Experimental system of distance learning for Kazakhstan's specialists

Fig. 5.16 Prof. Vladimir S. Chaplinskiy, one of the most skillful experts of the Space Systems Research Institute, teaching for the KazSat program

- Terminal facility for distance learning at Kazakhstan's center of space systems and electromagnetic compatibility of electronic systems in Astana, equipped with video conferencing systems for real-time video sessions.

On the basis of results of the joint methodological conferences on organization of education for Kazakhstan's specialists, it was decided to use teaching materials, audio and video aids (video lectures), and online seminars in "point-to-point" mode

Fig. 5.17 Teaching aids created in the Space Systems Research Institute for the experimental system of distance learning under the KazSat program

(Fig. 5.17). Without much delay a textbook was written and digitized entitled "The basics of building and operating a space communication and broadcast system." This envisioned a 320 h course of lectures by leading experts of the Research Institute of Space Systems (a division of the Khrunichev State Research and Production Space Center), Salyut Design Bureau, Khrunichev-Telecom, and GPKS (State Enterprise for Space Communication). The use of distance learning enabled Kazakh specialists to obtain basic theoretical knowledge of the on-Earth system of the space communication and monitoring system (KazSat). This was the basis of their successful practical training and work on standard systems. In this case the cost reached 30% of the predicted cost of traditional theoretical training.

The accrued expertise makes it possible not only to organize effective training of specialists under the KazSat program, but also optimize technologies and methods, which are supposed to be used not only in the IGMASS project but also in the framework of the distance learning system being created for the Russia-Belarus state alliance. The system is designed to train specialists and students in space technologies. Pavel P. Borodin, the secretary of the Allied State, confirmed the importance of this system for optimizing the principles and design decisions in organization of the World Space University. In his opinion, the distance learning program in space technologies in its Russian-Belorussian format will reduce by 30–35% the spending on training and retraining of specialists (relative to the cost of traditional training); provide continuity of training, retraining, instruction, and testing of staff of Russian and Belorussian dedicated enterprises, without

Fig. 5.18 Mr. Pavel P. Borodin, the Secretary of the Allied State "Belarus-Russia" during presentation of the distance learning system in the Space Systems Research Institute

discontinuance of the profile job; reduce training and retraining time, and distribute, using specialized training programs, the knowledge, skill, and expertise accrued in the space industry (Fig. 5.18).

5.2.2 Telemedicine and Emergency Medicine

Modern medicine tends to unite principle problems of critical, disaster, anesthesiology, and reanimation medicine in the framework of a single scientific and methodological basis – emergency care medicine (hereinafter referred to the environment surrounding the patient) or extreme medicine. The main unifying factor is a strict time limit and the range of engaged experts when the pace and level typical for ordinary therapeutic actions bring no result.

Contemporary medical science has accumulated experience in treating patients in extreme environmental conditions. However, the implementation of the available approaches under real conditions can be hindered by the absence of concrete real-time information about the situation. As information technologies spread across the globe the problem is being resolved through telecoms access of doctors to special information at the moment required for decision-making on various aspects of treatment and diagnosis. The efficiency of treating patients in emergency situations is considerably enhanced even without engaging additional doctors and material resources [7]. Besides, there are situations where patients and the injured have to be diagnosed and treated in hard-to-access areas where there are no doctors at all.

Such situations call for multidisciplinary tele consultations of doctors through modern powerful telecom networks.

The idea has long been of great interest to professional medical workers and was implemented in an applied scientific guideline – related to the design and use of remote medical assistance methods and specialized information exchange on the basis of modern telecom technologies – called telemedicine. Many countries pay priority attention to its development. Norway was the first to apply telemedicine because of numerous hard-to-access areas in its country where traditional medical assistance is lacking. France followed with a project for civilian and navy seafarers.

Telemedicine assumed wide scope in the United States as the country annually appropriates 13–14% of GDP for its development [8] both from the health and defense department budgets. The U.S. Department of Defense is very active in implementing telemedicine research results. Thus, in particular, a mobile telemedicine system has been designed and is operating with the use of digital channels of the INMARSAT space communications system, while the portable Medic-Cam telesystem with an 8-h self-sufficient operation on batteries (a full-color image from a 7-mm video camera lens is transmitted by an optical line to a ground space communications station which retransmits it to any part of the globe) allows one to receive consultations from the necessary doctors right on the battlefield.

Telemedicine services were widely used by U.S. and NATO troops during the war in the Balkans. At that time the center for perspective medical technologies of the U.S. Ground Forces launched an automatic telemedicine network for the 20-thousand-strong U.S. force. Medical establishments in Bosnia, Hungary, Germany, the United Kingdom, and the United States were linked into a single communication network. Portable computers equipped with a modem, text and graphic editors, and database software were used as terminals in hospitals. Internet or satellite communications channels were used to exchange data. Special devices were designed on the orders of the U.S. defense department that provided remote control of current coordinates (through GPS) of U.S. servicemen and their state of health via a personal medical card with a rigid-body chip, a biomedical belt with a set of sensors, and processor and communication means of urgent entry into a local telemedicine network.

The first major international experience of telemedicine and health use during the aftermath of natural and technogenic calamities was obtained over 20 years ago during the Spitak earthquake in Armenia (1988) and gas pipeline explosion near Ufa (1989) when TV linkups (audio, video, fax) were arranged between the calamity zones and leading U.S. medical centers under the auspices of the Soviet-U.S. commission for space biology and medicine [9]. Teleconsultations and video conferences involved experts from Moscow hospitals and major U.S. medical centers who provided consultations for burn, psychiatric, and other patients. Over 12 weeks the TV linkup held 34 four-hour video conferences in which 247 Soviet (from Armenia, Moscow, Bashkiria) and 175 U.S. experts participated. They considered a total of 209 clinical cases of 20 medical disciplines. As a result, numerous adjustments were introduced into diagnosis and treatment processes (in 33% of cases the diagnosis was changed, and in 21% – the treatment tactic), new treatment methods were introduced (in 10% of cases), and major volumes of medical information were transmitted to doctors and patients (in 46% of cases additional diagnostic and prophylactic measures were advised).

At present there is a series of telemedicine systems and a major number of various projects. The authors do not plan to consider them in detail, but will outline their common provisioning characteristics in terms of the use of space navigation and telecom resources. Thus, a typical telemedicine system comprises:

- Medical establishments with professional and educational information resources, medical diagnosis devices, databanks, as well as system users, etc.
- Sensors and other devices processing medical data into digital electric signals transmitted through communication channels.
- Telecom network access systems
- Communication links and network tools and forms of access to networks.
- As for the medical side of the issue, (information resources, databases, etc.) doctors have to provide all of them. Other components, i.e., technical implementation of telemedicine systems do not practically differ from remote training devices.
- Telemedicine terminals should provide [9]: reliable audio and video support for adopted decisions; express analysis of functional research data.
- Information and methodological back-up to medical staff, including availability of special databases; possibility of video consultations for patients, teletraining of medical staff, exchange of experience; video conferences to discuss new treatment methods, etc.

Thus, for example, intensive care monitors of vital functions that can be plugged into a parallel computer port are currently used to register and transmit physiological data of patients. They can, in particular, register and transmit ECG data, heart rate and rhythm indicators, hemoglobin saturation and oxygen in arterial blood. Video conferences use personal computers or special video linkup equipment, communication channels with protocols IP, ISDN, ATM, equipment of H.320 standard (narrowband phone), H.261 (video codec for audio and video services at a speed of 64 kb/s), H.231 (multichannel access to audio and video services), H.324 (general-purpose network video telephone), H.263 (video codec for low-speed communications), H.245 (video communication between multimedia terminals), H.323 (local network video communications), G.728 (coded speech at 16 kb/s speed), G.723 (audio communications at 5.3 kb/s and 6.4 kb/s speeds).

Therefore, the telecom infrastructure is a major component of the telemedicine system and a decisive factor in its efficient operation in extreme conditions. Satellite communications play a decisive role. Already in 1998 the international forum for the use of space technologies in the development of personal and mobile communications paid attention to the efficiency of VSAT technologies for telemedicine networks. In particular, it cited deployment data for telemedicine systems in scarcely populated regions: the TransTel Satellite Company (South Africa) deployed in regions with a low density of the population a ground fixed-line network of satellite telephony, as well as mobile communications based on unified platforms that can be transported in a railway carriage or an automobile. The station can service commercial clients (fax and telephony) and noncommercial users, mostly medical and educational establishments, and specific patients (access to Internet, databanks, diagnostic centers). Such a station cost 60,000 US dollars 12 years ago. The number of noncommercial queries comprised 1–2 thousand a week.

The telemedicine guideline directly promotes the operation of disaster medicine services – an independent organized structure intended to provide extreme medical assistance to civilian populations in case of natural and technogenic disasters, local armed conflicts and terrorist acts. Its main tasks include [10]:

- Events aimed at preventing, localizing, and eliminating the medical and sanitary aftermath of possible disasters, preventing and decreasing psychoneurological/emotional impact on the population, and promoting an early rehabilitation of people;
- Organization, preparation, and maintenance at a high readiness level of management bodies, establishments, and personnel for operations to eliminate medical and sanitary aftermath of emergency situations;
- Timely provision of medical assistance, evacuation and treatment of the affected people, their early health recovery and return to normal way of life, and maximum decrease of lethal outcomes and disability;
- Provision of sanitary wellbeing of the population in calamity areas, prevention of the emergence and spread of mass infectious diseases among the population in disaster areas and adjacent territories;
- Forensic examination of the killed, medico legal investigation of the injured to determine the degree of affection and workability forecast;
- Health preservation of the personnel during the elimination of medical and sanitary aftermath of emergency situations, provision of medical assistance to the personnel of rescue units.

The solution of the above-mentioned tasks can be achieved through a complex of events to create, equip, train, and maintain a high readiness level of the disaster medicine service; design and introduce into everyday practice theoretical, methodological, and organizational basics of medical assistance to the population in emergency situations; train the medical staff to work in extreme conditions and train the population to render first aid and know the rules of adequate conduct in conditions of natural calamities and technogenic disasters.

Elements of the orbital and ground infrastructure of the International Global Monitoring Aerospace System can be engaged and widely used by national disaster medicine services and similar units to determine the exact location of people affected by an emergency, for remote diagnosis and control of their physical condition, provision of first aid, pre-doctor and medical assistance including treatment (Fig. 5.19).

Potential users of the extreme medicine subsystem in the IGMASS structure can be divided into three groups. The first group comprises wounded and sick participants of extreme expeditions in any part of the globe (oceans, mountains, tropical forests, glaciers, extreme temperature regimes, etc.). The second group includes injured and affected in zones of natural calamities and technogenic

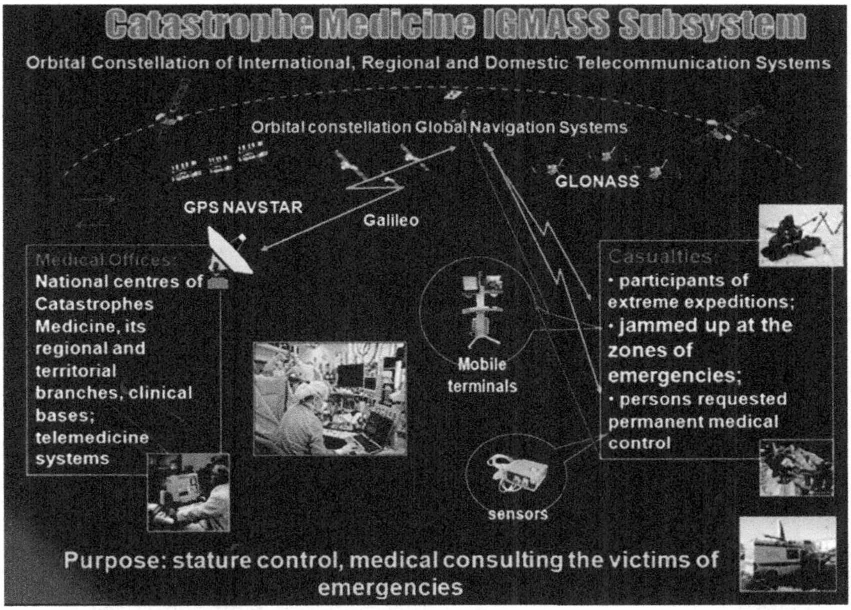

Fig. 5.19 IGMASS and medicine of catastrophes

disasters. The third group comprises people (including VIPs) in need of permanent control of vital functions for one or another reason.

Facilities and methods to monitor the health of people can change depending on the group. Thus, the first group can use light mobile terminals and individual sensors. It is characteristic for this group to use space telecom and navigation resources to establish and support reliable two-way data exchange traffic and to determine the location of the injured people. The second group of users can use mobile terminals with a wide range of diagnostic equipment that provides for a deep examination of the affected people. Mobile hospitals deployed in the emergency area have a full range of telemedicine services, including video conference communications and efficient two-way exchange of the necessary medical information. The third group of users shall be equipped with a set of sensors to control physical and medical parameters (current health indicators) that are critical for one or another disease. Such patients in remote areas or VIPs shall have access to space telecom (telematic) systems.

One of the prevention measures against technogenic disasters directly related to human activities can include remote control of the physical and emotional-psychological condition of operators of complex engineering, technical, and transportation facilities, and those involved in the production of potentially hazardous materials. A prototype of such solutions can be Russia-designed which is shown in Fig. 5.20.

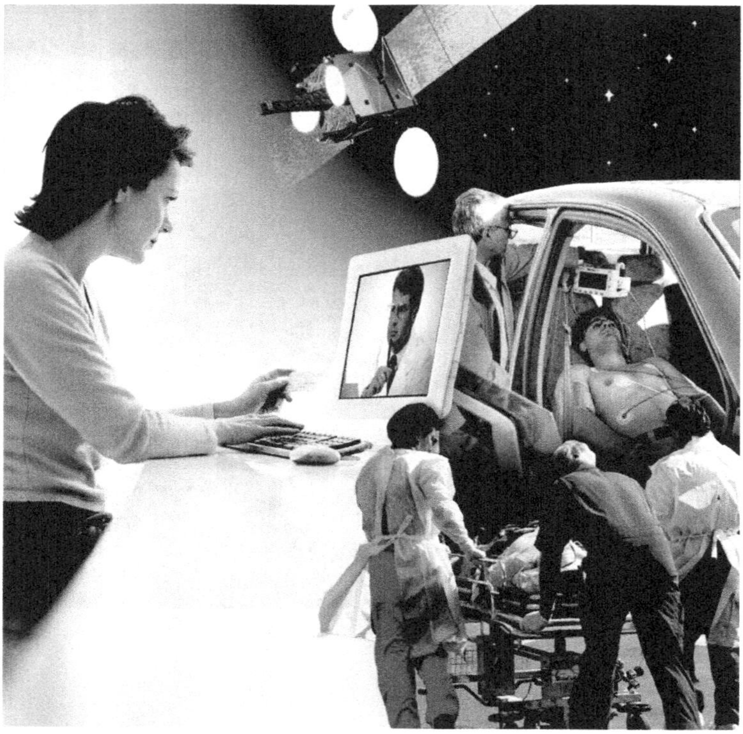

Fig. 5.20 Hardware-software complex for permanent remote control of human psychosomatic condition

Its control contour includes sensors of parameters of organs of interest, portable devices to control the sensors and for preliminary processing of the signals, interfaced mobile transmitters and receivers, stationary equipment to collect and process information. The complex is to provide everyday control of the physical, moral, and psychological state of the personnel operating public transport (international bus drivers, railway engineers, civil aviation pilots, etc.), and those involved in the production of potentially hazardous materials (nuclear power and chemical industries, energy generation) directly at work places. The obtained information is transmitted through various telecom networks to the interested customer for direct control or further use. The variety of engaged telecom subsystems will help in determining optimal ways of information collection, processing, and delivery. For hard-to-access and specifically sensitive objects direct satellite communication channels can be used. The methods and means of a system of permanent remote control of the psychosomatic human condition can be directly used in rendering assistance to participants in extreme expeditions, to the people affected by natural calamities and technogenic disasters, and patients requiring permanent control of vital functions.

To conclude our review of disaster telemedicine topics we shall note that human physiological parameter sensors and their miniaturization comprise a major component of all the above-mentioned remote control systems. Thus, in Denmark experts of the Electronic Path Consortium designed a sensor for long-term care of patients to control the cardiac muscle, body temperature, breathing, and oxygen concentration in the blood. The device is 3–4-mm thick and transmits all data to a mobile phone or personal computer of the patient and attending doctor both for self-control, early diagnosis, and electronic out-patient medical records. British scientists from the Royal University of Belfast designed efficient types of Wi-Fi antenna that protrude from the body surface by 5 mm, which allows one to create biosensors that can be carried for a long time to monitor the state of health of a patient. The achievements in the design and creation of sensors for remote control of physiological parameters allow predicting their early introduction into everyday life which will demand enhanced opportunities in the field of telecom and information-navigation systems to promote future emergence of the "global information security space."

5.3 The IGMASS Project as a Potential Tool for Evolving the "Planetary Security Information Field"

With the progress of astronautics human life is acquiring a new dimension. Man is beginning to better understand nature and gain control over it, unveiling its mysteries, both on Earth and in outer space. The pondering of the universe gave birth to no end of fantastic ideas, mystic notions and myths, urging man towards the study of periodic celestial phenomena and attempts to evolve a systematic knowledge of this world, albeit initially based on hypotheses and conjectures.

Philosophically, for ancient Greeks the term "space" meant the ultimate generalization of all that is visible and invisible. It was in this sense that under the influence of Plato, as early as I AD, Philo of Alexandria spoke of "the intelligible cosmos." We encounter this term used by Neo Platonians, for instance by Plotinus (III AD). The issue of space genesis was not raised in ancient times. It emerged only with the coming of Christianity when, together with the Bible, an idea of an individual creator surfaced. As for "the middle" of the universe, the geocentric concept had endured till the Renaissance era. The evolution of knowledge about the composition and structure of space went hand-in-hand with the advance of natural history in general. The study of the stellar universe, the subject matter of astronomy, rests on the achievements of mathematics and other natural sciences. As for methodological aspects they are based on theoretical conclusions of philosophic thinking. The aggregate of scientific knowledge about space testifies to the infinity of the universe and eternity of time. The progress of astronautics and space exploration will indisputably enrich science with new facts which will give ground for new hypotheses and concepts aimed at ensuring the steady and safe development of the planet.

In the 50 odd years that elapsed since the launch of the first Soviet artificial satellite, knowledge about the capabilities of astronautics has significantly increased. By astronautics we mean not only the assemblage of specific space vehicles, but also the rapidly growing production industry that resolves the fundamental social problem, the mastering of space and introduction of space-related technologies into human life. The solution of most of mankind's global problems is directly linked to astronautics, which is a concentrate of essential traits of revolution in science and technology, marking the establishment of a new technological basis of "a planet of people" and steadily promoting the buildup of mankind's intellectual and material capability. In it, we must recognize a new quality of the society's productive forces and ascent of humanity to an essentially new level. This can't but influence all that we call culture, which, essentially, is a complex of intransient material and spiritual values of the society attained in the course of its multifaceted development and is embodied in man. The logic of its progress generated the task known as space exploration and provided means of its fulfillment.

Meeting the essential needs of mankind in progress, astronautics refines them, thus contributing to the advance of human culture. In his time, K.E. Tsiolkovsky declared that the conquering of space must proceed simultaneously with improvement of relationship between peoples, the perfection of their life style across the globe (i.e., the progress of social culture), and the elimination of evil and suffering: "Living the life of the Universe you must be happy..." [11]. Such is the inevitable role of culture in a broad sense. This is its imperative. All this had been ripening over ages, gradually, initiating in the twenty-first century a rapid growth in material and spiritual achievements of man, due to which he dramatically changed his daily living and creative endeavors. As a result, man himself has changed. This can be understood fully only based on philosophy, sociology, and culture study since these sciences determine the general view of the world, and the principal attitude to it and provide ground for man's goal-oriented activity. Without them, the latter is like a ship without sails and compass. For K.E. Tsiolkovsky this was the basis of his research and engineering ventures.

The humanitarian aspects of astronautics became obvious from its incipience when the practical use of space was only a dream and a matter of only theoretical speculations. The latter is most graphically reflected in K.E. Tsiolkovsky's legacy, whose works contain, in addition to speculative notions, practical calculations on which he relied in his far-reaching theories.

Man has entered space by virtue of overpowering trends and objective laws of scientific, technological, industrial, and social progress. The role of an outstanding individual in human history (as in history of science) is well known – this is the role of a captain who steers the ship towards the goal. The particularly shrewd minds can perceive in a specific solution, invention, or discovery a promise for the future which will dramatically contribute to the advance of science, technology, and social life, bringing mankind, as K.E. Tsiolkovsky put it, "mountains of bread and no end of power." In the equation of a variable-weight body's jet-aided propulsion the great Russian scientist saw the possibility for mankind to free itself of the geo-centric life pattern and enter the galaxy without bounds. The launch of the first artificial satellite, the space flight of Yuri Gagarin, Neil Armstrong's landing on the Moon, and other stellar events marked the daring and highly risky attempts of mankind to realize that possibility.

We witness the steady and ever faster introduction of space technologies into purely human activities on Earth (the use of space systems for solution of problems on Earth). Simultaneously space is acquiring a human dimension, i.e., it becomes intensely used and mastered by man. This dual process, objective and inevitable, is a subject matter of such human activities as natural science (fundamental and applied) and engineering endeavors (in design, technology, and economics). In addition, it increasingly becomes impacted by fundamental humanitarian issues. "The queen of sciences," as philosophy is often referred to, is precisely the science of thinking and, in a sense, a theoretical reflection of the world in terms of definitive generalizations. This stipulates the initial ground for formulating a viewpoint on objective reality, evolution of public activity standards, determination of fundamental principles of socially significant creativity, and establishment of pedagogical (didactic, theoretic, and explorative) patterns of education. The definition "space" with reference to philosophy accentuates its essential feature in the present day – the understanding by man of his place in Space, and its impact on everything on Earth.

The birth of applied astronautics and direct instrumental exploration of near-Earth space signified a leap not only in natural science and technology, but also in philosophy, public opinion, and historical progress in general. Contributing to this were the preceding successes in natural science and technology. The decisive contribution was made by the works of the great Russian humanitarian scientists K.E. Tsiolkovsky, V.I. Vernadsky, and A.L. Chizhevsky. Among other things, their passion for exploration stemmed from the Russian cultural tradition of search for novelty. From the days of its inception, the Russian Academy of Astronautics named after Tsiolkovsky, one of the promoters of global aerospace monitoring, has been concerned with such space-related issues as perception of the galaxy and its reflection in terms of science, and the impact of space on the world's biological and sociological processes. Astronautics not only urged mankind to think and act on a planetary scale, but also launched geo-centric tendencies in a social stratum whose connection with space had been for some time rather weak. This confirms the forecasts of Russian space researchers in which, despite occasionally controversial conjectures and philosophically vulgar notions, the ideas of geniuses were absolutely obvious.

The establishment and development of applied astronautics not only in Russia but also in the rest of the world was determined by the historic progress and response to challenges of time arising independently of anyone's will or intention. Today, astronautics has become a new major element in the society's productive forces. Employing the best achievements in all branches of science and technology, it is the backbone of science and engineering, bringing about dramatic changes (leaps) in knowledge and practice. It is safe to say that man as a subject of historic creativity, is an embodiment of these processes. Astronautics gave birth to an essentially new type of man's productive activity for public good. The chief feature of this activity is not lack of knowledge about possible collisions in the future, hitherto unknown, and the extraordinariness of the emerging challenges, but also the extreme nature of physical and other conditions in which they are addressed, conditions not quite predictable and unavailable for reproduction or simulation on Earth. That is why to this day those processes are highly risky and fraught with hazard for mankind.

The capabilities of astronautics in the permanent global monitoring of all layers of the geo-sphere, geo-information technologies, and space facilities for rescue during emergencies are well known. These achievements, now in use across the globe, are hard to overestimate. Those include satellite geodesy, navigation, meteorology, communication and television, satellite relaying information, the Internet, distance learning and universal education. Without them it is hard to imagine today the daily life of "the planet of people." The combination of achievements in space and information technologies integrates the world community into "a global village" and promotes steady development. However, in order to avoid unpleasant surprises in this area in the form of unpredictable consequences, it is necessary to look into the future, for which purpose the services of sociologists and political analysts are unavoidable.

Space exploration today starts the socialization of space. With the growth of applied use of space, new questions arise spontaneously and new aspects become obvious in the system of social and political priorities. The classical political and economic problems come to be regarded in a new light and the economy comes to grips with new tasks. Thus, we have not yet learned how to adequately measure and plan the funding of space exploration, how to assess the ecological damage sustained from this and how to handle such issues in terms of economy and engineering. Also, we need to learn how to forecast the pricing of space products and services and how to calculate the economic effect of each space launch.

Today's multifaceted space exploration rests on a single foundation whereas its components supplement one another. The ultimate goal of space exploration is to serve mankind in general and each individual in particular, primarily in the sphere of security. The better understanding of Nature by man helps to enhance both environmental and social security. That is why space exploration issues have become vital today. Addressing them adequately and quickly is a top priority task for understanding the social genesis of progress of science, technology, and civilization in general. Without this it is impossible to map out the strategy and tactics of advance of science and technology, to improve forecasting, programming, planning, decision-making in any sphere of material and spiritual production, including applied astronautics, which from the theoretical point of view is today a humanistic value that ensures global security and means much more than the conventional interpretations. Thus, one of the urgently important tasks may become the formation of "the secure information space" as part of the above mentioned "information sphere" that thwarts global threats and reduces the risk of their emergence .

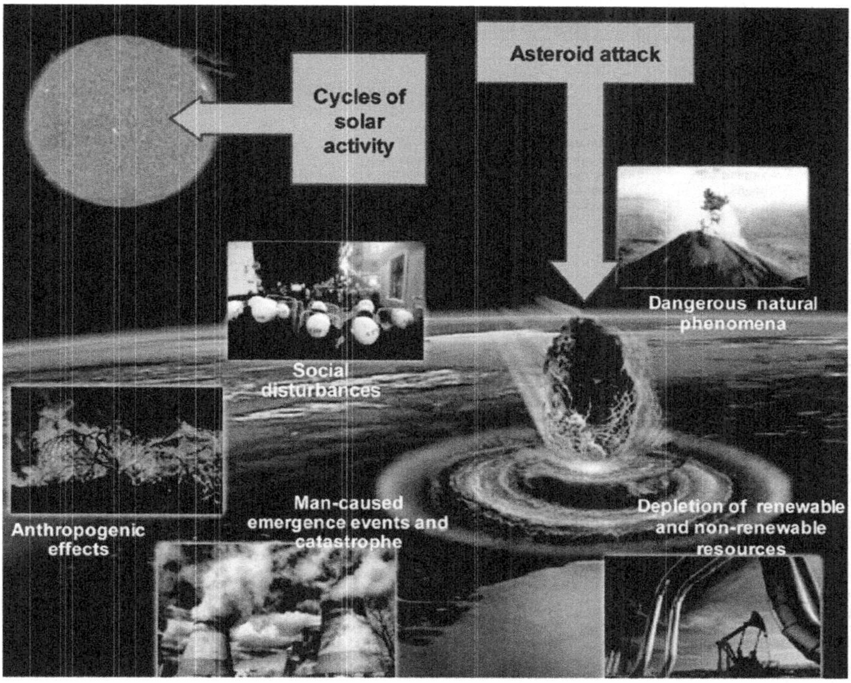

Fig. 5.21 Contemporary global risks and threats

Its common knowledge that mankind has always lived in zones of internal and external risks, which have become part of people's everyday life. As a source of tension and apprehension, such risks stimulate the progress of civilization. With the change of epochs and technologies some of them disappeared (softened), whereas others were handed down as a "legacy" to the next generations. Such risks often turned into new, more formidable ones (Fig. 5.21), which time and again materialized in the form of outer and inner threats to civilization.

Outer threats are those coming from outer space. So, the cycles of anomalous solar activity were followed by cycles of global climatic changes with glaciations, floods, and Earth's collisions with asteroids and comets, which led to global cataclysms. The inner threats are natural calamities and man-made disasters (accidents resulting from human activities). Dangerous natural phenomena are often triggered by man's actions whereas natural calamities, in their turn, often cause man-made catastrophes. In any case, such dangerous natural phenomena as the sinking and under flooding of terrain, landslides, tropical hurricanes, and even increased seismic activity may result from mankind's intensified attempts "to transform nature." At the turn of the century the risks from man's impacts came close to a crisis which may become a threat capable of irreversible processes with the most deleterious consequences for mankind (Fig. 5.22).

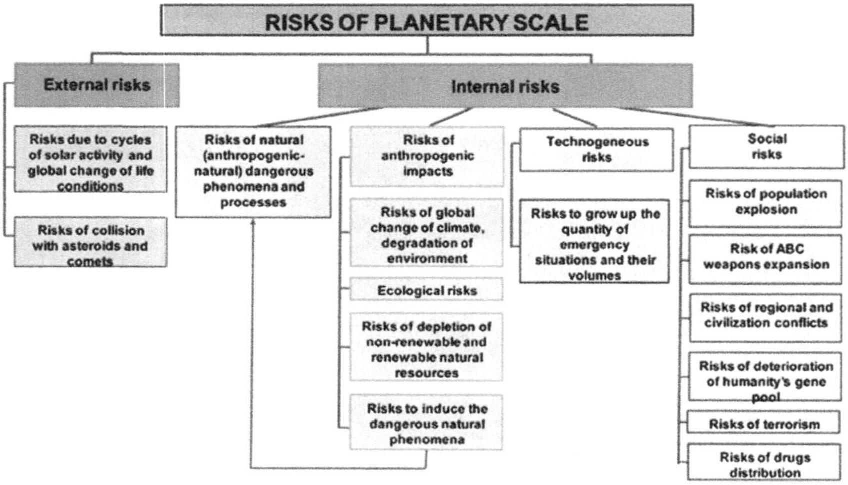

Fig. 5.22 Classification of global risks

The analysis of threats and risks of political emergencies are beyond the scope of consideration of this book. Nor is this what the international global aerospace monitoring system mainly deals with (this is the concern of dedicated systems of the military and intelligence services). It should be only noted that in today's multipole world it is necessary to elaborate such political decisions in the field of global security and stable development sought by all countries. Today's industrial production of most countries features a consumerism attitude to nature, which destroys environment and increasingly goes beyond their national borders (for instance as in the case of a recent ecological disaster in the Gulf of Mexico caused by a massive spill of millions of barrels of oil). Whereas the national interests of industrial giants in most cases continue to dominate over the planetary ones, the developing countries seek to reach at any cost such an economic, military, and political level which would make them figures to reckon with in addressing global issues. As a result, the entire mankind is globally at a loss.

Under the current conditions, the elaboration and realization of balanced decisions by authoritative international institutions (the UN being the main one), it is necessary to possess relevant information whose amount, the rate of delivery and update correspond to the dynamics of the global on-going processes. Such information could be obtained using the resources of space facilities belonging to national and transnational institutions. The latter, being territorially separated, could be united under the IGMASS project, which is being implemented under the auspices and supervision of the United Nations.

The current globalization gives any user of communication facilities a convenient opportunity, on the one hand, to access an aggregate information resource (which in some cases may be anonymous), and on the other, to distribute information freely among a vast number of users. In this case, the transnational information traffic in science, technology, economics, education, culture, business, advertizing, etc., eliminates the historically established value system, and levels the mentality, thus turning political borders into artificial obstacles. If brought to the extreme, this process may lead to an information war. The trend can be prevented by further development of purely humanitarian aspects of IGMASS by uniting telecommunication resources for addressing mankind's issues of spreading education and preserving cultural and moral values.

Thus, "the information space of global security" may be converted from a philosophic and futuristic notion to a practical one. For this purpose, the rightly organized power, which ordinarily supports direct and indirect information links, must draw up a compromise that would ensure a steady advance. This can only be achieved if all the operators have an adequate access to information that can be placed in the global security space created with the help of modern and future space systems of all countries.

References

1. Булгак В.Б., Варакин Л.Е. и др. Концепция развития связи РФ // Технологии электронных коммуникаций. – М., 1996 (in Russian).
2. Миркин Б.М., Наумова Л.Г. Экология России. – М.: Устойчивый мир, 2001 (in Russian).
3. Овсянников В.И., Густырь А.В. Введение в дистанционное образование. – М.: Министерство образования РФ. http://www.gdenet.ru (in Russian).
4. Полат Е.С., Петров А.Е., Аксенов Ю.В. Концепция дистанционного обучения на базе компьютерных телекоммуникаций. – М.: Российская академия образования, 2003 (in Russian).
5. Попова М. Дистанционное образование в мире и в России. http://www.intermba.ru/ (in Russian).
6. Технический комплекс Всемирного космического университета дистанционного обучения (шифр « ЭРУДИТ»): системный проект. Технический отчет. – г. Королев: МОРО РАКЦ, 2004 (Russian).
7. В.А. Меньшиков и др. "Система дистанционного обучения, подготовки и переподготовки специалистов на основе современных компьютерных и телекоммуникационных технологий.". \ "Современные проблемы науки и образования.". – 2006. – № 4.
8. Андреев А.А. Евразийская ассоциация дистанционного образования // Материалы IV международной конференции по дистанционному образованию (in Russian).
9. Медведев О.С., Столяров И.Н. Телемедицина: обзор современного состояния и перспективы развития в России. – http://www.rfbr.ru (in Russian).
10. Prof. Valery A. Menshikov, Dr. Vyachelav M. Klyuzhev, Dr. Sergey F. Goncharov Space and Global Security of Humanity (Book of Abstracts of the First Specialized International Symposium, Limassol, Cyprus, November 2–7, 2009), M., A. A. Maksimov Space Systems Research Institute (NIIKS), 2009. – 65 p.
11. Сенкевич В.П. Космонавтика: системный анализ, информация, прогнозы. – ЦНИИМаш, 2000. (in Russian).
12. Вержбицкий В.В., Власова Ю.Ю. Социальные условия развития системы профессионального дистанционного образования (основные результаты мониторинга за 1999–2002 гг.). – X всероссийская научно-методическая конференция « Телематика 2003». http://www.ict.edu.ru (in Russian).

13. В.А. Меньшиков, С.Р. Лысый. – "Система дистанционного обучения специалистов России и Беларуси с использованием современных компьютерных и телекоммуникационных технологий.". – Второй белорусский космический конгресс. – 25–27 октября 2005, Минск.
14. Menshikov V.A. Proposition concerning creation of international aerospace system for monitoring of global phenomena. – Acta Astronautica, v. 64, #1, 2009.
15. Васильков В.Г., Щукин В.С. Возможности использования телекоммуникационных технологий в медицине критических состояний// Вестник интенсивной терапии. – 1998. – № 1. (in Russian).
16. Меньшиков В.А., Вокин Г.Г., Макаров М.И. Концепция построения интегрированной системы государственного мониторинга важнейших технико-экономических объектов, источников природных ресурсов и социально-экономических процессов в России. Российская академия наук, Институт проблем управления им. В.А. Трапезникова, РАН, Государственный космический научно-производственный центр им. М.В. Хруничева. III научно-техническая конференция. – М.: 2003 (in Russian).
17. Измалков В.И., Измалков А.В. Техногенная и экологическая безопасность и управление риском. – М.: МЧС России, РАН, 1998 (in Russian).
18. Малинецкий Г.Г. и др. О национальной системе научного мониторинга // Вестник РАН. – 2005. – Т. 75 – № 7 (in Russian).
19. Воробьев Ю.Л. Безопасность жизнедеятельности. – М.: Деловой экспресс, 2005 (in Russian).
20. Ranganath Navalgund, Valery Menshikov, Joseph Akinyede Space-Based disaster Managment: the Weed for international cooperation IAA, Bengalor-Paris, 2010, 80 p. (Heads of space Agencies Summit on November 17, 2010 in Washington)

Conclusion

So, dear readers, you have turned the last page of the book.... We hope that despite the various technical approaches and styles of the materials you have enjoyed the book. In this book we have been trying to prove that contemporary humanity needs to look beyond its own internal issues of survival in this so unusual and speedy 21st century and create mechanisms for the sustainable development of humanity as a whole. One of them (according to the point of view of the authors' and their supporters around the World) is the IGMASS Project.

For more than 3 years since its appearance, IGMASS has evolved from a conceptual idea, presented in Dnepropetrovsk (Ukraine) in 2007 to a full-scale international project, supported by the International Academy of Astronautics (IAA). In 2008, at a meeting in Glasgow city the IGMASS project was for the first time formally presented to management and leading scientists of IAA as a promising design, which required strong and comprehensive interstate support. In 2009, at the next summit of the Academy in Paris, the IGMASS Project Manager from IAA was officially appointed and a working (study) group of experts from 12 countries was formed to assess the level of IGMASS development prospects. In autumn of the same year (2009), the results of the group were formally unveiled at the IAA Summit in Daejon (South Korea). The Group resolution recognized IGMASS as a system capable of being built in the future, and recommended that the IAA "...ask the UN and its core committees, departments, and programs to examine key aspects of the proposed creation and utilization of the system in the interest of the whole international society, to realize the proposed project in the frame of UN programs, and to submit it for examination by concerned UN committees and commissions" (Figs. C1 and C2).

The IGMASS concept was officially approved and supported by the participants of the First International Specialized symposium "Space and Global Security of Humanity" held in Limassol, Cyprus, on November 2–4, 2009. This representative scientific forum became the first stage in practical promotion of the initiative, which enjoyed active support from the IAA. The initiative passed a preliminary multistage scientific approbation at a series of international events and was approved by distinguished scientists and experts from leading space institutions, including the heads of four national space agencies. Information and organization supporting the IGMASS project is provided by the Russian Federal Space Agency and the ZNANIE International Association, which enjoys a general consultative status in

V.A. Menshikov et al., *Global Aerospace Monitoring and Disaster Management*,
DOI 10.1007/978-3-7091-0810-9, © Springer-Verlag/Wien 2012

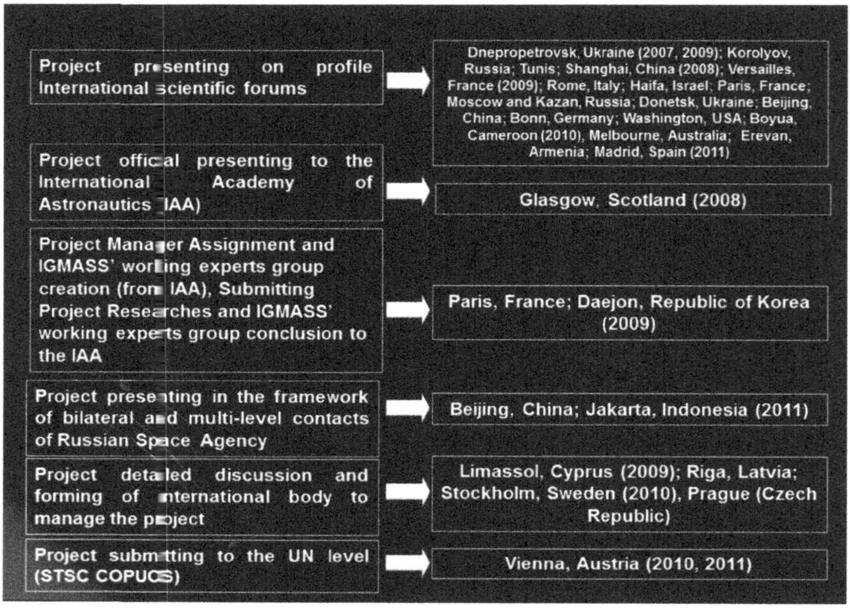

Fig. C1 Main steps of the IGMASS project's promotion

Fig. C2 Participants of the first international specialized symposium "Space and Global Security of Humanity," Limassol, Cyprus, November 2–4, 2009

ECOSOC and UNIDO. The Ministry of Foreign Affairs of the Russian Federation provides political support to the Project according to Russian foreign policy guidelines to promote the development of international cooperation in peaceful space exploration for global security and sustainable development (Fig. C3).

Fig. C3 The first presentation of the IGMASS project on UN level: 47th session of the scientific and technical subcommittee of the United Nations on the Peaceful Uses of Outer Space (COPUOS), Vienna, Austria, February 8–19, 2010

Over a period of 2 years, work to initiate the IGMASS Project achieved several major results: broad scientific and theoretical research has been carried out and involved foreign experts; the project was actively and rather successfully presented at the international level; full-scale R&D to create the multifunctional system of the Union State of Russia and Belarus as a prototype of key system segments was launched; small and microsatellites for the IGMASS orbital constellation were designed and the deployment of a specialized system of ground infrastructure for IGMASS to receive and process monitoring information from satellites started.

On the UN level IGMASS was first officially presented at the 47th session of the Scientific and Technical subcommittee of the United Nations on the Peaceful Uses of Outer Space held in Vienna on February 8–19, 2010 (UN document A/AC.105/958 of 11.03.2010). The session drew the attention of the world community to the IGMASS project, as well as prospects for cooperation in its implementation under the aegis of the United Nations and its involved institutions. There took place, in particular, a number of meetings and consultations with the leadership of UN-OOSA, GEOSS, ISPRS, as well as experts from the countries of all continents. Many of them at that time expressed readiness to join the "International Committee of the IGMASS Project Implementation" (ICPI), which had been established along with the adoption of a technical concept for the project at the

Second Specialized International Symposium "Space and Global Security of Humanity" in Riga (Latvia) and Stockholm (Sweden), on July 5–9, 2010. Besides the creation of the IGMASS Committee and the discussions of the special taskforce "research on elimination of natural calamity and disaster aftermath," the agenda of the 3-day symposium in Riga and Stockholm also included political, legal, organizational, and technical issues of project implementation and potential growth of the number of its member-states (Figs. C4 and C5).

Fig. C4 Presidium of the second international specialized symposium "Space and Global Security of Humanity," Riga, Latvia, July 5–7, 2010

Fig. C5 The first ICPI working session adopted some statutory documents for IGMASS project promotion (Prague, Czech Republic, September, 2010)

The main theme of the first working ICPI session on September, 27, 2010 held over the days of the 61st International Astronautic Congress, were the adoption of the Committee Charter and "Operation plan for the years 2010–2011." In accordance with a resolution of the Prague Committee session, the activity of the ICPI Executive Secretariat was project realization within the framework of the adopted ICPI Charter with the assistance of the scientific, political, and business community of different countries from all over the world. Although, the Executive Secretariat didn't get a chance to organize IGMASS research within the framework of wide international cooperation, project system design was started by the initiative group of Russian developers on the basis of research that had been done in 2010 by the Space Systems Research Institute and Russian Academy of Cosmonautics, named after K.E. Tsiolkovskiy (the volume of scientific materials exceeded a thousand pages). These results fully proved that milestone ideas of IGMASS could be realized (Fig. C6).

In regard to political promotion of the IGMASS project the Washington Summit of the Heads of Space Agencies, devoted to the IAA's Golden Jubilee, should be mentioned. On 17 November, 2010 the summit was attended by leaders of 29 national space agencies and equivalent organizations representing Argentina, Austria, Belarus, Brazil, Britain, China, Germany, European Union, India, Israel, Italy, Kazakhstan, Canada, China, Mexico, Netherlands, Nigeria, Norway, Russia, Romania, Saudi Arabia, Thailand, Ukraine, Czech Republic, Chile, France, South Korea, and Japan, as well as the Chairman of UN COPUOS. The summit demonstrated genuine interest in the project. One of its plenary lectures, delivered

Fig. C6 Plenary lecture of Dr. R. Navalgund, dedicated to the IGMASS project (IAA Washington Summit, November, 17, 2010)

by the Chairman of the IAA working group, director of the Space Facilities Center of Indian Space Research Organization (ISRO) R. Navalgund was almost entirely devoted to IGMASS. The Indian scientist emphasized the importance of this initiative aimed at reducing the devastating effects of global geophysical phenomena, but also to prevent man-made disasters caused by them. He pointed out the need to consider IGMASS in relation to complex problems such as asteroid danger and "extreme space weather," expanding orbital grouping for Earth remote sensing, primarily for the prediction of natural and man-made disasters, and development and verification of predictive mathematical modeling. It was particularly emphasized that the IGMASS project would promote international cooperation, without which effective action to anticipate and prevent threats of global character is impossible.

As the result of the Summit a declaration was adopted, which among other things provides for the development of forecasting and disaster prevention in "...refinement of remote sensing technology for disaster management through the transition from response methods to prevention methods, the strengthening of national, regional or international level links between stakeholders, including public and private agencies to prevent emergencies." Thus, at the highest international level the relevance and practical significance of the IGMASS project was proven once again as the heads of 30 space agencies and other profile institutions from all over the world solemnly gave positive assessments of the IGMASS profile ideas (Figs. C7 and C8).

Promotion of IGMASS on the UN level was associated with the presentation in Vienna, Austria at the 53rd COPUOS Session (June, 2010) and 48th session of COPUOS Scientific-technical Subcommittee (February, 2011). There the IGMASS

Fig. C7 The IGMASS project presentation at the plenary session of the IAA Washington summit (November, 17, 2010)

Fig. C8 Space dignitaries of the 2010 Washington summit

project was for the first time openly supported by the heads of the Argentinean and Nigerian delegations. They publicly appealed to their Subcommittee colleagues to "join the Project" and to the Subcommittee – "...to assess the IGMASS implementation in the frameworks of wide international cooperation." Moreover, a representative of UN OOSA joined the ICPI as an observer. During coming COPUOS sessions, ICPI is planning to submit a political declaration about "Consolidation of International Community Efforts on Using Aerospace Capabilities to Warn About Global Natural and Man-made Threats."

Unfortunately, because of the compressed mission timetable of the Executive Secretariat members this spring (Indonesia, Australia, and Armenia) ICPI didn't use an opportunity to present the IGMASS project at the last regular session of the COPUOS Legal Subcommittee. However, this gap was compensated by participation in the 1st IAA regional Conference « Small Satellites Programmes for Socio-Economic Benefits, and the Role of the IGMASS" (December 6–7, 2010, Boya, Cameroon). At this scientific forum, held for the first time in South-West Africa with participation of representatives from Great Britain, Gabon, Germany, Cameroon, Kenya, Nigeria, Russia, Sudan, France, and the Republic of South Africa, the ICPI Executive Secretariat gave presentations on the political and legal aspects of IGMASS project implementation. Moreover, Chinese experts expressed their readiness to take part both in IGMASS scientific research and system design.

On January 10–13, 2011, Committee members held negotiations in Beijing with representatives of the Chinese National Space Administration (CNSA) and profile organizations for their engagement in the IGMASS project. The Chinese side were particularly interested in joint research in the field of identifying the precursors of natural and man-made disasters, complex use of the monitoring information from Russian and Chinese satellites, the receiving and processing of this data by ground infrastructure facilities of both countries, in joint work in the field of small and microsatellites, as well as in creation of the next generation of space geophysical equipment. At the same time there was a working trip of the Committee Executive Secretary with a presentation of the IGMASS project in Indonesia. Three-day talks in Jakarta with senior management and leading specialists of the Indonesian National Institute of Aeronautics and Space (LAPAN) – Indonesian Space Agency – resulted in the signing of a Memorandum of Understanding (MOU) with LAPAN. A working meeting also took place in Jakarta with the ASEAN Secretariat, where the IGMASS Executive Secretary discussed the prospects of involving this well-known Asian regional organization in IGMASS activity via the signing of a MOU. The Secretariat disseminated a MOU draft among its members for their consideration for signing. In addition, during the visit to Indonesia the ICPI drew the attention of the Economic and Social Commission for Asia and the Pacific (ESCAP) to the project. In this regard, the ESCAP Disaster Risk Reduction Division requested via diplomatic channels that the Executive Secretariat make (together with representatives from China and India) a special presentation of the project during a regular session of the organization. All such options open up new prospects for political positioning of IGMASS and its Committee in the South-East Asia and Pacific region (Fig. C9).

Presentation of the IGMASS project also took place on the Australian continent – during the 14th International Aerospace Congress and Airshow "Avalon 2011" (Melbourne, February 29–March 4, 2011). During the Congress members of the Executive Secretariat held working discussions with representatives of Australian institutions, working on aerospace and disaster management, and also meetings and discussions with representatives of the Space Policy Unit, Ministry of Research, Science and Technology, of the Australian Government, and the administration of the Royal Melbourne University of Technology (Fig. C10).

Political activity of the ICPI among CIS countries was also noteworthy: for example, on April 18–20, 2011 the IGMASS project was presented in Yerevan (Armenia) during the "First Russian-Armenian Interstate Forum," which was held under the auspices of the Administration of the Russian President. At this Forum, in the presence of top leadership of the two countries the Executive Secretariat signed five MOUs with the National Academy of Sciences of Armenia and its specialized profile institutes: at the least, we sincerely hope to start special research on the IGMASS concept within the frameworks of the CIS scientific cooperation program with the participating Belarusian, Kazakhstan, and Ukrainian research organizations.

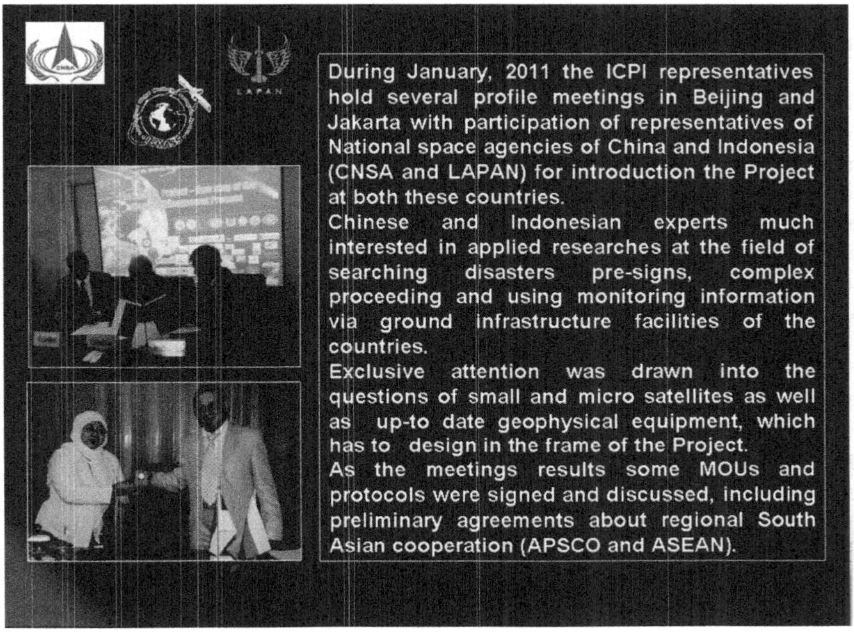

Fig. C9 ICPI activity in South Asia (Beijing, China and Jakarta, Indonesia, spring, 2011)

Fig. C10 Discussing of IGMASS project implementation during IAA working meeting, Paris, France, March 22, 2010

Therefore, the result of the IGMASS presentations at a number of scientific meetings and political events, where our representatives have offered to sign and signed documents in regard to cooperation, was the accession to the project of more than 30 countries and international organizations. The main achievement of the ICPI was beginning the system design of the project. The Committee is planning to release the IGMASS avant-project up to autumn 2011, coordinate it with all members of the Committee, and thereby start concrete work on developing the system. At the same time I'd like to note that the project developers have gradually come to the idea of beginning IGMASS implementation particularly in Russia, involving countries of the CIS, and possibly China. The positive experience of cooperation with our Belarusian colleagues in the framework of design of the Multifunctional Space System of Union State, unambiguous support for the project from the Ukrainian side, and the recent interest in IGMASS from the Academy of Sciences of Armenia gives us a reliable basis for such a conclusion.

The first (pilot) phase of the project would be limited to the mission of forecasting seismic danger in relation to two or three main precursors, reliably fixed from space. As far as Russia is concerned there exist a number of concerned state bodies (Federal Space Agency, Academy of Sciences, Emergency Ministry, Ministry of Natural Resources etc.), possessing the necessary space, air, and ground infrastructure, and activities could be started within the sphere of functional unification of existing technical facilities and software with a view to solving two or

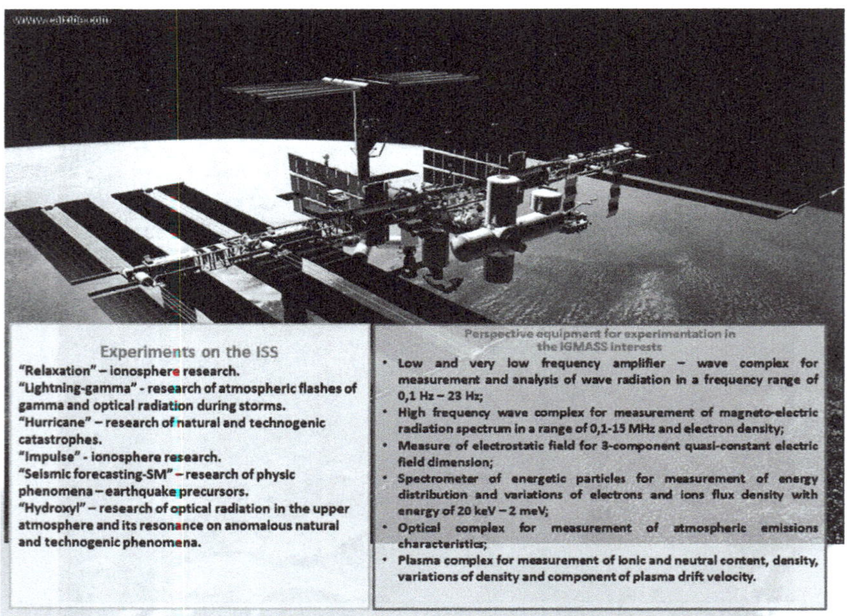

Fig. C11 Using the ISS facilities for IGMASS development

three specific forecasting tasks; integration of predictive monitoring data received both from space and ground sensor systems; realization of space experiments (including on-board the ISS) with ionosphere precursors of seismic events; development of IGMASS subsystems to provide consumers with prospective monitoring information about natural and technological disasters significant for Russia (volcanic eruptions, forest fires, pipeline accidents, etc.)

Thus, the concept of international global monitoring as an instrument to counter natural calamities and emergencies is gradually developing from general theory into a real project with clear implementation prospects in the foreseeable future. Every month it wins new supporters both in Russia and abroad. As IGMASS is supposed to be in charge of the complex solution of tasks to provide early and short-term forecasts of devastating natural phenomena and man-made disasters, it can be stated with a great deal of confidence that it is capable of developing into the backbone idea, which, if implemented, would signal the beginning of a new and common strategy of space exploration aimed at promoting environmentally safe and socially sustainable development of the whole world community on the basis of the eternal value of life preservation on our favorite planet – the Earth (Fig. C11).

Glossary

AIAA	American Institute of Aeronautics and Astronautics
AoA	Analysis of Alternatives
AU	Astronomical Unit (1 AU = the mean distance of the Earth from the Sun)
CSS	Catalina Sky Survey
CERN	European Center of Nuclear Researches
CIS	Commonwealth of Independent States
COPUOS	UN Committee on the Peaceful Uses of Outer Space
COMPASS	Complex Orbital Magneto–Plasma Autonomous Small Satellite
CRED	Center for Research on the Epidemiology of Disasters
CERES	Clouds and the Earth's Radiant Energy System
CNSA	Chinese National Space Administration
CPR	Cloudiness and Aerosol Meter Cloud Profiling Radar
CDV	Compressed Digital Video
CSA	Canadian Space Agency
DRDC	Defence Research and Development Canada
DCT	Discovery Channel Telescope
DB	Database
DD	Dilatancy Diffusion Model
DSO	Dangerous Space Objects
DVB-RCS	Digital Video Broadcasting – Return Channel via Satellite
ECC	Earth Crossing Comets
ELF-components	Extra Low Frequency components
ERS	Earth Remote Sensing
ETH	Extraterrestrial Hypothesis
ESA	European Space Agency
ECOSOC	UN's Economic and Social Council
ESCAP	Economic and Social Commission for Asia and the Pacific
ESRIN	European Space Research Institute
ESV	Earth Satellite Vehicle
FSC	Fixed Space Communication
GAIA	Global Astrometry Interferometer for Astrophysics
GCR	Galactic Cosmic Rays
GDP	Gross Domestic Product

GEO	Geostationary Orbit
GEOSS	Global the Earth Observation System of Systems
GLI	Global Imager
GLAS	Geosciences Laser Altimeter System
GMES	Global Monitoring for Environment and Security
GNI	Gross National Income
GPS	Global Position System
IAA	International Academy of Astronautics
IADC	Inter-Agency Space Debris Coordination Committee
IAU	International Astronomical Union
ICBM	Intercontinental Ballistic Missile
ICPI	International Committee of the IGMASS Project Implementation
IGMASS	International Global Aerospace Monitoring System
ILAS	Improved Limb Atmospheric Spectrometer
IM	Impact Module
IP	Internet Packet
ISDN	Integrated Services Digital Network
ISPRS	International Society for Photogrammetry and Remote Sensing
ISRO	Indian Space Research Organization
ISS	International Space Station
ITU	International Telecommunication Union
IEO	Inner Earth Objects
JAXA	Japan Aerospace Exploration Agency
JHU	Johns Hopkins University
JPL NASA's	Jet Propulsion Laboratory
JSGA	Japanese Space Guard Association
KSP	Kinetic Star-Shaped Penetrator
LAPAN	Indonesian National Institute of Aeronautics and Space
LaRC	Langley Research Center
LDEF	Long Duration Exposure Facility
LEO	Low Earth Orbit
LHB	Late Heavy bombardment
LINEAR	Lincoln Near-Earth Asteroid Research
LONEOS	Lowell Observatory Near-Earth Object Search
LSST	Large Aperture Synoptic Survey Telescope
MBLA	Multi-Beam Laser Altimeter
MIPAS	Michelson Interferometer for Passive Atmospheric Sounding
LPC	Long-Period Comet
M	Earthquake Magnitude
MIPAS	Michelson Interferometer for Passive Atmospheric Sounding
MIT	Massachusetts Institute of Technology
MoD	Ministry of Defence

MODIS	Moderate-Resolution Imaging Spectro-radiometer
MOPITT	Measurements of Pollution in the Troposphere
MOU	Memorandum of Understanding
MPC	Minor Planet Center
MCU	Multi Conference Unit
MT	Megatons of TNT
My	Million years
NASA	National Aeronautics and Space Agency
NEA	Near-Earth Asteroid
NEAR	Near-Earth Asteroid Rendezvous
NEAT	Near-Earth Asteroid Tracking
NED	Nuclear Explosive Devices
NEO	Near-Earth Object
NEODyS	Near-Earth Object Dynamics Site
NEOMAP	Near-Earth Object Mission Advisory Panel
NEOSS	Near-Earth Observation Surveillance Satellite
NGO	Non-Governmental Organization
NORAD	North American Aerospace Defense Command
OLR	Outgoing Long-wave Radiation
PA&E	Program Analysis and Evaluation
Pan-STARRS	Panoramic Survey Telescope and Rapid Response System
PHA	Potentially Hazardous Asteroid
PHC	Potentially Hazardous Long-Period Comet
POLDER	Polarization and Directionality of the Earth's Reflectance
PHO	Potentially Hazardous Object
RAKTS	Russian Academy of Cosmonautics n.a. K.E.Tsiolkovsky
R&D	Research and Development
SCIAMACHY	Scanning Imaging Absorption Spectrometer for Atmospheric Cartography
SDT	Science Definition Team
SPC	Short-Period Comet
SMTP	Simple Mail Transport Protocol
TAR	True-Amplitude Recovery
TOMS	Total Ozone Mapping Spectrometer
TNT	Trinitrotoluol Equivalent
TPS	Thermal Protection System
UAV	Unmanned Aerial Vehicle
UCF	Unsteady Crack Formation
UFO	Unidentified Flying Object
UN	United Nations
UNESCO	UN Educational Scientific and Cultural Organization
UN OOSA	UN Office for Outer Space Affairs
UNIDO	United Nations Industrial Development Organization
ULF-ELF-VLF	Ultra-Extra-Very Low Frequency

URL	Uniform Resource Locator
USI	Universal Space Interceptor
UV	Ultraviolet Radiation
VLF precursors	Very Low Frequency precursors
VSAT	Very Small Aperture Terminal
WSIS	World Summit on Information Society
WWW	World Wide Web
XPS	XUV Photometer System

The manufacturer's authorised representative in the EU is Springer Nature Customer Service Centre GmbH, Europaplatz 3, 69115 Heidelberg, Germany. If you have any concerns regarding our products, please contact ProductSafety@springernature.com

Printed and bound by CPI Group (UK) Ltd, Croydon, CR0 4YY
07/01/2026
02030929-0008